THE METHOD OF QUASI-REVERSIBILITY

Applications to Partial Differential Equations

MODERN ANALYTIC AND COMPUTATIONAL METHODS
IN SCIENCE AND MATHEMATICS

A Group of Monographs and Advanced Textbooks

Editor: Richard Bellman, University of Southern California

MODERN ANALYTIC AND COMPUTATIONAL METHODS IN SCIENCE AND MATHEMATICS

METHODES MODERNES D'ANALYSE ET DE COMPUTATION EN SCIENCE ET MATHÉMATIQUE

NEUE ANALYTISCHE UND NUMERISCHE METHODEN IN DER WISSENSCHAFT UND DER MATHEMATIK

НОВЫЕ АНАЛИТИЧЕСКИЕ И БЫЧИСЛИТЕЛЬНЫЕ МЕТЕОЛЫ В НАУКЕ И МАТЕМАТИКЕ

Editor

RICHARD BELLMAN, UNIVERSITY OF SOUTHERN CALIFORNIA

THE METHOD OF QUASI-REVERSIBILITY
Applications to Partial Differential Equations

by

R. Lattès
Société d'Informatique Appliquée, Paris

and

J.-L. Lions
Faculté des Sciences de l'Université de Paris
et École Polytechnique

Translated from the French edition and edited by Richard Bellman

AMERICAN ELSEVIER PUBLISHING COMPANY, INC.
New York 1969

244223

ORIGINALLY PUBLISHED AS
Méthode de Quasi-Réversibilité et Applications
French Edition Copyright © by Dunod, Paris, 1967

AMERICAN ELSEVIER PUBLISHING COMPANY, INC.
52 Vanderbilt Avenue
New York, N. Y. 10017

ELSEVIER PUBLISHING COMPANY
Barking, Essex, England

ELSEVIER PUBLISHING COMPANY
335 Jan Van Galenstraat, P. O. Box 211
Amsterdam, The Netherlands

Standard Book Number 444-00057-7

Library of Congress Card Number 71-79964

MANUFACTURED IN THE UNITED STATES OF AMERICA

TRANSLATOR'S PREFACE

One of the outstanding challenges to mathematical theory and ingenuity is the construction of algorithms which provide numerical answers to numerical questions. Much of the work of the masters of the seventeenth, eighteenth, and nineteenth centuries centered about this theme. Particular emphasis was placed upon those questions arising in the scientific domain. In the latter half of the nineteenth century and the early part of the twentieth, a strange cultural development took place. There was a bifurcation in the mathematical community. Not only did many mathematicians reject questions of computation as beneath them, but, in addition, by a simple extension of this idea they rejected all investigations connected with the real world. A peculiar snobbery arose in which those theories least related to engineering and science were esteemed the most. The word ''pure'' became an accolade, and ''applied'' a pejorative term.

Fortunately, this aberration is an unstable one. In the first place, the problems posed by the outside world are so varied and intriguing that they inevitably attract the attention of mathematicians. These investigations continually create new branches of mathematics and illuminate existing domains. Young minds automatically reject the orthodoxy of their elders—and fortunately so for human progress.

Secondly, mathematics has a strange gift of universality. A method constructed specifically for analytic number theory plays an important role in celestial mechanics; a topological theory constructed for the study of celestial mechanics becomes vital in the theory of scheduling. All mathematical theories share this trait of being so much more powerful than they seem at first construction. Thus, new fields are always borrowing from old classical domains. Mathematics is too vital for the monastery; it will not stay cloistered.

Finally, there is the digital computer. A device for doing arithmetic a billion, and a billion billion times faster than ever before has irrevocably altered the concept of algorithmic feasibility. Consequently, the kinds of mathematical questions we can profitably ask has expanded without limit. The game has become unbelievably more interesting.

As a result of the efforts of the last twenty years, it has become quite clear that effective use of the digital computer requires a combination of superior mathematical training and mathematical ingenuity. There were some, ten years ago, who predicted glumly that the development of the computer would mark the end of mathematical analysis. We now know that precisely the opposite is true.

The book by Lattès and Lions is a beautiful illustration of this combination required for modern analysis. The central motif is the numerical solution of fundamental problems in mathematical physics which are intrinsically unstable. These are improperly posed problems, in the terminology of Hadamard. Their solution requires both basic theory and experimentation and adaptation. A computational solution thus becomes a stochastic control process in which the results previously obtained become valuable clues as to the subsequent path that should be pursued. Their solution is never routine.

The book is written in a fashion that will soon be standard. An equation is presented, methods are proposed, analyzed rigorously, and then carried through computationally for a wide range of parameter values. We see a mixture of fundamental equations, important methods, and effective algorithms.

The volume will be of equal significance to those interested in the modern theory of partial differential equations, to those primarily interested in numerical analysis, and to those in the sciences who want numerical answers to specific questions.

Richard Bellman

University of Southern California
February 1969

CONTENTS

INTRODUCTION

1. The method of *Quasi-Reversibility* (abbreviated QR) which we introduce and study in this book has as its essential aim the numerical solution of *classes of boundary-value problems which are improperly posed* in the sense of Hadamard [1], problems whose theoretical and practical[1] importance have been repeatedly stressed.

Let us consider a *physical system evolving in time*: an essential problem consists of attaining at the end of time T (or in the neighborhood of T) a given objective. This can theoretically be achieved by appropriate initial conditions. Unfortunately the difficulty of perfect realization of these conditions introduces some perturbations (and therefore some deviations from the ideal conditions).

Two problems then arise:

— to understand the influence of these "deviations" on the solution, if the system is allowed to evolve freely by itself;
— to correct the evolution of the system, that is, to control the physical process. This means exerting influences between the times 0 and T, not only to compensate for the initial deviations, but also for other perturbations, random or not, which occur in the course of the process. These actions are all aimed at improving the quality of the desired objective (or at the conditions to attain it).

Of course the final situation can be defined in a complex fashion. In addition, there may be constraints on the initial conditions, on the controls in the course of the process and on the phenomena themselves.

---

[1]We encounter them more and more in numerous fields, for example: Reservoir engineering, seismology, meteorology, hydrodynamics, magnetohydrodynamics, etc.

Other basic problems arise—always considering a system in evolution—in the course of the determination of parameters, or coefficients characterizing the system, from a series of measured states.

A number of such problems are considered and solved in the present work, namely:

(I) Problems connected with the *backwards* solution of *irreversible* problems of evolution;

(II) Problems of the *control* of "systems" governed by partial differential equations, where the control appears in the *boundary conditions*;

(III) Boundary value problems (including *inverse problems*) with "overabundant" data on one part of the boundary and "insufficient" data on the remainder of the boundary.

The general idea of the method introduced in this book is to *modify suitably the partial differential operators arising in the problem* (this is of course an absolutely "universal" idea in Numerical Analysis!).

The modification of the operators is accomplished by the introduction of *differential* terms, which are:

— "small" (can formally tend to zero);

— "degenerate at the boundary" (for example, to "eliminate" boundary conditions which are mathematically inconvenient, or which are precisely the unknowns to be determined).

The operators thus modified are generally *of a different order* than the initial operator and of the same type (elliptic, etc.) *or not.*

To make this precise, let us now describe in more detail the problems approached in this book.

2. The simplest example of problems of class (I) consists in determining "optimally" the *initial temperature* ζ of a plate whose edges are maintained at zero temperature, in order that at a *given time* T the corresponding temperature $u_\zeta(T)$ approximate "optimally" a temperature given in advance.

In general, there exists no ζ such that $u_\zeta = \chi$: this is connected with the *irreversibility* of the problem. But on the other hand however small η we can always find ξ such that u_ξ "approximates"[1] χ to within η; this is connected with *the backwards uniqueness*.[2]

The problem is then to find *one* ξ yielding such an approximation (there are an infinite number), and the situation is "unstable in ξ." The problem in fact can be interpreted as a Sensitivity Analysis (of which the foregoing instability is a characteristic feature).

The system here is ruled by the heat operator $\partial/\partial t - \Delta$.

The "naive" method would consist of "solving" the problem

$$\left.\begin{array}{l} \left(\dfrac{\partial}{\partial t} - \Delta\right)u = 0 \\ \qquad u = 0 \text{ on the edges of the plate,} \\ u(x, T) = \text{final temperature} = \chi(x). \end{array}\right\} \tag{1}$$

But this problem is *improperly posed* and the general idea (stated in 1) consists here of replacing the heat operator $\partial/\partial t - \Delta$ by the operator

$$\frac{\partial}{\partial t} - \Delta - \varepsilon\Delta^2, \qquad \varepsilon > 0 \text{ "small."}$$

a "properly posed" operator in the backward sense. Then we replace (1) by the *properly posed problem*:

$$\left.\begin{array}{l} \left(\dfrac{\partial}{\partial t} - \Delta - \varepsilon\Delta^2\right)u = 0 \\ \qquad u_\varepsilon(x, T) = \chi(x) \\ u_\varepsilon \text{ and } \Delta u_\varepsilon \text{ zero on the edges of the plate.} \end{array}\right\} \tag{2}$$

It is essential to note that the operator (2) *is not* reversible, no more than the heat operator is. But one is "properly posed" in the sense of increasing t, the other in the sense of decreasing t, whence the terminology adapted: *quasi-reversibility* (abbreviated QR).

It is important to observe that there is *no* uniqueness to "the" QR method; in all the examples we have met, there are *always* an

---

[1] In Chapter 1 we make the topology precise.
[2] J. L. Lions—B. Malgrange [1].

infinite number of possible QR methods—all aspects of the same idea: we change the "system" in such a way that an improperly posed problem becomes properly posed.

Naturally the example which precedes calls forth, in applications, a certain number of generalizations, and variants. These are studied in Chapters 1 and 2, then in Chapter 6; more precisely:

(i) we can replace the heat operator by general parabolic operators of evolution of arbitrary order;

(ii) we can also replace the heat operator by integro-differential operators (such as the "transport equation"), by coupled parabolic-hyperbolic systems, etc.;

(iii) we can replace the "optimal" determination of the temperature χ at *time T* by a "mean" determination in the neighborhood of T (this corresponds to what is often required in applications, and is therefore more realistic).

(iv) the "system" can be represented by *nonlinear*[1] differential operators[2]; some examples are studied in Chapter 6, some others are studied in a separate article of M. Cicurel[3];

(v) there are constraints on the initial ξ.[4]

3. In the preceding example, the "control" is an *initial condition*. This is then, for practical applications, a quite particular *type* of control problem[5] for a process ruled by partial differential equations. On the other hand, another type, *absolutely fundamental*, uses *control in the boundary conditions*. These are problems of class (II). To fix our ideas, in the simplest example, we suppose that the initial temperature is fixed (let us say at zero) and we look for "a" *condition on the boundary values g so as to attain a final temperature* χ *"in an optimal fashion."*

---

[1]As a consequence of the nonlinearity and essentially open problem of controllability, the majority of the results in this case are basically heuristic.

[2]By "differential operator" we mean also "system of differential operators" (elasticity, magneto-hydrodynamics, etc.).

[3]We cannot, naturally, in this book present "all" the cases to which the QR method is applicable. . We try only, in attacking large classes of problems amenable to this method, to furnish means of adapting it without difficulty whenever possible (in particular for control processes, inverse and control problems, etc.).

[4]Naturally the constraints can pertain to other aspects: boundary conditions, the solution itself, a functional dependent on the solution, etc.

[5]To be precise, it is not control, that is to say, action over time, in the course of evolution of phenomenon.

In the parabolic cases, there is always "controllability" (this is connected with the uniqueness of the Cauchy problem), but this is not always true in the hyperbolic cases.

The QR method rests here on the same general ideas, but the technical details are quite different and more complicated than for problems of class (I). In particular, *we now change the nature of the operator*, a parabolic or hyperbolic one becomes "elliptic." This is studied in Chapter 3.

4. Problems of class (III) are studied in Chapters 4 and 5. These problems are connected with a very important class of *inverse* problems for differential operators, problems where the unknowns are *coefficients* of differential operators (or initial conditions, or boundary conditions, etc.). For example, in the elliptic case it is a question of calculating a coefficient appearing in the boundary conditions pertaining to a *part* of the boundary inaccessible to measurement, but *measurements of the Cauchy data* are accessible on *another* part of the boundary. These problems are linked therefore to the numerical solution of Cauchy problems which are *improperly posed* such as the Cauchy problem for elliptic operators—or again to "harmonic continuation" (and to analytic continuation).

Problems of analogous type present themselves (both theoretically and practically) for parabolic operators, the object of Chapter 5: for example, finding boundary conditions corresponding "optimally" to a succession of "measured states"; or also determination of the coefficients of a system when the solution is known at different points at different times.

The idea of the QR method is here to *increase the order* of the operator when the boundary data are of Cauchy type and to *lower* it when there are no (or not enough) data at the boundaries.[1]

5. In a general fashion, for any given problem to which the QR method can be applied, there is not *one* QR method, but rather an *infinite* number. For the choice of a method, we restrict ourselves most frequently to that which appears "simplest"[2]—or to one which is suceptible to a physical interpretation which can aid in the choice

---

[1] An analogous idea has been introduced independently by J. Kohn and L. Nirenberg [2].
[2] Notably from the point of view of numerical work.

of certain parameters—of course in a first analysis we must determine the nature and the properties of the terms which are essential to transform the problem into a well-posed problem.

Naturally the idea of modifying differential or integro-differential operators to "improve the situation" is not new. Let us note, in the same order of ideas:

5.1 The method of *pseudo-viscosity* introduced by Von Neumann-Richtmyer [1], which allows the replacement of a hyperbolic problem with a parabolic problem (whence suppression of discontinuities and uniqueness of solution). Modifications and variants are due to O. A. Oleinik [1] and Lax-Wendroff [1].

5.2 The *elliptic regularization*, which permits an approach to a *nonelliptic* problem by a family of elliptic problems (cf. J. L. Lions [2], O. A. Oleinik [2], J. Kohn-L. Nirenbeeg [1], M. S. Baquendi [1]).

5.3 The entire theory of singular perturbations (cf. K. O. Friedrichs [1], D. Huet [1], I. M. Visik-Liousternik [1]).

5.4 The method of *penalty function* in the calculus of variations (cf. R. Courant [1]).

5.5 The method of stabilization of Tychonnoff [1], [2], [3] (cf. also for integral equations, B. L. Phillips [1]) whose connections with our QR method are studied in Chapter 4, Section 8 for the case of elliptic equations.

5.6 The method called "auxiliary domain" (where the operator *and* the domain are changed), cf. J. L. Lions [4], Sauliev [1], A. Mignot [1].

If the problems considered in Chapters 1, 2, 3, 5 do not seem to have been previously studied in a systematic fashion, the same cannot be said for those of Chapter 4 which have given rise to numerous developments (in one particular case, a method close to ours has been proposed by M. M. Lavrentiev [3]). Let us note, in particular, in addition to the work already cited of M. M. Lavrentiev, the work [1] by the same author and Garabedian, Lieberstein [1], F. John [1], C. Pucci [1], [2], Douglas-Gallie [1] and P. Henrici [1].

For improperly posed problems in general, see M. M. Lavrentiev [2] B. Lago [1].

In a recent work Douglas [1, 2] uses the methods of mathematical programming (cf. also the bibliography of Douglas [2]). A systematic comparison of our method with the preceding will probably allow the efficiency of each to be improved; for example, it seems plausible that our QR method can furnish useful and usable information for the application of the method of Douglas.

6. From the point of view of *numerical application* of the QR method, we have in general followed the following program:

(i) We chose the "simplest" QR method, with one or more "small" coefficients ε_i to be determined (we have not in general devoted our efforts to a study of possible variants obtained by means of different quasi-reversibility terms).

(ii) We determine these parameters by an error calculation.

(iii) We solve numerically the QR system chosen using difference approximations (we have not studied these methods in detail; generally it was a matter of standard systems except in some situations where the difference methods are presented in full detail).

(iv) To verify the agreement between "theory" and "practice," we have in most cases, not only carried out the calculations with many choices of "small" parameters, but also reintegrated the "direct" problems, once the unknowns had been determined with the aid of the QR method (initial conditions, boundary conditions, coefficients, etc.).

The numerical calculations were carried out on the CDC 3600 of the Society of Applied Information (S.I.A.), some with a contract between the D.R.M.E.[1] and the SEMA[2] (for the Control and Analysis aspects of sensitivity), others within the context of a contract between the D.G.R.S.T.[3] and the SEMA (for inverse problems and harmonic continuation).

---

[1] Direction des Recherches et Moyens d'Essais.
[2] Société d'Economie et de Mathématique Appliquees.
[3] Délégation Générale à la Recherche Scientifique and Technique.

We are most grateful to D.R.M.E. and D.G.R.S.T. for having permitted us through these contracts to carry out this work, notably the indispensable numerical examples.

The carrying out of the method and the programming for the numerous examples in the book was done in collaboration with Mrs. Bosset (Engineer in Chief at SEMA), Mrs. Guignot and Richard (Engineers at SEMA) and Mr. Cicurel (Principal Engineer at SEMA). By many apt comments they led us to many improvements of the QR method and we wish particularly to thank them for this.

Chapter 1

QUASI-REVERSIBILITY
AND EQUATIONS
OF PARABOLIC TYPE

DETAILED OUTLINE

1. AN EXAMPLE

In this section we present a *simple example* of the problem which we will study in this chapter.

1.1 Heat Equation

Let Ω be an open domain of $\mathbf{R}^n$,[1] with the generic point $x = \{x_1, ..., x_n\}$. Let t denote the time variable. Let $u(x, t) = u$ be the solution of

$$\frac{\partial u}{\partial t} - \Delta u(x, t) = 0, \quad x \in \Omega, \quad t > 0, \tag{1.1}$$

or

$$\left(\Delta u(x, t) = \sum_{i=1}^{n} \frac{\partial^2}{\partial x_i^2} u(x, t)\right),$$

with the *initial condition*

$$u(x, 0) = \xi(x), \tag{1.2}$$

ξ a given function in Ω, and the *boundary condition*

$$u(x, t) = 0 \text{ if } x \in \Gamma = \text{boundary of } \Omega, \, t > 0. \tag{1.3}$$

It is a classic result that conditions (1.1), (1.2), and (1.3) determine u under suitable hypotheses on Γ and ξ.

The operator $\partial/\partial t - \Delta$ is the *heat operator*. In suitable units $u(x, t)$ denotes the temperature a time t at a point x in the plate Ω, where the initial temperature is $\xi(x)$ and Γ is maintained at temperature zero.

---

[1] $n = 1, 2, 3$ in applications.

1.2 First Formulation of the Problem

Let $T > 0$ be a given value and let χ be a given function in Ω. To each function ξ corresponds the solution of (1.1), (1.2), (1.3) denoted by

$$u = u(x, t ; \xi) .$$

Therefore the integral

$$J = \int_\Omega \left| u(x,T; \xi) - \chi(x) \right|^2 dx \qquad (^1)$$

is a *function* of ξ. We set

$$J(\xi) = \int_\Omega \left| u(x,T; \xi) - \chi(x) \right|^2 dx , \qquad (1.4)$$

which may also be written

$$J(\xi) = \int_\Omega \left| u(x, T) - \chi(x) \right|^2 dx . \qquad (1.4')$$

The first formulation of our problem is as follows: *minimize* $J(\xi)$ *where* ξ *varies without constraints.*$(^2)$ $(^3)$ ∎

The first result (under hypotheses which will be made precise in what follows), is

$$\mathop{\text{Inf}}_{\xi} J(\xi) = 0 . \qquad (^4) \qquad (1.5)$$

This leads to a definitive formulation.

$(^1)$We suppose that the integral exists. Further on, sufficient conditions for this to be so will be given.

$(^2)$The case where ξ varies subject to constraints will be discussed in Chapter 6.

$(^3)$The sign ∎ indicates the end of a demonstration, remark, etc.

$(^4)$An analogous result in a more general setting will be demonstrated in Section 4.

1.3 Formulation of the Problem and Various Remarks

Problem 1.1: *For u given $\eta > 0$, find $\xi_\eta = \xi_\eta(x)$ such that*

$$J(\xi_\eta) \leqslant \eta .\tag{1.6}$$

Remark 1.1. It follows from (1.5) that a function ξ_η exists. However, the demonstration of (1.5) which we will give *is not constructive.* Our problem reduces therefore to giving a constructive solution of (1.5) or (1.6)—one that is *numerically practical.* ∎

Remark 1.2. Evidently there exist an infinite number of solution functions ξ_η. The class of possible solutions will be reduced *subsequently* by constraints on ξ;[1] for the moment, no constraint is imposed on ξ. It is therefore only a question of constructing *particular* functions ξ_η answering the question. ∎

Remark 1.3. Observe that the unknown is ξ and *not* the associated solution $u(x, t; \xi)$. ∎

Remark 1.4. An "obvious" solution would seem to be the following: we can choose ξ so that $J(\xi) = 0$, namely, $u(x, T) = \chi(x)$. Therefore, we consider u as the "solution" of

$$\left.\begin{aligned} &\frac{\partial u}{\partial t} - \Delta u = 0 \\ &u(x, T) = \chi(x), \\ &u = 0 \quad \text{on} \quad \Sigma = \Gamma \times (0, T) \end{aligned}\right\} \tag{1.7}$$

and we take

$$\xi(x) = u(x, 0) .\tag{1.8}$$

But, the "solution" of (1.7) *does not exist in general.* By this we mean that the solution of (1.7) exists only if χ satisfies extremely restrictive conditions which are not easy to make explicit. Furthermore, the solution is of a *very unstable* type, blocking any simple numerical solution.[2] ∎

---

[1] It is even possible that no solution will exist (controllability).
[2] This does not mean that this type of problem (*improperly posed problem*), is unapproachable; cf. F. John [1], J. Douglas, Jr. [1]. Our method can be considered as a *contribution to the study of improperly posed problems.*

Remark 1.5. Another more "natural" solution is as follows: let ξ_1, ξ_2, ..., ξ_N, be N *linearly independent functions*. Set

$$u_j(x, t) = u(x, t; \xi_j).\tag{1.9}$$

The mapping $\xi \rightarrow u(x, t; \xi)$ is linear:

$$u\left(x, t; \sum_{j=1}^{N} \lambda_j \xi_j\right) = \sum_{j=1}^{N} \lambda_j u_j, \qquad \lambda_j \in \mathbf{R},$$

and the λ_j can be determined so that

$$\int_{\Omega} \left| \sum_{j=1}^{N} \lambda_j u_j(x, T) - \chi(x) \right|^2 dx\tag{1.10}$$

is a *minimum*. But the difficulties are the following:

1. How can one *choose* the ξ_j and N so as to be assured that the minimum in (1.10) is less than or equal to $\leq \eta$?
2. From the practical point of view, we will be required to solve N heat equations numerically. As soon as we encounter multidimensional geometries, serious problems of calculating time and computer storage capacity will be encountered. ∎

In the following section we present a *heuristic* version of the method we propose. In subsequent sections the method will be generalized and made more precise.

2. HEURISTIC PRESENTATION OF THE METHOD OF QUASI-REVERSIBILITY

2.1 General Ideas

The general idea is the following: we are going to look for an operator P_ε "near" the heat operator $\partial/\partial t - \Delta$ for which the "obvious" solution of Remark 1.4 *is possible*. Therefore it is necessary that

i) P_ε is an evolution operator

ii) there exists one and only one solution of

$$P_\varepsilon u_\varepsilon = 0, \qquad x \in \Omega, \qquad t \in (0, T) \tag{2.1}$$

$$u_\varepsilon(x, T) = \chi(x), \tag{2.2}$$

$$u_\varepsilon = 0 \quad \text{on} \quad \Sigma, \tag{2.3}$$

with possibly some *other conditions at the boundaries*, satisfied, if possible, by u, the solution of (1.1), (1.2) and (1.3).

Supposing that these conditions are satisfied, we take:

$$\xi(x) = u_\varepsilon(x, 0). \ \blacksquare \tag{2.4}$$

The problems are therefore:

1) Find P_ε;

2) Choose P_ε so that for the choice in (2.4) we have $J(\xi) \leqslant \eta$;

3) Integrate (2.1), (2.2) and (2.3) numerically, possibly with some additional conditions.

Remark 2.1. A *verification* of the adequacy of the proposed solution ξ is possible by means of a *numerical integration* of (1.1), (1.2) and (1.3) and calculation of $J(\xi)$. $\blacksquare$

Remark 2.2. As we will see, there are an *infinite number of possible choices of P_ε*. In each particular case, we restrict ourselves to taking "the" P_ε leading to the simplest calculations.

Remark 2.3. We often express the fact that problem (1.7) is *improperly posed* (cf. Remark 1.4), by saying that the *heat operator cannot be time-inverted (irreversibility)*. Our method consists in finding an *operator P_ε* "near" the heat operator for which the *backward problem*, (2.1), (2.2), (2.3), with possibly supplementary conditions, is *well posed*. This is the origin of the terminology for our method: Method of Quasi-Reversibility (QR).

FUNDAMENTAL REMARK: This does not at all signify that the operator P_ε is reversible.

2.2 Presentation of the Method

We will show farther on (and in a more general context), that the following problem is *properly posed*:

$$\frac{\partial u_\varepsilon}{\partial t} - \Delta u_\varepsilon - \varepsilon \, \Delta^2 u_\varepsilon = 0 \,, \qquad x \in \Omega \,, \qquad t < T \,, \qquad \varepsilon > 0 \,, \qquad \text{(2.5)}$$

$$u_\varepsilon(x, T) = \chi(x) \,, \qquad \text{(2.6)}$$

$$u_\varepsilon = 0 \quad \text{on} \quad \Sigma \,, \qquad \text{(2.7)}$$

$$\Delta u_\varepsilon = 0 \quad \text{on} \quad \Sigma \,. \qquad \text{(2.8)}$$

The condition 1) of (2.1) holds, thus ii). If u is a sufficiently regular solution of (1.1), (1.2), (1.3) then

$$\Delta u = 0 \quad \text{on} \quad \Sigma \,.$$

In effect $\Delta u = \partial u / \partial t$ and $u \mid_\Sigma = 0$ ensure $\partial u / \partial t \mid_\Sigma = 0$. Therefore, (2.8) is compatible with the conditions satisfied by u.

Remark 2.4. Having made the choice (2.5), (2.6), (2.7), (2.8), we take

$$\xi(x) = u_\varepsilon(x, 0) \,.$$

When $\varepsilon \to 0$, u_ε *does not converge*, in general. The important point (which we demonstrate further on), is the following: let U_ε be *the* solution of (direct integration of the initial value problem)

$$\left. \begin{aligned} &\frac{\partial U_\varepsilon}{\partial t} - \Delta U_\varepsilon = 0 \,, \\[2mm] &U_\varepsilon(x, 0) = u_\varepsilon(x, 0) \\[2mm] &U_\varepsilon \mid_\Sigma = 0 \,. \end{aligned} \right\} \qquad \text{(2.9)}$$

Then, in a sense which we will make precise

$$U_\varepsilon(x, T) \to \chi(x) \quad \text{when} \quad \varepsilon \to 0 \,. \blacksquare \qquad \text{(2.10)}$$

Remark 2.5. Cauchy Problem. Let us examine the situation in more detail in the case where $\Omega = \mathbf{R}^n$, the Cauchy problem. We use here the *Fourier transform* in x, supposing that the given functions are *tempered* (cf. L. Schwartz [1]). Let

$$\hat{f}(y) = \int_{\mathbb{R}^n} e^{-ix.y} f(x) \, dx \,, \qquad xy = x_1 y_1 + \cdots + x_n y_n$$

(an integral in the distribution sense; cf. L. Schwartz, *loc. cit.*). Put

$$\hat{u}_\varepsilon(y, t) = \int_{\mathbb{R}^n} e^{-ix.y} u_\varepsilon(x, t) \, dx$$

then in (2.5), (2.6), ... , (2.8), the conditions (2.7), (2.8) disappear.[1] We obtain

$$\frac{d\hat{u}_\varepsilon}{dt} + y^2 \, \hat{u}_\varepsilon - \varepsilon y^4 \, \hat{u}_\varepsilon = 0 \,, \qquad y^2 = y_1^2 + \cdots + y_n^2 \,, \tag{2.11}$$

$$\hat{u}_\varepsilon(y, T) = \hat{x}(y) \,, \tag{2.12}$$

and thus

$$\hat{u}_\varepsilon(y) = \hat{u}_\varepsilon(y, 0) = e^{-(\varepsilon y^4 - y^2)T} \, \hat{\chi}(y) \tag{2.13}$$

Hence, we make the choice

$$\hat{\xi}(y) = \hat{u}_\varepsilon(y, 0) = e^{-(\varepsilon y^4 - y^2)T} \, \hat{\chi}(y) \,. \tag{2.14}$$

Note that if si $\chi \in \mathscr{S}'$ (that space of tempered distributions), then $\xi(y)$ given by (2.4) is still in $\mathscr{S}'$ (the distributions are written as functions). On the other hand, it is impossible to take $\varepsilon = 0$ for an arbitrary χ in $\mathscr{S}'$. ∎

Let us verify Remark 2.4 in the present case. We have:

$$\left. \begin{aligned} &\frac{d\hat{U}_\varepsilon}{dt} + y^2 \, \hat{U}_\varepsilon = 0 \\ &\hat{U}_\varepsilon(y, 0) = \hat{u}_\varepsilon(y, 0) = \hat{\xi}(y) \qquad \text{(given in \quad (2.4))} \end{aligned} \right\} \tag{2.15}$$

from which follows

$$\hat{U}_\varepsilon(y, t) = e^{-y^2 t} \, \hat{\xi}(y)$$

---

[1] They are replaced by *conditions of growth at infinity.*

therefore

$$\hat{U}_\varepsilon(y, t) = e^{-\varepsilon y^4 T}\, e^{y^2(T-t)}\, \hat{\chi}(y)\,, \tag{2.16}$$

and

$$\hat{U}_\varepsilon(y, T) = e^{-\varepsilon y^4 T}\, \hat{\chi}(y) \tag{2.17}$$

and finally

$$\left. \begin{aligned} & U_\varepsilon(x, T) \to \chi \quad \text{in} \quad \mathscr{S}'(\text{or in} \quad L_2(\mathbf{R}^n)) \quad \text{when} \\ & \varepsilon \to 0,\, \chi \quad \text{given in} \quad \mathscr{S}'(\text{or in} \quad L_2(\mathbf{R}^n))\,. \end{aligned} \right\} \tag{2.18}$$

Note that, for $t < T$, U_ε *does not converge in general* when $\varepsilon \to 0$. ∎

Remark 2.6. The same results will hold, at least from the theoretical point of view, if we consider, in place of (2.5), . . . , (2.8), the problem

$$\frac{\partial u_\varepsilon}{\partial t} - \Delta u_\varepsilon - \varepsilon(-1)^m\, \Delta^m u_\varepsilon = 0, \qquad x \in \Omega\,, \qquad t < T\,, \tag{2.19}$$

$$u_\varepsilon(x; T) = \chi(x)\,, \tag{2.20}$$

$$\left. \begin{aligned} & u_\varepsilon = 0 \quad \text{sur} \quad \Sigma \\ & \Delta u_\varepsilon = 0 \quad \text{sur} \quad \Sigma \\ & \cdots\cdots\cdots\cdots\cdots \\ & \Delta^{m-1} u_\varepsilon = 0 \quad \text{sur} \quad \Sigma\,. \end{aligned} \right\} \tag{2.21}$$

The simplest case numerically corresponds to $m = 2$. ∎

Orientation. We are going to pose the problem in a more general and rigorous fashion. The following section gives the background required for this.

3. REVIEW OF EQUATIONS OF EVOLUTION

3.1 Function Spaces

Let $\mathscr{H}$ be a Hilbert space. To simplify a bit, we suppose that $\mathscr{H}$ is *real*, as are *all* the spaces introduced subsequently. We also suppose that it is *separable*, which is to say there exists a *denumerable* set of elements dense in $\mathscr{H}$. This is the case for all of the spaces occurring in the examples given in this book.

Let $L_2(0, T ; \mathcal{H})$ denote the space of functions which are square summable over $(0, T)$ with values in $\mathcal{H}$; in other words, if $f \in L_2(0, T ; \mathcal{H})$, then:

i) f is *measurable with values in* $\mathcal{H}$; by this we mean $\forall h \in \mathcal{H}$, the *scalar* function, $t \rightarrow (f(t), h)_{\mathcal{H}}$,[1] is measurable in $(0, T)$, with the Lebesgue measure dt;

ii) $\left(\int_0^T \| f(t) \|_{\mathcal{H}}^2 \ dt \right)^{1/2} = \| f \|_{L_2(0,T,\mathcal{H})} < \infty .$

In fact, $L_2(0, T, \mathcal{H})$ is more precisely a space of *classes* of functions f_1 and f_2 being *identified* if $f_1(t) = f_2(t)$ almost everywhere (a.e.). Then, given the *norm* $\| f \|_{L_2(0,T, \mathcal{H})}$, the space $L_2(0, T ; \mathcal{H})$ *is a Hilbert space.* ∎

Let $\mathcal{H}$ and $\mathcal{K}$ denote two spaces of Hilbert, with

$$\mathcal{H} \subset \mathcal{K} ,$$

the sign $\subset$ denotes here and in what follows both algebraic and topological inclusion. This means that the identity mapping of $\mathcal{H}$ in $\mathcal{K}$ is continuous. We will consider functions $v \in L_2(0, T ; \mathcal{H})$ such that $v' = dv/dt \in L_2(0, T ; \mathcal{K})$; here, the derivative dv/dt is taken in the sense of *distributions* on $]0, T[$, with values in $\mathcal{H}$; cf., for this notion L. Schwartz [2]. This means that when v is given in $L_2(0,T ; \mathcal{H})$, there exists a function v_1 in $L_2(0, T ; \mathcal{K})$, and *only one*, such that

$$-\int_0^T v(t) \, \varphi'(t) \, dt = \int_0^T v_1(t) \, \varphi(t) \, dt , \qquad (3.1)$$

$\forall \varphi \in D(]0, T[) =$ space of functions C^∞,[2] with compact support on $]0, T[$. In (3.1), the integrals have values in $\mathcal{K}$. ∎

3.2 The Operators $A(t)$

Let V and H be two Hilbert spaces with

$$V \subset H , \quad V \text{ dense in } H. \qquad (3.2)$$

---

[1] $(h_1, h_2)_{\mathcal{H}}$ denotes the scalar product in $\mathcal{H}$ and $\| h \|_{\mathcal{H}} = ((h, h)_{\mathcal{H}})^{1/2}$

[2] This is to say, infinitely differentiable. A function C^k will be k times continuously differentiable, k a nonnegative integer.

We denote by $((,))$ (respectively $(,)$) and $\| \ \|$ (respectively $| \ |$) the scalar product in V (respectively H) and the norm on V (respectively H). Let V' be the dual of V; we identify H with its dual so that

$$V \subset H \subset V' ; \tag{3.3}$$

If $f \in V'$, $v \in V$, (f, v) denotes their scalar product; if $f \in H$, it coincides with the scalar product in H.

For each $t \in (0, T)$, we are given a continuous bilinear form $a(t ; u, v)$ on V, having the following properties:

$$\forall u, v \in V, \text{ the function } t \to a(t ; u, v) \text{ is measurable and} \atop | a(t ; u, v) | \leqslant M \|u\| \ \|v\| , M = \text{constant independent of } t, u, v. \tag{3.4}$$

For fixed u in V, the linear form

$$v \to a(t ; u, v)$$

is continuous on V; therefore it can be written

$$a(t ; u, v) = (A(t) u, v), \qquad A(t) u \in V' . \tag{3.5}$$

We deduce also from (3.4) that

$$\| A(t) u \|_* \leqslant M \| u \| , \qquad \forall u \in V \tag{3.6}$$

where $\| \ \|_* =$ dual norm of $\| \ \|.$[1]

The family of operators $A(t) \in \mathscr{L}(V ; V')$[2] is said to be *coercive*[3] if:

$$\text{there exists } \lambda \text{ and } \alpha > 0 \text{ such that} \atop a(t ; v, v) + \lambda | v |^2 \geqslant \alpha \| v \|^2, \ \forall v \in V. \ \blacksquare \tag{3.7}$$

3.3 Equations of Evolution

Let us recall the following theorem for the demonstration, see J. L. Lions [1].

---

[1] $\| f \|_* = \sup\limits_{v \in V} \dfrac{| (f, v) |}{\| v \|}, \quad f \in V'$.

[2] Space of continuous linear mapping from V onto V'.

[3] Or elliptic.

Theorem 3.1: *Let us suppose that* (3.4) *and* (3.7) *hold. Let f and* ξ *be given functions with*:

$$f \in L_2(0, T; V'),\qquad\qquad (3.8)$$

$$\xi \in H.\qquad\qquad (3.9)$$

There exists one function u, and only one, satisfying

$$u \in L_2(0, T; V'),\qquad\qquad (3.10)$$

$$u' = \frac{du}{dt} \in L_2(0, T; V'),\qquad\qquad (3.11)$$

$$A(t)\, u + u' = f,\qquad\qquad (3.12)$$

$$u(0) = \xi.\qquad\qquad (3.13)$$

Remark 3.1. It can be shown (cf. Lions [1] and [2]) that if u satisfies (3.10) and (3.11), then it is, after possible modification on a set of measure zero, continuous from $[0. T] \to H$. Therefore, in particular, (3.13) *is meaningful.* ∎

3.4 First Example

We will subsequently give some much more general examples. Here we wish simply to indicate how the heat equation (1.1), (1.2), (1.3) fits into the preceding format.

An important supplementary notion is indispensable:

Sobolev space $H^1(\Omega)$ (cf. S. L. Sobolev [1]).[1]

Let $H^1(\Omega)$ denote the space of (classes of) functions $v \in L_2(\Omega)$,[2] such that

$$\frac{\partial v}{\partial x_i} \in L_2(\Omega) \qquad \forall i = 1, ..., n.$$

[1] Ω will *always* denote an open set of $\mathbf{R}^n$.
[2] $L_2(\Omega)$= space of functions which are square summable on Ω.

The derivatives $\partial v/\partial x_i$ are taken in the sense of distributions on Ω. Provided with the norm

$$\| v \|_{H^1(\Omega)} = \left(\int_\Omega | v(x) |^2 \, dx + \sum \int_\Omega \left(\frac{dv}{\partial x_i} \right)^2 dx \right)^{1/2} , \quad dx = dx_1, ..., dx_n , \quad \textbf{(3.14)}$$

$H^1(\Omega)$ is a Hilbert space (cf. S. L. Sobolev [1], Magenes-Stampacchia [1]).

Space $H_0^1(\Omega)$.

Let $H_0^1(\Omega)$ denote the closure of $D(\Omega)$[1] in $H^1(\Omega)$. If we suppose that the boundary Γ of Ω is a variety C^1, with Ω on only one side of Γ, then for every $v \in H^1(\Omega)$ we can define in a unique fashion, *the trace of v on Γ:* $v|_\Gamma \in L_2(\Gamma)$, where $L_2(\Gamma)$ = space of functions of square summable on Γ, with the measure of the surface Γ.

In addition, we have:

$$H_0^1(\Omega) = \{ v \mid v \in H^1(\Omega), v|_\Gamma = 0 \} . \quad \textbf{(3.16)}$$

Therefore $H_0^1(\Omega)$ *is the sub-space of $H^1(\Omega)$ consisting of functions which are zero on Γ.* ■

Example of an Equation of Evolution

Take

$$V = H_0^1(\Omega), \qquad H = L_2(\Omega) ,$$

following the notation of 3.2 and 3.3. Then

V' = space of distributions on Ω of the form

$$f = f_0 + \sum_{i=1}^n \frac{\partial}{\partial x_i} f_i , \qquad f_0, f_i \in L_2(\Omega) .$$

We often set: $V' = H^{-1}(\Omega)$. ■

---

[1] $D(\Omega)$ = space of functions C^∞ and with compact support in Ω.

For $u, v \in V$, set

$$a(u, v) = \sum_{i=1}^{n} \int_{\Omega} \frac{\partial u}{\partial x_i} \frac{\partial v}{\partial x_i} \, dx = \sum_{i=1}^{n} \left(\frac{\partial u}{\partial x_i}, \frac{\partial v}{\partial x_i} \right). \tag{3.17}$$

Finally set

$$a(t \,; u, v) = a(u, v) \qquad \forall t \,.$$

It is easy to verify that (3.7) holds (even with $\lambda = 0$); (3.4) is trivial. The operator A is given by

$$Au = -\Delta u = -\sum_{i=1}^{n} \frac{\partial^2 u}{\partial x_i^2}. \tag{3.18}$$

Theorem 3.1 is applicable and gives:

There exists a unique u such that

$$u \in L_2(0, T; H_0^1(\Omega)), \tag{3.19}$$

$$u' \in L_2(0, T; H^{-1}(\Omega)), \tag{3.20}$$

$$\frac{\partial u}{\partial t} - \Delta u = f, \tag{3.21}$$

$$u(0) = \xi \,. \tag{3.22}$$

Note that $u(0)$ is a function $x \to u(x, 0)$; in general, $u(t)$ is the function $x \to u(x, t)$. Therefore (3.21) and (3.22) may be written

$$\frac{\partial}{\partial t} u(x, t) - \Delta u(x, t) = f(x, t), \quad x \in \Omega, \quad t > 0,$$

$$u(x, 0) = \xi(x), \qquad x \in \Omega \,.$$

The condition (3.19) is *equivalent* to

$$u \in L_2(0, T; L_2(\Omega)), \tag{3.19}_o$$

$$\frac{\partial u}{\partial x_i} \in L_2(0, T; L_2(\Omega)), \qquad \forall i, \tag{3.19}_i$$

which is to say

$$\int_0^T \int_\Omega (u(x, t))^2 \, dx \, dt < \infty$$

$$\int_0^T \int_\Omega \left(\frac{\partial u}{\partial x_i}(x, t)\right)^2 \, dx \, dt < \infty, \qquad \forall i$$

and

$$u(x, t)|_\Gamma = 0 \quad \textit{for almost all } t. \tag{3.19)i}$$

We see therefore that (3.21), in the particular case $f = 0$, (3.22) and (3.19) are respectively analogous to (1.1), (1.2), (1.3).

4. GENERAL PROBLEM

4.1 The Functional $J(\xi)$.

Let the family $A(t)$ be given along with (3.4), (3.7). Then to each corresponds a *unique* function $u = u(t) = u(t; \xi)$ satisfying:

$$u' + A(t)u = 0, \tag{4.1}$$

$$u(0) = \xi, \tag{4.2}$$

$$u \in L_2(0, T; V), \qquad u' \in L_2(0, T; V'). \tag{4.3}$$

cf. theorem (3.1). Then if χ is *given* in H, set

$$J(\xi) = |u(T; \xi) - \chi|^2 . \blacksquare \tag{4.4}$$

Example 4.1.

In the case of example 3.4, we find

$$J(\xi) = \int_\Omega |u(x, T; \xi) - \chi(x)|^2 \, dx ;$$

the functional in (1.4). $\blacksquare$

4.2 The Lower Bound for $J(\xi)$.

We have the following conjecture:

Conjecture 4.1.

Under the hypotheses of theorem 3.1, we have

$$\operatorname*{Inf}_{\xi \in H} J(\xi) = 0 . \tag{4.5}$$

We are going in what follows to demonstrate the conjecture in a particular case and then make a contribution to possible demonstration of the conjecture.

Theorem 4.1: *We suppose that*

$$\left.\begin{array}{l} a(t\,;u,v) = a(u,v) \text{ is independant of } t \text{ and} \\[6pt] a(u,v) = a(v,u) \qquad \forall u,v \in V \qquad (^1)\,; \end{array}\right\} \tag{4.6}$$

then **(4.5)** *is true*.

Proof.

1) Under the hypotehsis (4.6) the operator A is *self-adjoint*. We can then find a measurable hilbertian sum

$$\mathfrak{h} = \int_{\alpha_0}^{\infty} \mathfrak{h}(\lambda)\, d\mu(\lambda) , \qquad (^2)$$

$\lambda \to \mathfrak{h}(\lambda)$ is a measurable field of Hilbert spaces, and a *unitary* operator $\mathscr{U}$ of H on $\mathfrak{h}$ such that

$$\mathscr{U} V = \{ f \mid \lambda\, \mathscr{U} f \in \mathfrak{h} \}$$

and

$$\mathscr{U} A v = \lambda \mathscr{U} v .$$

$(^1)$We suppose that the hypotheses of Theorem 3.- are valid, and thus that

$$a(v,v) + \lambda\,|v|^2 \geqslant \alpha\,\|v\|^2 , \qquad \alpha > 0 , \qquad \forall v \in V .$$

$(^2)$It suffices to change u into $e^{kt}\,u$, $k > \lambda$, to suppose that $a(v,v) \geqslant \alpha_0\,\|v\|^2$, $\alpha_0 > 0$.

2) Then, setting

$$\mathscr{U}u(t) = \hat{u}(\lambda, t) \,,$$

the problem (4.1), (4.2) is equivalent to

$$\frac{\partial \hat{u}}{\partial t} + \lambda \hat{u} = 0 \,,$$

$$\hat{u}(\lambda, 0) = \hat{\xi}(\lambda) \,, \quad \text{where} \quad \hat{\xi} = \mathscr{U}\xi.$$

Therefore

$$\hat{u}(\lambda, T) = e^{-\lambda T} \, \hat{\xi}(\lambda) \,. \tag{4.7}$$

3) Everything reduces to showing that when ξ spans H, the space spanned by $\hat{u}(\lambda, T)$ given by (4.7) is *dense in* $\mathfrak{h}$. Since $\hat{\xi}$ spans $\mathfrak{h}$, the result is immediate. ∎

Contribution to general case.

1) In the general case, it is necessary to show that the space spanned by $u(T; \hat{\xi})$ is dense in H. Let us suppose therefore that $\psi \in H$, with

$$(u(T; \xi), \psi) = 0 \qquad \forall \xi \in H \,. \tag{4.8}$$

2) We introduce the *adjoint family* $A^*(t)$ of $A(t)$ by

$$(u(T; \xi), \psi) = 0 \qquad \forall \xi \in H \,. \tag{4.9}$$

Then, using a simple variant of Theorem 3.1 (essentially changing t into $-t$), there exists a unique $v = v(t)$ such that

$$- v' + A^*(t) v = 0 \,, \tag{4.10}$$

$$v(T) = \psi \,, \tag{4.11}$$

$$v \in L_2(0, T; V) \,, \quad v' \in L_2(0, T; V') \,. \tag{4.12}$$

3) Consider now the integral

$$\int_0^T [(u', v) + (u, v')] \, dt \,;$$

then([1]):

([1])This formula of integration by parts for "weak" derivatives can be rigorously justified.

$$\int_0^T \frac{\mathrm{d}}{\mathrm{d}t} \left(u(t), v(t)\right) \mathrm{d}t = \left(u(T), v(T)\right) - \left(u(0), v(0)\right) =$$

$$= \left(u(T; \xi), \psi\right) - \left(\xi, v(0)\right) = - \left(\xi, v(0)\right).$$

In addition, replacing u' by $-A(t) u$ and v' by $A^*(t) v$, it becomes

$$\int_0^T \left[- \left(A(t) u, v\right) + \left(u, A(t) v\right)\right] \mathrm{d}t$$

From (4.9) this is zero. Therefore

$$\left(\xi, v(0)\right) = 0 \qquad \forall \xi \in H$$

which is to say

$$v(0) = 0. \tag{4.13}$$

4) To obtain the usual case, change t into $T - t$ and set

$$v(T - t) = w(t).$$

Then:

$$\left.\begin{array}{r} w'(t) + A^*(T - t) w(t) = 0, \\ w(0) = \psi, \end{array}\right\} \tag{4.14}$$

and

$$w(T) = 0, \tag{4.15}$$

with

$$w \in L_2(0, T; V), \qquad w' \in L_2(0, T; V'). \tag{4.16}$$

The problem is then that of *backwards uniqueness*: *do the hypotheses* (4.14), (4.15), (4.16) *entail* $w \equiv 0$?

If the answer is affirmative, $\psi = 0$ and (4.5) is demonstrated.

5) The response is affirmative in the case of Theorem (4.1). In effect, we then have

$$\hat{w}(\lambda, t) = \mathrm{e}^{-\lambda t} \hat{\psi}(\lambda)$$

and if (4.15) holds, then $\psi = 0$.

6) In the general case, the problem of backwards uniqueness is resolved in the affirmative, with the aid of some additional hypotheses of regularity on $A(t)$[1] and on $w,$[2] in J. L. Lions–B. Malgrange [1]. ∎

4.3 The General Problem

Assuming the truth of (4.5) in the following, consider the following problem (generalization of problem 1.1):

Problem 4.1:   For $\eta > 0$ given, find $\xi_\eta \in H$ [3] such that

$$J(\xi_\eta) \leqslant \eta \cdot \blacksquare \tag{4.17}$$

Remark 4.1. The principle of the demonstration given under the heading "contribution to the general case" *is not constructive*.

As for the demonstration of Theorem 4.1, it is constructive, but not practical numerically. In effect, it depends upon the *diagonalization* of A. Let us suppose (a case frequently arising in applications), that *the identity mapping of V in H is compact*. Then, there exists a denumerable sequence of characteristic values λ_i and characteristic vectors w_i such that

$$Aw_i = \lambda_i w_i, \quad \lambda_i > 0, \quad \lambda_i \nearrow +\infty, \quad w_i \in V \quad (^4), \tag{4.18}$$

and (4.7) becomes

$$u(T) = \sum_{i=1}^{\infty} e^{-\lambda_i T}(\xi, w_i)\, w_i. \tag{4.19}$$

The problem is therefore that of *choosing the scalars ξ_i and the integer N* so that

$$\sum_{i=1}^{N} e^{-\lambda_i T}\, \xi_i\, w_i$$

[1] Probably essential.
[2] Very probably unnecessary.
[3] Therefore, with no further constraint than belonging to H.
[4] The w_i are normalized by $(w_i, w_j) = \delta_i^j$.

is near[3] χ, then of choosing ξ in H satisfying

$$(\xi, w_i) = \xi_i, \qquad 1 \leqslant i \leqslant N \qquad [4].$$

However, except in very special cases, the λ_i and w_i are unknown—and out of reach except for N "small."

We are going to give in the following section the QR method for the resolution of Problem 4.1, a method which generalizes and makes precise the heuristic considerations of Section 2. ∎

5. THE QR METHOD

5.1 The Spaces $D(A(t))$, $D(A^*(t))$.

Set

$$D(A(t)) = \{ v \mid v \in V, A(t) v \in H \} ; \tag{5.1}$$

this is therefore the *domain* of the operator $A(t)$ considered as an *unbounded operator* in H (see M. H. Stone [1], for this concept). Provided with the norm

$$\| v \|_{D(A(t))} = (\| v \|^2 + | A(t) v |^2)^{1/2} ,$$

it is a Hilbert space. In the same way $D(A^*(t))$ is defined. ∎

The space $D(A(t))$ [respectively $D(A^*(t))$], is *dense* in H. We can therefore consider the dual spaces as super-spaces of H; thus

$$D(A(t)) \subset H \subset D(A(t))' . \tag{5.2}$$

The operator $A(t)$ is linear continuous from $D(A(t)) \to H$, therefore by passage to the adjoint:

$$A^*(t) \in \mathscr{L}(H ; D(A(t))') . \tag{5.3}$$

We deduce from this

$$A^*(t) A(t) \in \mathscr{L}(D(A(t)) ; D(A(t))') . ∎ \tag{5.4}$$

---

[1] In the sense that the distance in H to χ be less than η.
[2] Evidently, ξ is not unique.

5.2 The QR Method

We always assume the conditions of Theorem 3.1. Consider, for each $\varepsilon > 0$, the following problem: find u_ε satisfying:

$$u_\varepsilon' + A(t)\, u_\varepsilon - \varepsilon A^*(t)\, A(t)\, u_\varepsilon = 0 \,, \tag{5.5}$$

$$u_\varepsilon(T) = \chi \,, \tag{5.6}$$

$$\left. \begin{array}{l} u_\varepsilon \in L_2(0,\, T;\, D(A(t)))\,, \\ u_\varepsilon' \in L_2(0,\, T;\, D(A(t))')\,. \end{array} \right\} \tag{5.7}$$

The first condition (5.7) signifies that: $u_\varepsilon \in L_2(0,\, T;\, V)$, $u_\varepsilon(t) \in D(A(t))$ for almost all t and $A(t)\, u_\varepsilon(t) \in L_2(0,\, T;\, H)$. Finally, we identify $L_2(0,\, T;\, D(A(t))')$ with $(L_2(0,\, T;\, D(A(t))))'$.

We will show subsequently, under suitable hypotheses, that *the problem of* (5.5), (5.6), (5.7) *has a unique solution.*

We then take:

$$\xi = u_\varepsilon(0) \,. \blacksquare \tag{5.8}$$

Example 5.1

Let us take the example treated in 3.4. Then

$$A(t) = A = -\Delta \,,$$

$$D(A) = \{\, v \mid v \in H_0^1(\Omega),\, \Delta v \in L_2(\Omega) \,\} \,.$$

If the boundary Γ of Ω is C^1, then $D(A) \subset H^2(\Omega)$, where $H^2(\Omega)$, the Sobolev space of order 2, denotes the space

$$H^2(\Omega) = \left\{\, v \mid v \in H^1(\Omega),\, \frac{\partial^2 v}{\partial x_i\, \partial x_j} \in L_2(\Omega) \quad \forall i, j \,\right\} ;$$

then $A^* = A$, and if u_ε is the solution of (5.5), (5.6), (5.7), we have:

$$\frac{\mathrm{d}}{\mathrm{d}t} \left(u_\varepsilon(t),\, v\right) + a(u_\varepsilon(t),\, v) - \varepsilon(Au_\varepsilon(t),\, Av) = 0 \qquad \forall v \in D(A) \,. \tag{5.9}$$

In the sense of distributions, we have

$$\frac{\partial u_\varepsilon}{\partial t} - \Delta u_\varepsilon - \varepsilon \Delta^2 u_\varepsilon = 0 ; \tag{5.10}$$

since $u_\varepsilon \in L_2(0, T; V)$, $V = H_0^1(\Omega)$, we have:

$$u_\varepsilon = 0 \text{ on } \Sigma. \tag{5.11}$$

Let us multiply (5.10) by $v \in D(A)$ and integrate over Ω. Using Green's formula and taking account of (5.10), the result is

$$\int_\Gamma \Delta u_\varepsilon(x, t) \frac{\partial v}{\partial n}(x) \, d\Gamma = 0 \qquad \forall v \in D(A),$$

and therefore

$$\Delta u_\varepsilon = 0 \text{ on } \Sigma. \tag{5.12}$$

We recognize the method given in 2.2. ∎

STATE OF PROBLEM

It is now necessary:

i) To see if the QR problem (5.5), (5.6), (5.7) admits a unique solution.
ii) With ζ chosen as in (5.8), and U_ε the solution (cf. Remark 2.4) of

$$\left.\begin{aligned}
&\frac{\partial U_\varepsilon}{\partial t} + A(t)\, U_\varepsilon = 0, \\
&U_\varepsilon \in L_2(0, T; V), \qquad U_\varepsilon' \in L_2(0, T; V'), \\
&U_\varepsilon(0) = u_\varepsilon(0).
\end{aligned}\right\} \tag{5.13}$$

to see if $U_\varepsilon(T) \to \chi$ when $\varepsilon \to 0$.
iii) To choose ε so that $J(u_\varepsilon(0)) \leqslant \eta$.

The questions i), ii), iii) are studied respectively in 5.3, 5.4, 5.5. ∎

5.3 Solution of the Problem (5.5), (5.6), (5.7)

We are first going to demonstrate:

Theorem 5.1: *Let the hypotheses of* 3.1 *hold. In addition, suppose that*

$$D(A(t)) = V_2 = \text{a space independent of } t. \; (^1) \; (^2) \qquad (5.14)$$

Then the problem (5.5), (5.6), (5.7) *possess a unique solution.*

Proof. Let us show that the theorem is a consequence of Theorem 3.1.

1) For $u, v \in V_2$, set:

$$\pi(t \, ; u, v) = (A(t) \, u, v) - \varepsilon(A(t) \, u, A(t) \, v).$$

This is a continuous bilinear form on V_2, and by virtue of Theorem 3.1 we have the result:

$$- \pi(t \, ; v, v) + \lambda \, | \, v \, |^2 \geqslant \alpha_2 \, \| \, v \, \|^2_{V_2}, \quad \forall v \in V_2, \quad \alpha_2 > 0 \; (^3) \qquad (5.15)$$

2) Now

$$- \pi(t \, ; v, v) = \varepsilon \, | \, A(t) \, v \, |^2 - (A(t) \, v, v),$$

$$\geqslant \varepsilon \, | \, A(t) \, v \, |^2 - \frac{\varepsilon}{2} \, | \, A(t) \, v \, |^2 - \frac{1}{2 \, \varepsilon} | \, v \, |^2.$$

from which follows (5.5), naturally with α_2 dependent on ε. ∎

Remark 5.1. The hypothesis in (5.14) is not indispensable to ensure that (5.5), (5.6), (5.7) possess a unique solution.

Let us show first that *existence always* holds, under the hypothesis of 3.1. Replace, as is permissible, Eq. (5.5) by

$$u'_\varepsilon + A(t) \, u_\varepsilon - \varepsilon A^*(t) \, A(t) \, u_\varepsilon - k u_\varepsilon = 0 \qquad (5.16)$$

(changing u_ε into $\exp(- k(t - T)) \, u_\varepsilon$, $k \in \mathbf{R}$). Choose, for example,

$$k > \frac{1}{2 \, \varepsilon}.$$

Set

$$b(v, \varphi) = \int_0^T \left[- (v, \varphi') + (A(t) \, v, \varphi) - \varepsilon(A(t) \, v, A(t) \, \varphi) - k(v(t), \varphi(t)) \right] dt \, ; \qquad (5.17)$$

(^{1})This can occur with $A(t)$ dependent on t.
(^{2})Provide V_2 with the norm of $D(A(0))$ to fix ideas.
(^{3})Replace π by $- \pi$ because we integrate here *in the backwards sense.*

for

$$v \in L_2(0, T; D(A(t))),$$

$$\varphi \in L_2(0, T; D(A(t))), \qquad \varphi' \in L_2(0, T; H), \qquad \varphi(0) = 0. \qquad \textbf{(5.18)}$$

We readily verify that the problem in (5.16), (5.6), (5.7) is equivalent to finding $u_\varepsilon \in L_2(0, T; D(A(t)))$, such that

$$b(u_\varepsilon, \varphi) = (\chi, \varphi(T)) \qquad \forall \varphi \qquad \textbf{(5.19)}$$

which satisfy (5.18). Now

$$b(\varphi, \varphi) = -\frac{1}{2} |\varphi(t)|^2 + \int_0^T ((A(t)\,\varphi, \varphi) - \varepsilon\,|\,A(t)\,\varphi\,|^2 - k\,|\,\varphi(t)\,|^2)\,dt$$

and (as in the demonstration of Theorem 5.1, 2), we deduce, choosing $k > 1/2\,\varepsilon$), that

$$- b(\varphi, \varphi) \geqslant \int_0^T \left(\frac{\varepsilon}{2}\,|\,A(t)\,\varphi\,|^2 + \left(k - \frac{1}{2\,\varepsilon} \right) |\,\varphi(t)\,|^2 \right) dt\,,$$

which shows the *existence* of u_ε (cf. Lions [1], chapter III).

We have established the uniqueness only by means of an additional hypothesis:

Theorem 5.2: *Let the hypotheses of Theorem 3.1 hold. Suppose in addition*([1]) *that for every function v which satisfies*

$$v \in L_2(0, T; D(A(t))), \quad v' \in L_2(0, T; D(A(t))'), \quad v(T) = 0 \qquad \textbf{(5.20)}$$

we have

$$\int_0^T (v', v)\,dt = -\frac{1}{2} |\,v(0)\,|^2\,. \qquad \textbf{(5.21)}$$

Then the problem in (5.5), (5.6), (5.7) *possesses a unique solution.*

Proof. From what has preceded, it remains only to demonstrate uniqueness. Suppose that (5.16) holds, with $k > 1/2\,\varepsilon$ and $u_\varepsilon(T) = 0$. Taking the scalar product of (5.16) with u_ε and *using* (5.21), we deduce:

$$-\frac{1}{2} |\,u_\varepsilon(0)\,|^2 + \int_0^T ((A(t)\,u_\varepsilon, u_\varepsilon) - \varepsilon\,|\,A(t)\,u_\varepsilon\,|^2 - k\,|\,u_\varepsilon(t)\,|^2)\,dt = 0\,,$$

([1])In reality, this *may* not be an *additional* hypothesis.

then

$$\frac{1}{2}|u_\varepsilon(0)|^2 + \int_0^T \left(\frac{\varepsilon}{2}|A(t)\,u_\varepsilon|^2 + \left(k - \frac{1}{2\,\varepsilon}\right)|u_\varepsilon|^2\right)dt = 0\,;$$

and then

$$u_\varepsilon = 0\,. \blacksquare$$

Remark 5.2. As we have already said, we do not know if (5.21) is *always* satisfied under the hypotheses of Theorem 3.1. It can be shown that this hypothesis is satisfied if $D(A(t))$ "depends sufficiently regularly" on t (cf. Lions [1]); it is easy to verify this, in particular, in the case of Theorem 5.1. See also the works of C. Baiocchi [1]. $\blacksquare$

5.4 The Problem of Convergence

Let us now consider problem ii) of 5.2. Let U_ε be the solution of (5.13). Does $U_\varepsilon(T) \to \chi$ when $\varepsilon \to 0$? We can demonstrate:

Theorem 5.3: *Under the hypotheses of Theorem 4.1, we have $U_\varepsilon(T) \to \chi$ when $\varepsilon \to 0$, where U_ε is defined by (5.13).*

Fundamental Remark: It is not true that $U_\varepsilon(t)$ converges when $\varepsilon \to 0$ for $t < T$.

Proof. Let us use the notation of the proof of Theorem 4.1. Let

$$\hat{U}_\varepsilon(\lambda, t) = \mathcal{U}U_\varepsilon(t)$$

$$\hat{u}_\varepsilon(\lambda, t) = \mathcal{U}u_\varepsilon(t)\,.$$

We have

$$\left.\begin{aligned}\frac{\partial \hat{u}_\varepsilon}{\partial t} + \lambda\hat{u}_\varepsilon - \varepsilon\lambda^2\,\hat{u}_\varepsilon = 0\,, \qquad (^1)\\[2mm] \hat{u}_\varepsilon(T) = \chi \end{aligned}\right\}$$

(¹) We can suppose $\lambda \geqslant \alpha_0 > 0$ by a preliminary change of A into $A + kI$, where k is conveniently chosen.

and

$$
\left.
\begin{aligned}
\frac{\partial \hat{U}_\varepsilon}{\partial t} + \lambda \hat{U}_\varepsilon &= 0 \\
\hat{U}_\varepsilon(0) &= \hat{u}_\varepsilon(0) .
\end{aligned}
\right\}
$$

We find easily (compare the calculations in Remark 2.5)

$$
\hat{u}_\varepsilon(\lambda, t) = e^{-(\varepsilon\lambda^2 - \lambda)(T-t)} \hat{\chi}(\lambda)
$$

from this follows

$$
\hat{u}_\varepsilon(\lambda, 0) = e^{-(\varepsilon\lambda^2 - \lambda)T} \hat{\chi}(\lambda) .
$$

Then

$$
\hat{U}_\varepsilon(\lambda, t) = e^{-\varepsilon\lambda^2 T} e^{\lambda(T-t)} \hat{\chi}(\lambda)
$$

and therefore

$$
\hat{U}_\varepsilon(\lambda, T) = e^{-\varepsilon\lambda^2 T} \hat{\chi}(\lambda) . \tag{5.22}
$$

Now, when $\varepsilon \to 0$, $e^{-\varepsilon\lambda^2 T} \hat{\chi}(\lambda) \to \hat{\chi}(\lambda)$ in $\mathfrak{h}$, from whence the result. ∎

Remark 5.3. In the general case, which is to say under the hypotheses of Theorem 3.1, we can only *conjecture* that $U_\varepsilon(T) \to \chi$ in H when $\varepsilon \to 0$. ∎

5.5 The Choice of ε.

Theorem 5.4: ([1]) *Under the hypotheses of Theorem 4.1, we have*

$$
| U_\varepsilon(T) - \chi |^2 = J(u_\varepsilon(0)) \leqslant \eta
$$

if

$$
(1 - e^{-\varepsilon\lambda_0^2 T}) | \chi | \leqslant \eta^{1/2} \tag{5.23}
$$

where

$$
\lambda_0 = \inf \lambda , \quad \lambda \in \text{spectrum of } A .
$$

([1])We suppose $A > 0$, as can be arranged.

Proof. This is an immediate consequence of (5.22). In effect

$$| U_\varepsilon(T) - \chi |^2 = \int_{\alpha_0}^{\infty} (1 - e^{-\varepsilon\lambda_0^2 T})^2 \, | \, \hat{\chi}(\lambda) \, |^2_{\mathfrak{H}(\lambda)} \, d\mu(\lambda)$$

$$\leqslant (1 - e^{-\varepsilon\alpha_0^2 T})^2 \, | \, \chi \, |^2$$

from which the result follows, taking account of the fact that we can take $\alpha_0 = \lambda_0$.

In the general case, *we adapt the following practical rule.* Let

$$\left. \begin{array}{l} \lambda_0(t) = \inf \text{Re } \lambda \,, \, \lambda \in \text{spectrum of } A(t), \quad \text{Re } \lambda = \text{real part of } \lambda \\ \lambda_0 = \inf \, \lambda_0(t) \,, \qquad t \in (0, T) \,. \end{array} \right\} \quad (5.25)$$

We choose $\varepsilon(^1)$ *so that*

$$(1 - e^{-\varepsilon\lambda_0^2 T}) | \, \chi \, | \leqslant \eta^{1/2} \,. \tag{5.26}$$

Remark 5.4. In fact, it is necessary to take ε such that

$$(1 - e^{-\varepsilon\lambda_0^2 T}) | \, \chi \, | \leqslant \theta\eta^{1/2} \,, \qquad \theta < 1 \,,$$

in order to be able to take account of the error introduced in the numerical resolution of the QR problem.

Remark 5.5. (Compare with 2.6.) There is no *uniqueness* for the QR-operator which we associate with $\frac{\partial}{\partial t} + A(t)$. *For example*, we can consider in place of (5.5) the equation

$$u'_\varepsilon + A(t) u_\varepsilon + (-1)^{m-1} \varepsilon (A^*(t) A(t))^m u_\varepsilon(t) = 0 \,. \tag{5.27}$$

From the practical point of view, the following remark is most important:

Remark 5.6. All that has been said holds if (5.5) is replaced by

$$u'_\varepsilon + A(t) u_\varepsilon - \varepsilon B^*(t) B(t) u_\varepsilon = 0 \,, \tag{5.28}$$

(^{1})A heuristic rule.

where the operators $B(t)$ are a family of operators equivalent to $A(t)$ in the sense:

$$D(B(t)) \quad (^1) \; = D(A(t)), \\ |B(t)\,v| \geqslant c\,|A(t)\,v| \qquad \forall v \in D(A(t)).$$ \qquad (5.29)

It would even be sufficient to have $D(B(t)) \subset D(A(t))$, with the inequality (5.29). For example, if $D(A(t))$ does not depend on t, we can take $B(t) = B$, *independent of* t.

In the general case, $B(t)$ *should be chosen in the "simplest possible way" consistent with* (5.29). ∎

Remark 5.7. Taking account of (5.29), (5.23) is replaced by

$$(1 - e^{-\varepsilon c^2 \lambda_0^2 T}) \,|\,\chi\,| \leqslant \eta^{1/2} \,. \quad ∎ \qquad (5.30)$$

Remark 5.8. The entire theory carries over to the case where (in place of 4.1), we have

$$u' + A(t)\,u = f\,. \quad f \text{ a } given \text{ function in } L_2(0,\,T;\,H)\,(^2) \qquad (5.30)$$

the conditions (4.2) and (4.3) are unchanged, that is to say

$$u(0) = \xi\,, \qquad (5.31)$$

$$u \in L_2(0,\,T;\,V)\,, \qquad u' \in L_2(0,\,T;\,V')\,. \qquad (5.32)$$

We still consider

$$J(\xi) = |\,u(T,\,\xi) - \chi\,|^2\,.$$

Naturally, if u_0 denotes the solution of (5.30), . . . , (5.32) with $f = 0$, and if ψ denotes the solution of

$$\psi' + A(t)\,\psi = f\,, \qquad (5.33)$$

$$\psi(0) = 0\,, \qquad (5.34)$$

we have

$$u = u_0 + \psi$$

$(^1)$Domain of $B(t)$.
$(^2)$Or even in $L_2(0,\,T;\,V')$.

and

$$J(\xi) = |\, u_0(T; \xi) - (\chi - \psi(T))\,|^2 \,.$$

We are thus led back to the preceding system with χ replaced by $\chi - \psi(T)$.

From the practical point of view, for the QR-method we will introduce u_ε as the solution of

$$u'_\varepsilon + A(t)\, u_\varepsilon - \varepsilon A^*(t)\, A(t)\, u_\varepsilon = f \quad \text{(in place of 0)} \tag{5.35}$$

with all other considerations unchanged. ∎

Remark 5.9. For a given η, (5.26) gives a bound of the type $\varepsilon \leqslant \bar{\varepsilon}$.

In order to modify the equations as little as possible, it is natural to wish to take ε as small as possible. But, by the nature of the problems under consideration, we can expect numerical instability as $\varepsilon \to 0$. Therefore, we can expect that *for each problem there is an optimal value ε_0 of ε. This remark is valid for the greater part of the problems considered in this book.* We are content to show this in several cases (cf., for example, Remark 11.8).

6. EXAMPLES

6.1 Example 1

Let $a_{ij}(x, t)$ be a family of functions defined in $\Omega \times (0, T)$, such that

$$a_{ij} \text{ is measurable and bounded in } \Omega \times (0, T)\, (^1) \tag{6.1}$$

and

$$\left. \begin{array}{c} \displaystyle\sum_{i,j=1}^{n} a_{ij}(x, t)\, \xi_i\, \xi_j \geqslant \alpha(\xi_1^2 + \cdots + \xi_n^2)\,, \qquad \alpha > 0\,, \\[2ex] \text{for almost all} \quad x, t\,. \end{array} \right\} \tag{6.2}$$

We take $V = H_0^1(\Omega)$ (cf. 3.4).
For $u, v \in H_0^1(\Omega)$, we set

$(^1)$ We write: $a_{ij} \in L_\infty(\Omega \times (0, T))$.

$$a(t ; u, v) = \sum_{i,j=1}^{n} \int_{\Omega} a_{ij}(x, t) \frac{\partial u}{\partial x_j} \frac{\partial v}{\partial x_i} \, dx . \tag{6.3}$$

Then, the operator $A(t)$ is given by [cf. (3.2)],

$$A(t) u = - \sum_{i,j=1}^{n} \frac{\partial}{\partial x_i} \left(a_{ij}(x, t) \frac{\partial u}{\partial x_j} \right). \tag{6.4}$$

The problem (4.1), (4.2), (4.3) becomes:

$$\left.\begin{aligned}
&\frac{\partial u}{\partial t} + A(t) u = 0 , \\[2mm]
&u(x, 0) = \xi(x) , \qquad \xi \quad \text{given in} \quad L_2(\Omega) , \\[2mm]
&u|_{\Sigma} = 0 .
\end{aligned}\right\} \tag{6.5}$$

For the QR-method. we can take $B = - \Delta$ (cf. Remark 5.6). Then, to (6.5) we associate the QR-problem

$$\left.\begin{aligned}
&\frac{\partial u_\varepsilon}{\partial t} + A(t) u_\varepsilon - \varepsilon \Delta^2 u_\varepsilon = 0 , \\[2mm]
&u_\varepsilon(x, T) = \chi(x) , \\[2mm]
&u_\varepsilon|_{\Sigma} = 0 , \qquad \Delta u_\varepsilon|_{\Sigma} = 0 \qquad (^1) . \quad \blacksquare
\end{aligned}\right\} \tag{6.6}$$

6.2 Example 2. Problem of Neumann

Notation of 6.1; *we now take* $V = H^1(\Omega)$; same choice of $a(t ; u, v)$. The problem (4.1), (4.2), (4.3) becomes:

$$\left.\begin{aligned}
&\frac{\partial u}{\partial t} + A(t) u = 0 , \\[2mm]
&u(x, 0) = \xi(x) , \qquad x \in \Omega , \\[2mm]
&\frac{\partial u}{\partial v_A}(x, t) = \sum_{i,j=1}^{n} a_{ij}(x, t) \frac{\partial u}{\partial x_j} \cos(v, x_i) = 0 \quad \text{on} \quad \Sigma \qquad (^2) .
\end{aligned}\right\} \tag{6.7}$$

$(^1)$The condition $\Delta u_\varepsilon|_{\Sigma} = 0$ is not satisfied by u. We will therefore lose precision (in the neighborhood of Σ in particular), with the choice

$$\left.\begin{aligned}
&\frac{\partial u_\varepsilon}{\partial t} + A(t) u_\varepsilon - \varepsilon A^*(t) A(t) u_\varepsilon = 0 , \\[2mm]
&u_\varepsilon(x, T) = \chi(x) , \\[2mm]
&u_\varepsilon|_{\Sigma} = 0 , \qquad A(t) u_\varepsilon|_{\Sigma} = 0 .
\end{aligned}\right\}$$

$(^2) v = $ normal to Γ exterior to Ω.

In this case, $D(A(t))$ depends—in general—on t. As the QR-problem, we take

$$\left.\begin{array}{l} \dfrac{\partial u_\varepsilon}{\partial t} + A(t)\,u_\varepsilon - \varepsilon A^*(t)\,A(t)\,u_\varepsilon = 0\,, \\[2mm] u_\varepsilon(x, T) = \chi(x)\,, \\[2mm] \dfrac{\partial u_\varepsilon}{\partial \nu_A} = 0 \ \text{ on } \ \Sigma\,, \qquad \dfrac{\partial}{\partial \nu_A} A(t)\,u_\varepsilon = 0 \ \text{ on } \ \Sigma\,. \ \blacksquare \end{array}\right\} \qquad (6.8)$$

6.3 Example 3. Higher Order

The examples so far given correspond to operators of the form, $\dfrac{\partial}{\partial t} + A(t)$ *parabolic of the second order.*

Analogous results hold for parabolic operators *of order superior to 2.*[1] Let us restrict ourselves to an example. Consider u defined by

$$\left.\begin{array}{l} \dfrac{\partial u}{\partial t} + \Delta^2 u = 0\,, \qquad x \in \Omega\,, \qquad t > 0\,, \\[2mm] u(x, 0) = \zeta(x)\,, \qquad x \in \Omega\,, \\[2mm] u\,|_\Sigma = 0\,, \qquad \dfrac{\partial u}{\partial \nu}\Big|_\Sigma = 0\,. \end{array}\right\} \qquad (6.9)$$

This problem falls into the category of Section 3 in the following fashion. Recall that

$$H^2(\Omega) = \left\{ v \mid v, \dfrac{\partial v}{\partial x_i}, \dfrac{\partial^2 v}{\partial x_i\,\partial x_j} \in L_2(\Omega),\, \forall i, j \right\}, \qquad (6.10)$$

is a Hilbert space with the norm

$$\left(|v|^2 + \sum_{i=1}^{n} \left| \dfrac{\partial v}{\partial x_i} \right|^2 + \sum_{i,j=1}^{n} \left| \dfrac{\partial^2 v}{\partial x_i\,\partial x_j} \right|^2 \right)^{1/2} \qquad (^2)\,.$$

Let $H_0^2(\Omega) = $ adherence of $D(\Omega)$ in $H^2(\Omega)$.

---

[1] The case of certain nonparabolic operators is studied in Chapter 2.

[2] $|f|^2 = \displaystyle\int_\Omega f(x)^2 \, dx\,.$

Following the notation of Section 3, take:

$$V = H_0^2(\Omega),$$

$$a(t; u, v) = a(u, v) = (\Delta u, \Delta v) \qquad \left(= \int_\Omega (\Delta u)(\Delta v) \, dx \right).$$

With these choices, the general problem (4.1), (4.2), (4.3) corresponds to (6.9); note that for:

$$H_0^2(\Omega) = \left\{ v \mid v \in H^2(\Omega), \ v\big|_\Gamma = 0, \ \frac{\partial v}{\partial v}\bigg|_\Gamma = 0 \right\}.$$

The QR-method then gives:

$$\left.\begin{aligned}
&\frac{\partial u_\varepsilon}{\partial t} + \Delta^2 u_\varepsilon - \varepsilon \Delta^4 u_\varepsilon = 0, \\
&u_\varepsilon(x, T) = \chi(x), \\
&u_\varepsilon\big|_\Sigma = 0, \qquad \frac{\partial u_\varepsilon}{\partial v}\bigg|_\Sigma = 0, \\
&\Delta^2 u_\varepsilon\big|_\Sigma = 0, \qquad \frac{\partial}{\partial v} \Delta^2 u_\varepsilon\big|_\Sigma = 0. \ \blacksquare
\end{aligned}\right\} \qquad (6.11)$$

Remark 6.1. The examples can be varied in an infinite number of ways. The same types of methods can be applied to all the examples treated by Lions [1], Chapters IV, V, VI, VII,[1] and in Lions–Magenes [1].

Let us note in particular that all that has been said holds if A is an elliptic *system* as, for example, *the system of elasticity.* $\blacksquare$

7. NUMERICAL APPLICATIONS FOR ONE-DIMENSIONAL EXAMPLES

7.1 Formulation of the Problem

7.1.1 THE INITIAL PROBLEM

We are looking for the function $\xi(x)$ defined in $[0, 1]$ such that, with $\chi(x)$ and η given, if $u(x, t)$ verifies:

---

[1] Add to Chapter VII the work of T. Kato–H. Tanabe [1], J. L. Lions [3]. The QR-method applied to those problems is probably correct, but still largely heuristic. See also the works of Baiocchi [1].

$$\frac{\partial u}{\partial t} - \frac{\partial}{\partial x}\left[a(x)\frac{\partial u}{\partial x}\right] = 0, \tag{7.1}$$

$$u(0, t) = u(1, t) = 0, \tag{7.2}$$

$$u(x, 0) = \xi(x), \tag{7.3}$$

(I)′

then we have

$$\left[\int_0^1 (u(x, T) - \chi(x))^2 \, dx\right]^{1/2} \leqslant \eta, \tag{7.4}$$

where η is a "small" number, given a priori.

7. 1. 2 THE APPROXIMATE PROBLEM

At the first stage, then, we resolve the problem[1]

$$\frac{\partial U}{\partial t} - \frac{\partial}{\partial x}\left[a(x)\frac{\partial U}{\partial x}\right] - \varepsilon\frac{\partial^4 U}{\partial x^4} = 0, \tag{7.5}$$

$$U(0, t) = U(1, t) = 0,$$
$$\frac{\partial}{\partial x}\left[a(x)\frac{\partial U}{\partial x}\right](0, t) = \frac{\partial}{\partial x}\left[a(x)\frac{\partial U}{\partial x}\right](1, t) = 0, \tag{7.6}$$

$$U(x, T) = \chi(x), \tag{7.7}$$

(II)

and we deduce from this:

$$U(x, 0) = \xi(x).$$

At the second stage, at the end of the verifications, we integrate system (I) with $\xi(x)$, the solution of (II), as initial value and we see whether the constraint of (7.4) is satisfied.

For convenience, we perform a change of variable in (II) from t to $T - t$, so that we integrate according to *increasing* t.

[1] We have set $u_\varepsilon = U$.

We are therefore looking for $v(x, T)$ such that:

$$v(x, t) = U(x, T - t)$$

satisfies

$$\frac{\partial v}{\partial t} + \frac{\partial}{\partial x}\left[a(x)\frac{\partial v}{\partial x}\right] + \varepsilon \frac{\partial^4 v}{\partial x^4} = 0, \tag{7.8}$$

$$v(0, t) = v(1, t) = 0,$$

$$\frac{\partial}{\partial x}\left[a(x)\frac{\partial v}{\partial x}\right](0, t) = \frac{\partial}{\partial x}\left[a(x)\frac{\partial v}{\partial x}\right](1, t) = 0, \tag{7.9}$$

$$v(x, 0) = \chi(x). \tag{7.10}$$

(III)

Remark 7.1. In reality, in order to have greater flexibility in the utilization of the program, we considered:

$$v(0, t) = f_0(t),$$
$$v(1, t) = f_1(t). \tag{7.11}$$

Remark 7.2. For analogous reasons, we took:

$$\frac{\partial}{\partial x}\left[a(x)\frac{\partial v}{\partial x}\right](0, t) = d_1,$$

$$\frac{\partial}{\partial x}\left[a(x)\frac{\partial v}{\partial x}\right](1, t) = d_2, \tag{7.12}$$

where d_1 and d_2 are arbitrary constants.

7.1.3 CHOICE OF ε AND DISCRETIZATION ERROR IN THE RESOLUTION OF THE SYSTEM

We should have

$$[1 - e^{-\varepsilon c^2 \lambda_0^2 T}]|\chi| \leqslant \eta, \tag{7.13}$$

$(|\chi| = \text{norm of } \chi \text{ in } L_2(0, 1));$

where c is a constant, which for (II) is equal to:

$$c = \inf_u \left(\frac{|u''|}{|a(x)\,u'' + u'|} \right) \tag{7.14}$$

and λ_0 is given by:

$$\lambda_0 = \inf_u \frac{\displaystyle\int_0^1 au'^2\,\mathrm{d}x}{\displaystyle\int_0^1 u^2\,\mathrm{d}x}. \tag{7.15}$$

If therefore we find $\bar{\lambda}_0$ and $\bar{c}$ such that

$$\lambda_0 \leqslant \bar{\lambda}_0 \qquad c \leqslant \bar{c}, \tag{7.16}$$

then we can take

$$\varepsilon = \varepsilon_* = \frac{-\log\left(1 - \eta/\|\chi\|\right)}{\bar{c}^2\,\bar{\lambda}_0^2\,T}. \tag{7.17}$$

Example.

Let $x \in [0, 1]$ and $a(x) = 1 + x$. Then

$$\lambda_0 \leqslant 2 \inf \frac{\displaystyle\int_0^1 u'^2\,\mathrm{d}x}{\displaystyle\int_0^1 u^2\,\mathrm{d}x} \leqslant 2\,\pi^2, \tag{7.18}$$

since π^2 is the smallest characteristic value of $(-\,\mathrm{d}^2/\mathrm{d}x^2)$ on $(0, 1)$ with the boundary conditions of (7.2). This bound can easily be improved, by considering in (7.15), for example, the function

$$u(x) = x(1 - x)\,;$$

we find then:

$$\lambda_0 \leqslant 15. \tag{7.19}$$

Calculation of $\bar{c}$ for $a(x) = 1 + x$.

It is a question of obtaining an inequality of the type:

$$| u'' | \geqslant c | (1 + x) u'' + u' | . \tag{7.20}$$

Let us first verify that an inequality of this nature *exists*. We have:

$$| (1 + x) u'' + u' | \leqslant 2 | u'' | + | u' | ;$$

then, taking account of $u(0) = u(1) = 0$:

$$(- u'', u) = \int_0^1 (- u'') u \, dx = | u' |^2 \leqslant | u | | u'' |$$

and, from (7.18), $\dfrac{| u' |^2}{| u |^2} \geqslant \pi^2$, therefore:

$$| u' |^2 \leqslant \frac{1}{\pi} | u' | | u'' | ;$$

thus,

$$| u' | \leqslant \frac{1}{\pi} | u'' |$$

and,

$$| (1 + x) u'' + u' | \leqslant \left(2 + \frac{1}{\pi} \right) | u'' | .$$

Hence

$$| u'' | \geqslant \frac{\pi}{2 \pi + 1} | (1 + x) u'' + u' | . \tag{7.21}$$

Consequently, an inequality of type (7.20) exists and

$$c \geqslant \frac{\pi}{2 \pi + 1} . \tag{7.22}$$

It remains to obtain an *upper* bound for c.

If we define the best constant c by

$$c = \inf_u \frac{| u'' |}{| (1 + x) u'' + u' |} ,$$

then

$$c \leqslant \frac{|\varphi''|}{|(1+x)\varphi'' + \varphi'|}$$

for a particular φ. Let us take $\varphi = x(1 - x)$. Then $\varphi'' = -2$ and $|\varphi''| = 2; |(1 + x)\varphi'' + \varphi'| = \sqrt{31/3}$. Therefore:

$$c \leqslant \frac{2}{\sqrt{31/3}}.$$

We can then take [a reasonable choice in view of (7.22)]:

$$\bar{c} = \frac{2\sqrt{3}}{\sqrt{31}}. \tag{7.23}$$

Naturally, this estimate can be considerably improved, but that is of no matter here.

Discretization Error of the System

Let $U(x, t)$ be the exact solution and $\tilde{U}(x, t)$ the approximate solution (calculated) of the system (II); let us suppose the discretization step has been chosen so that

$$|\tilde{U}(x, 0) - U(x, 0)| \leqslant \eta_1.$$

Let $u(x, t)$ be the exact solution of the system (I) with $\xi(x) = U(x, 0)$ as the initial condition and $\tilde{u}(x, t)$ the solution of the same system with $\tilde{\xi}(x) = \tilde{U}(x, 0)$ as the initial condition.

Then $w = u - \tilde{u}$ satisfies:

$$\left.\begin{aligned}
&\frac{\partial w}{\partial t} - \frac{\partial}{\partial x}\left(a(x)\frac{\partial w}{\partial x}\right) = 0, \\
&w(0, t) = w(1, t) = 0, \\
&w(x, 0) = \xi(x) - \tilde{\xi}(x).
\end{aligned}\right\} \tag{7.24}$$

From this we deduce:

$$\frac{\mathrm{d}}{\mathrm{d}t}|w(t)|^2 + 2\int_0^1 a(x)\left[\frac{\partial w}{\partial x}\right]^2 \mathrm{d}x = 0;$$

now:

$$\int_0^1 a(x) \left[\frac{\partial w}{\partial x} \right]^2 dx \geq \lambda_0 \, | \, w(t) \, |^2 \, ,$$

whence

$$\frac{d}{dt} \, | \, w(t) \, |^2 + 2 \, \lambda_0 \, | \, w(t) \, |^2 \leq 0 \, ,$$

and therefore

$$| \, w(t) \, |^2 \leq | \, \xi - \tilde{\xi} \, |^2 \, e^{-2\lambda_0 t} \, ,$$

from which finally:

$$| \, u(T) - \tilde{u}(T) \, | \leq | \, \xi - \tilde{\xi} \, | \, e^{-\lambda_0 T} \, .$$

If therefore we choose:

$$\xi(x) = \tilde{U}(x, 0) \, ,$$

then:

$$| \, \tilde{u}(T) - \chi \, | \leq | \, \tilde{u}(T) - u(T) \, | + | \, u(T) - \chi \, |$$
$$\leq \eta_1 \, e^{-\lambda_0 T} + | \, u(T) - \chi \, | \, .$$

We want:

$$\eta_1 \, e^{-\lambda_0 T} + | \, u(T) - \chi \, | \leq \eta \, . \qquad (7.25)$$

It is then natural to choose ε (cf. Remark 5.4) so that:

$$| \, u(T) - \chi \, | \leq \theta \eta \, , \qquad 0 < \theta < 1 \, ,$$

then the discretization step such that

$$\eta_1 \, e^{-\lambda_0 T} \leq (1 - \theta) \, \eta \, .$$

This last choice requires the determination of λ_0 and a suitable choice of Δx and Δt since η is a function of Δx, Δt *and of T*.

7.2 Difference Equations

7. 2. 1 CALCULATION OF DERIVATIVES

The numerical applications have been made using a symmetric discretization according to a Crank–Nicolson type scheme.

Remark 7.3. In numerical applications, we have never devoted any effort to the improvement of results by a precise investigation of methods for the numerical solution of the QR system. We show in fact that *once the QR method has been chosen, we are led to the usual problems.* ∎

Let

$$v_j^n = v(j\,\Delta x, n\,\Delta t)\,, \qquad\qquad 0 \le j \le J\,,$$
$$0 \le n \le N.$$

For the *x*-derivatives, take:

$$\frac{\partial v_j^n}{\partial x} \cong \frac{1}{\Delta x}\,(v_{j+1/2}^n - v_{j-1/2}^n)\,,$$

$$\frac{\partial^2 v_j^n}{\partial x^2} \cong \frac{1}{\Delta x^2}\,(v_{j+1}^n - 2\,v_j^n + v_{j-1}^n)\,,$$

$$\frac{\partial^3 v_j^n}{\partial x^3} \cong \frac{1}{\Delta x^3}\,(v_{j+3/2}^n - 3\,v_{j+1/2}^n + 3\,v_{j-1/2}^n - v_{j-3/2}^n)\,,$$

$$\frac{\partial^4 v_j^n}{\partial x^4} \cong \frac{1}{\Delta x^4}\,(v_{j+2}^n - 4\,v_{j+1}^n + 6\,v_j^n - 4\,v_{j-1}^n + v_{j-2}^n)\,.$$

Then

$$\frac{\partial}{\partial x}\left[a(x)\,\frac{\partial v}{\partial x}\right] \cong \frac{1}{\Delta x^2}\left[a_{j+1/2}[v_{j+1}^n - v_j^n] - a_{j-1/2}[v_j^n - v_{j-1}^n]\right].$$

For the *t*-derivatives, take:

$$\frac{\partial v_j^n}{\partial t} \cong \frac{1}{\Delta t}\,[v_j^{n+1} - v_j^n]\,.$$

7.2.2 MATRIX REPRESENTATION OF THE "BACKWARDS" SYSTEM

The system (III) [equations (7.8) to (7.10)] then becomes, with the aid of the foregoing formulas (after multiplying all terms by $2\,\Delta x^2$)

$$v_j^{n+1}\left[\frac{2\,\Delta x^2}{\Delta t} - a_{j+1/2} - a_{j-1/2} + \frac{6\,\varepsilon}{\Delta x^2}\right] + v_{j+1}^{n+1}\left[a_{j+1/2} - \frac{4\,\varepsilon}{\Delta x^2}\right] +$$

$$+ v_{j-1}^{n+1}\left[a_{j-1/2} - \frac{4\,\varepsilon}{\Delta x^2}\right] + \frac{\varepsilon}{\Delta x^2}\left[v_{j+2}^{n+1} + v_{j-2}^{n+1}\right]$$

$$= v_j^n\left[\frac{2\,\Delta x^2}{\Delta t} + a_{j+1/2} + a_{j-1/2} - \frac{6\,\varepsilon}{\Delta x^2}\right] - v_{j+1}^n\left[a_{j+1/2} - \frac{4\,\varepsilon}{\Delta x^2}\right]$$

$$- v_{j-1}^n\left[a_{j-1/2} - \frac{4\,\varepsilon}{\Delta x^2}\right] - \frac{\varepsilon}{\Delta x^2}\left[v_{j+2}^n + v_{j-2}^n\right]$$

for

$$1 \leqslant j \leqslant J - 1 \qquad J\,\Delta x = 1\,, \tag{7.26}$$

$$v_0^n = v_J^n = 0 \qquad \forall n \geqslant 0\,, \tag{7.27} \quad \text{(IV)}$$

$$v_j^0 = \chi(j.\Delta x)\,, \tag{7.28}$$

with the boundary conditions:

$$\frac{\partial}{\partial x}\left(a\,\frac{\partial v}{\partial x}\right) = \begin{cases} d_1 \text{ at } x = 0\,, \\ d_2 \text{ at } x = 1\,; \end{cases} \tag{7.29}$$

in general $d_1 = d_2 = 0$ which gives

at $x = 0$: $(j = 0)$,

$$\tfrac{1}{2}(a_{1/2} - a_{-1/2})(v_1 - v_{-1}) + a_0(v_1 + v_{-1} - 2\,v_0) = d_1\,\Delta x^2$$

at $x = 1$: $(j = J)$,

$$\tfrac{1}{2}(a_{J+1/2} - a_{J-1/2})(v_{J+1} - v_{J-1}) + a_J(v_{J+1} + v_{J-1} - 2\,v_J) = d_2\,\Delta x^2\,,$$

whence

$$v_{-1} = \frac{1}{a_0 + \dfrac{a_{-1/2} - a_{1/2}}{2}}\left[d_1\,\frac{\Delta x^2}{\Delta t} + 2\,a_0\,v_0 + v_1\left(\frac{a_{-1/2} - a_{1/2}}{2} - a_0\right)\right] \tag{7.30}$$

$$v_{J+1} = \frac{1}{a_J + \dfrac{a_{J+1/2} - a_{J-1/2}}{2}}\left[d_2\,\frac{\Delta x^2}{\Delta t} + 2\,a_J v_J + v_{J-1}\left(\frac{a_{J+1/2} - a_{J-1/2}}{2} + a_J\right) \tag{7.31}$$

Remark 7.4. Outside of the interval [0, 1], the function $a(x)$ may or may not keep the same analytic form as in the interior; $a(x)$ can always be continued by a different form. ∎

Expression of (7.26) in particular cases

It is worthwhile to write the particular equations corresponding to $j = 1, 2, J - 2, J - 1$, in which v_0 and v_J enter. In (IV) in the cases where $i = 2$ and $J = 2$, v_{-1} and v_{J+1} do not enter. We have:

$$v_2^{n+1} \left[\frac{2\,\Delta x^2}{\Delta t} - (a_{5/2} + a_{3/2}) + \frac{6\,\varepsilon}{\Delta x^2} \right] + v_1^{n+1} \left[a_{3/2} - \frac{4\,\varepsilon}{\Delta x^2} \right] +$$

$$+ v_3^{n+1} \left[a_{5/2} - \frac{4\,\varepsilon}{\Delta x^2} \right] + [v_4^{n+1} + v_0^{n+1}] \frac{2}{\Delta x^2} =$$

$$= v_2^{n} \left[\frac{2\,\Delta x^2}{\Delta t} + (a_{5/2} + a_{3/2}) - \frac{6\,\varepsilon}{\Delta x^2} \right] - v_3^{n} \left[a_{5/2} - \frac{4\,\varepsilon}{\Delta x^2} \right]$$

$$- v_1^{n} \left[a_{3/2} - \frac{4\,\varepsilon}{\Delta x^2} \right] - [v_4^{n} + v_0^{n}] \frac{\varepsilon}{\Delta x^2} .$$

with the analogous result for $j = J - 2$.

In the case $j = 1$ and $J - 1$, with the aid of (7.30) and (7.31), we write (IV) in the form, for $j = 1$:

$$v_1^{n+1} \left[\frac{2\,\Delta x^2}{\Delta t} - a_{3/2} - a_{1/2} + \frac{6\,\varepsilon}{\Delta x^2} - \frac{\varepsilon}{\Delta x^2} \left[\frac{\dfrac{a_{-1/2} - a_{1/2}}{2} - a_0}{\dfrac{a_{-1/2} - a_{1/2}}{2} + a_0} \right] \right] +$$

$$+ v_2^{n+1} \left[a_{3/2} - \frac{4\,\varepsilon}{\Delta x^2} \right] + v_3^{n+1} \frac{\varepsilon}{\Delta x^2} =$$

$$= v_1^{n} \left[\frac{2\,\Delta x^2}{\Delta t} + a_{3/2} + a_{1/2} - \frac{6\,\varepsilon}{\Delta x^2} - \frac{\varepsilon}{\Delta x^2} \left[\frac{\dfrac{a_{-1/2} - a_{1/2}}{2} - a_0}{\dfrac{a_{-1/2} - a_{1/2}}{2} + a_0} \right] \right] -$$

$$- v_2^{n} \left[a_{3/2} - \frac{4\,\varepsilon}{\Delta x^2} \right] - v_3 \frac{\varepsilon}{\Delta x^2} -$$

$$- (v_0^{n+1} + v_0^{n}) \left[a_{1/2} - \frac{4\,\varepsilon}{\Delta x^2} - \frac{2\,a_0\varepsilon/\Delta x^2}{a_0 + \dfrac{a_{-1/2} - a_{1/2}}{2}} \right] - \frac{2\,\varepsilon\,d_1}{a_0 + \dfrac{a_{-1/2} - a_{1/2}}{2}}$$

There are analogous formulas for $j = J - 1$.

Finally we can write v_0 and v_J assumed known:

$$Av^{n+1} = Bv^n + C \tag{7.32}$$

where A and B are two symmetric pentadiagonal matrices and C is a vector with the components C_j, zero except for $j = 1, 2, J - 1$, $J - 2$.

Remark 7.5. It is easy to verify that A and B are symmetric and further that whatever column j is chosen, the matrix B can be written as a function of the jth column of the matrix A. ∎

From the system $Av^{n+1} = Bv^n + C$ considered, we find $\xi(x) = v(x, T)$. Then we pass on to the second stage, *stage of verification*.

7. 2. 3 MATRIX REPRESENTATION OF THE DIRECT PROBLEM

Taking for our discretization, the same scheme of Crank-Nicolson type, we write the equations of system (I) [equations (7.1) to (7.3)]:

$$
\left.
\begin{aligned}
&\frac{1}{\Delta t}\left[u_j^{n+1} - u_j^n\right] - \\
&\qquad - \frac{1}{2\,\Delta x^2}\left[a_{j+1/2}(u_{j+1}^{n+1} + u_{j+1}^n - u_j^{n+1} - u_j^n)\right. \\
&\qquad \left. - a_{j-1/2}(u_j^{n+1} + u_j^n - u_{j-1}^{n+1} - u_{j-1}^n)\right] = 0, \\
&\qquad\qquad u_0^n = u_J^n = 0 \qquad \forall n,
\end{aligned}
\right\}
\begin{aligned}
(7.33) \\
\\
\\
(7.34)
\end{aligned}
\quad \text{(I)}
$$

$$u_j^0 = \xi(j\,\Delta x) = v_j \text{ solution in } x_j \text{ found in first stage} \quad (7.35)$$

We find u_j^{n+1} at time T by resolving successively for $t_n = n.\Delta t$ the system, starting with $u_j^0 = v_j$, found previously:

$$
u_j^{n+1}\left[\frac{1}{\Delta t} + \frac{a_{j+1/2} + a_{j-1/2}}{2\,\Delta x^2}\right] - u_{j+1}^{n+1}\frac{a_{j+1/2}}{2\,\Delta x^2} - u_{j-1}^{n+1}\frac{a_{j-1/2}}{2\,\Delta x^2} =
$$

$$
= u_j^n\left[\frac{1}{\Delta t} - \frac{a_{j+1/2} + a_{j-1/2}}{2\,\Delta x^2}\right] + u_{j+1}^n\frac{a_{j+1/2}}{2\,\Delta x^2} + u_{j-1}^n\frac{a_{j-1/2}}{2\,\Delta x^2}
$$

for $j = 1$, we have

$$u_1^{n+1} \left[\frac{1}{\Delta t} + \frac{a_{3/2} + a_{1/2}}{2 \, \Delta x^2} \right] - u_2^{n+1} \frac{a_{3/2}}{2 \, \Delta x^2} =$$

$$= u_1^n \left[\frac{1}{\Delta t} - \frac{a_{3/2} + a_{1/2}}{2 \, \Delta x^2} \right] + u_2^n \frac{a_{3/2}}{2 \, \Delta x^2} + [u_0^{n+1} + u_0^n] \cdot \frac{a_{1/2}}{2 \, \Delta x^2} .$$

Remark 7.6. In the calculations made at the end of the verification, we do not necessarily *follow the same grid* as for the retrograde system. ∎

7.2.4 CALCULATION OF THE MEAN SQUARE ERROR

Starting with $\chi(x)$ the *approximate solution* of the system (III) [equations (7.5) to (7.7)] furnishes $\xi(x) = \tilde{U}(x, 0)$. The exact solution of the system (I) with $\xi(x)$ for initial condition is a function $\tilde{u}(x, T)$ and we want:

$$| \tilde{u}(x, T) - \chi(x) | < \eta .$$

Now the verification, starting with the solution $\tilde{\xi}(x)$ of the system (II) can only be obtained by integrating system (I) numerically. Let η_2 be the quadratic error in this procedure. We can then verify, denoting by $\chi^*(x)$ the numerical approximation to $\tilde{u}(x, T)$, that

$$I = | \chi^*(x) - \chi(x) | < \eta + \eta_2 . \tag{7.36}$$

For the first term, we calculate the integral which appears by the trapazoidal method, obtaining

$$I = \sqrt{\frac{\Delta x}{2}} \left[(\chi_0^* - \chi_0)^2 + (\chi_J^* - \chi_J)^2 + 2 \sum_{j=1}^{J-1} (\chi_j^* - \chi_j)^2 \right]^{1/2} .$$

8. NUMERICAL RESULTS

8.1 Generalities

The system (III) is solved for various forms of the condition (7.10):

$$v(x, 0) = \chi(x) .$$

Two series of calculations have been carried out taking respectively

$$a(x) = \text{constant} = k$$

and

$$a(x) = 1 + x .$$

In the first case:

$$\bar{\lambda}_0 = k\pi^2 , \qquad C_0 = 1/k .$$

In the second case, we take the values (7.19) and (7.23).

8.2 Case Where $a(x) = 1$.

We have to solve

$$\frac{\partial v}{\partial t} + \frac{\partial^2 v}{\partial x^2} + \varepsilon \frac{\partial^4 v}{\partial x} = 0 , \qquad 0 \leqslant x \leqslant 1 ,$$

where

$$\left. \begin{aligned} v(0, t) &= v(1, t) = 0 , \\ \frac{\partial^2 v}{\partial x^2}(0, t) &= \frac{\partial^2 v}{\partial x^2}(1, t) = 0 , \\ v(x, 0) &= \chi(x) . \end{aligned} \right\}$$

8. 2. 1 USE OF FOURIER SERIES

If $\chi(x) = \sum_1^\infty A_m \sin m\pi x$ (a series convergent in $L_2(0, 1)$), the exact solution is given by

$$v(x, T) = \sum_1^\infty A_m \, e^{m^2\pi^2(1 - \varepsilon m^2\pi^2)T} \sin m\pi x.$$

In this case, the quantity to make less than $\theta\eta$ is

$$\sum_1^\infty A_m \, e^{-\varepsilon m^4\pi^4 T} \sin m\pi x .$$

We verify in particular that in the case $\chi(x) = \sin \pi x$, the numerical results are very good for values of T not exceeding $2/10$. For values of T larger than $2/10$, we meet numerical difficulties connected with stability (cf. Section 9).

For $\chi(x) = \sin 5 \pi x$, ε must be chosen so that

$$25 \, \varepsilon \pi^4 \, T \quad \text{is ``small.''}$$

8. 2. 2 OTHER FUNCTIONS $\chi(x)$.

$$\chi(x) = x(1 - x). \tag{a}$$

We take 50 intervals on the x-axis and $T = 1/10$, $\Delta t = T/100$.

The results are given in the Graphs 1 and 1 bis;[1] ε is given by

$$\varepsilon \simeq \frac{\theta \eta \, |\chi|}{\pi^4 T} \simeq \frac{\theta \eta \, |\chi|}{10} .$$

We take $\theta \eta \, |\chi| = 1/10$ and thus

$$\varepsilon = 10^{-2} .$$

The results are considerably improved by choosing $\varepsilon = 5.10^{-3}$ and 10^{-3}. For values of ε less than 10^{-3}, the scheme becomes much too unstable on $(0, 1/10)$. ∎

$$\chi(x) = \begin{cases} x & 0 \leqslant x \leqslant \tfrac{1}{2}, \\ 1 - x & \tfrac{1}{2} \leqslant x \leqslant 1. \end{cases} \tag{b}$$

Taking the relative theoretical error $\theta \eta / |\chi| = 1/10$, we have

$$\varepsilon = 10^{-2} .$$

Here again, the results are better with $\varepsilon = 10^{-3}$ but the procedure is unstable for $\varepsilon < 10^{-3}$.

The results are given in Graphs 2 and 2 bis.

[1] The graphs may be found at the end of each chapter.

If now we smooth the function:

$$\chi(x) = \begin{cases} x & \text{on} & (0; \ 0{,}5), \\ 1 - x & \text{on} & (0{,}5; \ 1), \end{cases}$$

by

$$\chi(x) = 0{,}4 + \frac{0{,}2}{\pi} \cos\left(\pi \frac{x - 0{,}5}{0{,}2}\right) \qquad \text{on} \qquad (0{,}4; \ 0{,}6),$$

we see (Graph 2 ter) that the result obtained is improved.([1])

Remark 8.1. This principle of smoothing of discontinuous functions, or those with discontinuous derivatives, is general. It is often a source of considerable improvement from the numerical point of view. ■

8. 2. 3 A STEP FUNCTION $\chi(x)$

$$\chi(x) = \begin{cases} 0 & x = 0, \\ 0{,}5 & 0 < x < 1, \\ 1 & x = 1. \end{cases} \qquad \text{(a)}$$

We obtain the Graphs 3 and 3 *bis*.

Smoothing the function at the endpoints of $(0,1)$, we obtain some results which are slightly better (Graphs 4 and 4 *bis*).

$$\chi(x) = \begin{cases} 0 & 0 \leqslant x < 0{,}5, \\ 0{,}5 & x = 0{,}5, \\ 1 & 0{,}5 < x \leqslant 1. \end{cases} \qquad \text{(b)}$$

The discontinuity at 0.5 (Graph 5) has not been taken account of in a satisfactory manner.

Smoothing the function $\chi(x)$ improves the results, but this improvement is not significant. Let us note, however, that it allows

([1])In the sense of difference with the more physically admissible function considered. We could improve further by smoothing in a smaller neighborhood of singularity.

a reduction of instability since the curves representative of χ^* (Graph 6) all pass through the point (0.5, 0.5), center of symmetry of the system.

We tried a series of numerical experiments with this function, modifying the step Δx and Δt for a given value of ε. We always obtain the same function χ^*. What is important is therefore a choice of a QR method and a choice of ε; in return the numerical integration of the QR system is standard.

8.3 Case Where $a(x) = 1 + x$.

In the case $a(x) = 1 + x$, the numerical results are not in general very different from the case $a = 1$ in quality. The Graphs 7 and 7 *bis* give respectively the results for $\chi^*(x)$ and the control ξ when

$$\chi = x(1 - x) .$$

Remark 8.2. Some of the preceding results, although satisfactory, can be improved by refining certain purely technical points (smoothing, integration schemes, optimization in ε, acceleration of convergence, other operators leading to Quasi-Reversibility, schemes with variable step (particularly in the neighborhood of singularities), etc.), or by means of standard procedures of approximation theory (cf. the following section).

8.4 Improvement of Results

8.4.1 Given a function $\chi(x)$, we have seen that it is possible to find $\xi_\varepsilon(x)$ so that if $u(x, t)$ is the solution of:

$$\left.\begin{array}{ll} \dfrac{\partial u}{\partial t} - \dfrac{\partial}{\partial x}\left(a(x)\dfrac{\partial u}{\partial x}\right) = 0 & x \in (0, 1), \qquad t \in (0, T) \\[2mm] u(0, t) = 0 \\[2mm] u(1, t) = 0 \\[2mm] u(x, 0) = \xi_\varepsilon(x) \end{array}\right\} \tag{8.1}$$

---

[1]What follows is naturally valid for the entire book, *in the linear cases.*

then

$$|u(x, T) - \chi(x)| < \eta.$$ (8.2)

To each value of ε, we thus make correspond a function $\xi_\varepsilon(x)$. Let us suppose that these calculations have been effected for n values of ε: $\varepsilon_1, \varepsilon_2, ..., \varepsilon_n$; write $\xi_{\varepsilon_j} = \varphi_j$; denote by L the linear mapping which allows us to go by means of integration of (8.1) from φ_j to $\chi_j = u_{\varepsilon_j}(x, T)$. If $\xi = \sum_{j=1}^n \lambda_j \varphi_j$, $\lambda_j \in \mathbf{R}$, we have

$$L\xi = \sum_{j=1}^n \lambda_j \chi_j.$$

It is natural to seek to determine the λ_j so that $|L\xi - \chi|$ is a minimum. This leads to the linear system:

$$\sum_{i=1}^n \lambda_i(L\varphi_i, L\varphi_i) = (L\varphi_i, \chi_j) \quad j = 1, ..., n.$$ (8.3)

This method has been used in Example 8.2.2b, taking these values for ε, $\varepsilon_1 = 10^{-2}$, $\varepsilon_2 = 5.10^{-3}$, $\varepsilon_3 = 10^{-3}$. The Graph 2 quarto shows the effective improvement obtained by application of this standard idea.

8.4.2 We can equally well use an acceleration process such as:

$$\xi_{ij} = \frac{\varepsilon_i \xi_i - \varepsilon_i \xi_j}{\varepsilon_j - \varepsilon_i}.$$ (8.4)

9. STABILITY OF NUMERICAL INTEGRATION SCHEMES FOR THE BACKWARD PARABOLIC SYSTEM

9.1 General Case. Energy Method

Let us consider the system (5.5), (5.6), having inverted the sense of time. Suppress the index ε to simplify the notation. We must then integrate:

$$u'(t) - A(t) u(t) + \varepsilon A^*(t) A(t) u(t) = 0$$ (9.1)

$$u(0) = \chi \quad \text{given} \tag{9.2}$$

(with u subject to suitable boundary conditions).

Let us simplify further by supposing:

$$A(t) = A \text{ independent of } t. \tag{9.3}$$

Remark 9.1. What follows carries over to the case where $A(t)$ depends on t as in Raviart [1].

Consider the Hilbert space $D(A)$, with the norm$(| u |^2 + | Au |^2)^{1/2})$; for $u, v \in D(A)$, we put:

$$b(u, v) = (Au, Av), \tag{9.4}$$

$$c(u, v) = (Au, v). \tag{9.5}$$

Equation (9.1) can be written in the equivalent form

$$(u'(t), v) - c(u(t), v) + \varepsilon b(u(t), v) = 0 \qquad \forall v \in D(A). \tag{9.6}$$

To simplify the notation, set

$$||| u ||| = \text{norm of } u \text{ in } D(A), \tag{9.7}$$

and suppose, without loss of generality, that

$$b(u, u) \geqslant \beta_0 ||| u |||^2. \tag{9.8}$$

We have

$$| b(u, v) | \leqslant \beta ||| u ||| \, ||| v ||| \tag{9.9}$$

and

$$| c(u, v) | \leqslant \gamma ||| u ||| \, | v |. \; \blacksquare \tag{9.10}$$

Consider now a *discretization* of (9.6). Let H_h be a family of spaces of finite dimension, depending on a parameter h (a vector in $\mathbf{R}^n$ eventually tending to zero). We suppose known (cf. J. P. Aubin [1]), for each h an operator p_h (''extension'') having the following properties:

$$p_h \in \mathscr{L}(H_h; D(A)), \qquad p_h \text{ one-to-one}. \tag{9.11}$$

Let H_h be provided with *two norms*:

$$| u_h |_h = | p_h u_h |, \qquad ||| u_h |||_h = ||| p_h u_h |||. \qquad (9.12)$$

Since H_h is of finite dimension, these two norms are *equivalent* and therefore

$$||| u_h |||_h \leqslant S(h) | u_h |_h. \qquad (9.13)$$

Remark 9.2. In the examples, H_h is the euclidean space of vectors defined at the grid-points of the step $h = \{h_1, ..., h_n\}$ which are "at the interior" of the domain Ω where the problem is posed. The operator p_h is an extension operator.

If $A(t)$ is an *elliptic differential operator of order* $2\,m$, then

$$S(h) = O\left(\frac{1}{| h |^{2m}} \right), \qquad | h | = (h_1^2 + \cdots + h_n^2)^{1/2}. \; \blacksquare \qquad (9.14)$$

Remark 9.3. Note that in (9.13) we *always* have $S(h) \to \infty$ if $h \to 0$ [except for the trivial case where $A(t)$ is an operator *restricted* in H]. $\blacksquare$

Remark 9.4. We can also avoid the operators p_h using some related notions which are technically different (cf. J. Cea [1], P. A. Raviart [1]). $\blacksquare$

EXPLICIT SCHEME

Introduce:

$$\Delta t = \text{step in time } t,$$
$$U_h^n = \text{approximation of } u(n\,\Delta t) = \text{«} x \to u(x, n\,\Delta t)\text{»}.$$

Set

$$(u_h, v_h)_h = (p_h u_h, p_h v_h),$$

and define u_h^{n+1} in terms of u_h^n by

$$\frac{1}{\Delta t}(u_h^{n+1} - u_h^n, v_h)_h - c_h(u_h^n, v_h) + \varepsilon b_h(u_h^n, v_h) = 0 \qquad \forall v_h \in H_h \qquad (9.15)$$

with

$$u_h^0 | = \text{approximation of } \chi \text{ in } H_h, \qquad (9.16)$$

and where the forms b_h and c_h are approximations of the forms b and c.

The scheme (9.15) is an *explicit* scheme whose stability will be studied, in a sense which will be made precise.

ENERGY INEQUALITY

Replace first v_h by u_h^n in (9.15). Then

$$\frac{1}{2\,\Delta t} \left(|\, u_h^{n+1} \,|_h^2 - |\, u_h^n \,|_h^2 \right) - \frac{\Delta t}{2} \left| \frac{u_h^{n+1} - u_h^n}{\Delta t} \right|_h^2 + \varepsilon b_h(u_h^n, u_h^n) = c_h(v_h^n, u_h^n) \,,$$

whence

$$\left. \begin{array}{l} \dfrac{1}{2\,\Delta t} \left(|\, u_h^{n+1} \,|_h^2 - |\, u_h^n \,|_h^2 \right) + \varepsilon \beta_0 \, ||| \, u_h^n \, |||_h^2 \leqslant \\[2mm] \qquad\qquad \leqslant \dfrac{\Delta t}{2} \left| \dfrac{u_h^{n+1} - u_h^n}{\Delta t} \right|_h^2 + \gamma \, ||| \, u_h^n \, |||_h \, |\, u_h^n \,|_h \,. \end{array} \right\} \qquad (9.17)$$

The last term on the right is majorized, *for example*, by

$$\frac{\varepsilon \beta_0}{2} \, ||| \, u_h^n \, |||_h^2 + \frac{\gamma^2}{2\,\varepsilon\beta_0} \, |\, u_h^n \,|_h^2 \,,$$

and we deduce then from (9.17):

$$\left. \begin{array}{l} \dfrac{1}{\Delta t} \left(|\, u_h^{n+1} \,|_h^2 - |\, u_h^n \,|_h^2 \right) + \varepsilon \beta_0 \, ||| \, u_h^n \, |||_h^2 \leqslant \\[2mm] \qquad\qquad \leqslant \Delta t \left| \dfrac{u_h^{n+1} - u_h^n}{\Delta t} \right|_h^2 + \dfrac{\gamma^2}{2\,\varepsilon\beta_0} \, |\, u_h^n \,|_h^2 \,. \end{array} \right\} \qquad (9.18)$$

Replace next v_h by $\dfrac{1}{\Delta t} (u_h^{n+1} - u_h^n)$ in (9.15). We have

$$\left| \frac{1}{\Delta t} (u_h^{n+1} - u_h^n) \right|_h^2 = c_h \left(u_h^n, \frac{1}{\Delta t} (u_h^{n+1} - u_h^n) \right) - \varepsilon b_h \left(u_h^n, \frac{u_h^{n+1} - u_h^n}{\Delta t} \right) \leqslant$$

$$\leqslant \gamma \, ||| \, u_h^n \, |||_h \left| \frac{u_h^{n+1} - u_h^n}{\Delta t} \right|_h + \varepsilon \beta \, ||| \, u_h^n \, |||_h \, \left|\left|\left| \frac{u_h^{n+1} - u_h^n}{\Delta t} \right|\right|\right|_h,$$

and thanks to (9.13), we deduce:

$$\left| \frac{u_h^{n+1} - u_h^n}{\Delta t} \right|_h \leqslant (\gamma + \varepsilon \beta S(h)) \, \| u_h^n \|_h \; ; \tag{9.19}$$

using (9.19) in (9.18), we have:

$$\frac{1}{\Delta t} \left(| u_h^{n+1} |_h^2 - | u_h^n |_h^2 \right) + \varepsilon \beta_0 \, \| u_h^n \|_h^2 \leqslant$$

$$\leqslant \Delta t (\gamma + \varepsilon \beta S(h))^2 \, \| u_h^n \|_h^2 + \frac{\gamma^2}{2 \, \varepsilon \beta_0} | u_h^n |_h^2 \, ,$$

whence finally

$$\left. \begin{aligned} \frac{1}{\Delta t} \left(| u_h^{n+1} |_h^2 - | u_h^n |_h^2 \right) + \left(\varepsilon \beta_0 - \Delta t (\gamma + \varepsilon \beta S(h))^2 \right) \| u_h^n \|_h^2 \leqslant \\ \leqslant \frac{\gamma^2}{2 \, \varepsilon \beta_0} | u_h^n |_h^2 \, . \end{aligned} \right\} \tag{9.20}$$

We make the stability hypothesis:

$$\Delta t (\gamma + \varepsilon \beta S(h))^2 = (1 - \delta) \, \varepsilon \beta_0, \quad \delta > 0 \text{ arbitrarily small.} \tag{9.21}$$

Inequality (9.20) then gives:

$$\frac{1}{\Delta t} \left(| u_h^{n+1} |_h^2 - | u_h^n |_h^2 \right) + (1 - \delta) \, \varepsilon \beta_0 \, \| u_h^n \|_h^2 \leqslant \frac{\gamma^2}{2 \, \varepsilon \beta_0} | u_h^n |_h^2 \, . \tag{9.22}$$

Sum this inequality over n from 0 to m, with

$$m \, \Delta t = T. \tag{9.23}$$

We have

$$| u_h^{m+1} |_h^2 + (1 - \delta) \, \varepsilon \beta_0 \sum_{n=0}^{m} \Delta t \, \| u_h^n \|_h^2 \leqslant$$

$$\leqslant | u_h^0 |_h^2 + \frac{\gamma^2}{2 \, \varepsilon \beta_0} \sum_{n=0}^{m} \Delta t \, | u_h^n |_h^2 \, . \tag{9.24}$$

CONSEQUENCE

If the stability condition of (9.21) *holds, then*

$$\sum_{n=0}^{m} \Delta t \, \| u_h^n \|_h^2 \leqslant \text{constant (independent of } h \text{ and } \Delta t) \tag{9.25}$$

Further, the discrete form of the Bellman–Gronwall inequality yields

$$| u_h^n |_h^2 \leqslant | u_h^0 |_h^2 \exp\left(\frac{T\gamma^2}{2\,\varepsilon\beta_0}\right).$$
(9.26)

Condition (9.21) will be realized if, *for example*:

(i) Δt is "sufficiently small" in the sense

$$\Delta t \leqslant (1 - \delta)\frac{\varepsilon\beta_0}{2\,\gamma^2}\,;$$

(ii) $\varepsilon\,\Delta t S(h)^2 \leqslant \dfrac{(1 - \delta)\,\beta_0}{2\,\beta^2}.$ ∎

We can improve this and rederive the stability condition of Example 9.2 using P. A. Raviart [2].

9.2   Study of A Particular Case By the von Neumann Method

Consider the particular one-dimensional case where $A = - a\dfrac{\mathrm{d}^2}{\mathrm{d}x^2}$ $(a > 0)$. The problem (9.1), (9.2), may then be written

$$\frac{\partial u}{\partial t} + a\frac{\partial^2 u}{\partial x^2} + \varepsilon\frac{\partial^4 u}{\partial x^4} = 0 \qquad 0 \leqslant x \leqslant 1, \qquad 0 \leqslant t \leqslant T;$$
(9.27)

$$u(x, 0) = \chi(x) \text{ given}$$
(9.28)

$$u(0, t) = u(1, t) = 0.$$
(9.29)

If we discretize (9.27) according to the *explicit scheme*

$$\frac{u_j^{n+1} - u_j^n}{\Delta t} + a\frac{u_{j+1}^n - 2u_j^n + u_{j-1}^n}{\Delta x^2} + $$
$$+ \varepsilon\frac{u_{j+2}^n - 4u_{j+1}^n + 6u_j^n - 4u_{j-1}^n + u_{j-2}^n}{\Delta x^4} = 0,$$
(9.30)

the amplification matrix—here reduced to a single element—associated with (9.30) is, after all calculations have been made

$$G(\Delta t, \Delta n, j) = 1 + 4 \sin^2 \frac{\beta}{2} \left(a \frac{\Delta t}{\Delta x^2} - 4 \frac{\Delta t}{\Delta x^4} \sin^2 \frac{\beta}{2} \right), \qquad (9.31)$$

where we have set

$$\beta = j \, \Delta x \,.$$

A *sufficient* condition for stability in a sense "close" to that of (9.26) is (cf. Richtmyer [1])

$$|G| \leqslant 1 + O(\Delta t) \,. \qquad (9.32)$$

The amplification factor G has the form

$$G = 1 + 4 y \left(\frac{a \, \Delta t}{\Delta x^2} - \sigma y \right), \qquad 0 \leqslant y \leqslant 1 \,, \qquad (9.33)$$

with

$$\sigma = \frac{4 \, \varepsilon \, \Delta t}{\Delta x^4} \,. \qquad (9.34)$$

The maximum of $G(y)$ is

$$1 + a^2 \frac{\Delta t}{4 \, \varepsilon} = 1 + O(\Delta t) \,,$$

therefore to satisfy (9.32) it is sufficient to write

$$G(y) \geqslant - 1 + O(\Delta t) \qquad \text{for} \qquad 0 \leqslant y \leqslant 1 \,.$$

Since $G(0) = 1$ and the curve is concave it suffices to write

$$G(1) \geqslant - 1 + O(\Delta t) \,,$$

which is to say

$$1 + 4 \frac{\Delta t}{\Delta x^2} \left[a - \frac{4 \, \varepsilon}{\Delta x^2} \right] \geqslant - 1 + O(\Delta t) \,. \qquad (9.35)$$

Since $\frac{\Delta t}{\Delta x^2} \left[a - \frac{4 \, \varepsilon}{\Delta x^2} \right]$ cannot be made $O(\Delta t)$ as $\Delta x \to 0$, we write (9.35) under the stricter form

$$1 + 4 \frac{\Delta t}{\Delta x^2} \left[a - \frac{4\varepsilon}{\Delta x^2} \right] \geqslant - 1 ,$$

or

$$\varepsilon \frac{\Delta t}{\Delta x^4} - \frac{a}{4} \frac{\Delta t}{\Delta x^2} \leqslant \frac{1}{8} , \tag{9.36}$$

a condition which is automatically satisfied if (cf. the end of point 9.1):

$$\varepsilon \frac{\Delta t}{\Delta x^4} \leqslant \frac{1}{8} . \ \blacksquare \tag{9.37}$$

Remark 9.5. The maximum $G_+ = 1 + a^2 \frac{\Delta t}{4\varepsilon}$ is attained for certain harmonics of the solution, which gives at the end of n steps, the amplification factor

$$\left(1 + \frac{a^2}{4\varepsilon} \Delta t \right)^n \simeq \exp\left(\frac{a^2}{4\varepsilon} n \Delta t \right).$$

Consequently, if we wish to reach $t = T$ starting from $t = 0$, certain harmonics of the solution can be amplified (in quadrate norm) by $\left(\frac{a^2}{4\varepsilon} T \right)$. This explains the form of $\tilde{\xi}(x)$ actually obtained (cf. Graph 8). In order that a factor such as this be "reasonable" it is necessary that $1/\varepsilon$ not be too "large." Since the choice of ε depends on η and T, it follows that the problem is not necessarily soluble for *every* pair $\{T, \eta\}$ (or $\{T, \varepsilon\}$). Here are some numerical results in this direction ($a = 1$):

T	ε	$\exp\left(\frac{a^2}{4\varepsilon} T \right)$	Observed factor of amplification	
$\frac{1}{10}$	10^{-2}	$\simeq 10$	$\simeq 3$	
$\frac{1}{10}$	5.10^{-3}	$\simeq 100$	$\simeq 40$	$\chi(x)$ can be found directly
$\frac{1}{10}$	10^{-3}	$\simeq 10^{10}$	$\simeq 10^9$	
$\frac{1}{10}$	5.10^{-4}	$\simeq 10^{20}$	$\simeq 10^{19}$	$\chi(x)$ cannot be determined

On the other hand, in a divergent case, we can calculate the time T_0 where the amplification factor attains the maximum value permitted by the computer being used (10^{300} for the CDC 3600). There still, the numerical values recorded are of the order which had been predicted.

10.   TWO-DIMENSIONAL EXAMPLES

10.1 The Problem

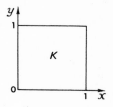

Figure 1

The domain considered is $K \times (0, T)$, where K is the square of side 1 in the (x, y)-plane, as indicated in Figure 1.

Given a function $\chi(x, y)$, we look for a function $\xi(x, y)$ such that $u(x, y, t)$ is the solution of:

$$\frac{\partial u}{\partial t} - \frac{\partial}{\partial x}\left(a(x, y)\frac{\partial u}{\partial x}\right) - \frac{\partial}{\partial y}\left(a(x, y)\frac{\partial u}{\partial y}\right) = 0, \qquad (10.1)$$

$$u(x, y, 0) = \xi(x, y), \qquad (10.2)$$

$$u(x, 0, t) = u(x, 1, t) = u(0, y, t) = u(1, y, t) = 0 ; \qquad (10.3)$$

(I)

then

$$|u(x, y, T) - \chi(x, y)| \leqslant \eta . \qquad (10.4)$$

Having performed a change of variable from t to $T - t$, we must solve:

$$\frac{\partial v}{\partial t} + \frac{\partial}{\partial x}\left(a(x, y)\frac{\partial v}{\partial x}\right) + \frac{\partial}{\partial y}\left(a(x, y)\frac{\partial v}{\partial y}\right) +$$

$$+ \varepsilon\left[\frac{\partial^4 v}{\partial x^4} + 2\frac{\partial^4 v}{\partial x^2 \partial y^2} + \frac{\partial^4 v}{\partial y^4}\right] = 0, \qquad (10.5)$$

$$v(x, y, 0) = \chi(x, y), \qquad (10.6)$$

$$v(x, 0, t) = v(x, 1, t) = v(0, y, t) = v(1, y, t) = 0, \qquad (10.7)$$

$$\frac{\partial}{\partial x}\left(a\frac{\partial v}{\partial x}\right)_{\substack{x=0\\x=1}} = 0, \qquad \frac{\partial}{\partial y}\left(a\frac{\partial v}{\partial y}\right)_{\substack{y=0\\y=1}} = 0 ; \qquad (10.8)$$

(II)

and then take

$$\xi(x, y) = v(x, y, T).$$ (10.9)

10.2 Matrix Representation of the Problem

10.2.1 DISCRETIZATION

We discretize in a symmetric fashion, according to a scheme Crank–Nicolson type.

Remark 10.1. We do not attempt to utilize refined methods of numerical integration. Naturally, this would be useful, and even essential, in a three-dimensional space and for more complicated systems. We could, using our QR equations, use, for example, the method of alternating directions or "splitting-up"; see Douglas [1], Marchuk [1], Peaceman–Rachford [1], R. Teman [1, 2], Yaneko [1]. Set:

$$x = i\,\Delta x \qquad i \text{ going from 0 to } I, \text{ with } I\,\Delta x = 1;$$

$$y = j\,\Delta y \qquad j \text{ going from 0 to } J, \text{ with } J\,\Delta y = 1;$$

$$t = n\,\Delta t \qquad n \text{ going from 0 to } N, \text{ with } N\,\Delta t = T;$$

$$v_{ij}^n = v(i\,\Delta x, j\,\Delta y, n\,\Delta t),$$

$$a_{ij} = a(i\,\Delta x, j\,\Delta y).$$

In what follows, we will call v^n the vector whose components are v_{ij}^n. For the derivatives in x and y, we then take, at the time $n\,\Delta t$:

$$\frac{\partial v_{ij}}{\partial x} \cong \frac{1}{\Delta x}\,[v_{i+1/2,j} - v_{i-1/2,j}],$$

$$\frac{\partial v_{ij}}{\partial y} \cong \frac{1}{\Delta y}\,[v_{i,j+1/2} - v_{i,j-1/2}] \text{ etc.,}$$

$$\frac{\partial}{\partial x}\left[a(x, y)\,\frac{\partial v_{ij}}{\partial x}\right] \cong \frac{1}{\Delta x^2}\,[a_{i+1/2,j}(v_{i+1,j} - v_{ij}) - a_{i-1/2,j}(v_{ij} - v_{i-1,j})].$$

For the derivative with respect to t:

$$\frac{\partial v_{ij}}{\partial t} \cong \frac{1}{\Delta t}\,[v_{ij}^{n+1} - v_{ij}^n].$$

10. 2. 2 Matrix Representation of the System (II)

We proceed in all matters in a fashion completely analogous to the one-dimensional case (cf. 7.2.2).
We have

$$Av^{n+1} = Bv^n + C,\qquad(10.10)$$

where A and B are matrices of rank $(I-1)(J-1)$ and C is a vector of the same rank which takes account of the boundary conditions. The matrices A and B are symmetric 13 diagonals. If I denotes the unit matrix,

$$B = -A + \frac{2}{\Delta t}I.\qquad(10.11)$$

Naturally, some precautions must be taken for the equations and the formulas corresponding to the particular cases

$$i = 1, 2, I-2, I-1,$$
$$j = 1, 2, J-2, J-1,$$

which introduce points situated on the boundary or outside of the net. The system in (10.10) allows us to determine

$$\xi(x, y) = v(x, y; T).\qquad(10.12)$$

10. 2. 3 Matrix Representation of the Direct Problem (I)

At the end of the verification, we must integrate the direct initial problem; with an analogous discretization scheme, system (I) may be written:

$$\left.\begin{aligned}
\frac{u_{i,j}^{n+1} - u_{i,j}^n}{\Delta t} &- \frac{1}{\Delta x^2}\left[a_{i+1/2,j}(u_{i+1,j}^{n+1} - u_{i,j}^{n+1} + u_{i+1,j}^n - u_{i,j}^n) - \right.\\
&\left.\qquad - a_{i-1/2,j}(u_{i,j}^{n+1} - u_{i-1,j}^{n+1} + u_{i,j}^n - u_{i-1,j}^n)\right]\\
&- \frac{1}{\Delta y^2}\left[a_{i,j+1/2}(u_{i,j+1}^{n+1} - u_{i,j}^{n+1} + u_{i,j+1}^n - u_{i,j}^n) - \right.\\
&\left.\qquad - a_{i,j-1/2}(u_{i,j}^{n+1} - u_{i,j-1}^{n+1} + u_{i,j}^n - u_{i,j-1}^n)\right] = 0
\end{aligned}\right\}\qquad(10.12)$$

with

$$u_{ij}^0 = \xi(i\,\Delta x, j\,\Delta y) \tag{10.13}$$

a solution for x_i, y_j found in the retrograde solution (cf. 10.12) and:

$$u_{i0}^n = u_{iJ}^n = {}^n_{0j} = u_{Ij}^n = 0 \tag{10.14}$$

and we seek u_{ij}^{n+1} at the time $T (= N\,\Delta t$ if we take equal time steps), which is to say

$$u(i\,\Delta x, j\,\Delta y, N\,\Delta t)\,.$$

The equation (10.12) may be written in matrix form:

$$Au^{n+1} = Bu^n + C\,, \tag{10.15}$$

where u is of dimension $(I-1)(J-1)$, and where $B = -A + \dfrac{2}{\Delta t} I$: A and B are two pentadiagonal symmetric matrices of dimension $(I-1)\cdot(J-1)$.

10.3 Numerical Results

10. 3. 1 CALCULATION OF ε WHEN $a(x, y) = 1$.

We evaluate ε using:

$$\varepsilon = -\frac{\log\,(1 - \eta/|\chi|)}{\lambda_0\,C_0\,T}\,.$$

If $a(x, y) = 1$, then $C_0 = 1$; λ_0 is the smallest characteristic value of the operator Δ in the square K with boundary values zero; this characteristic value therefore satisfies:

$$\frac{\partial^2 u}{\partial x^2} + \frac{\partial^2 u}{\partial y^2} + \lambda_0\,u = 0\,, \tag{10.16}$$

$$u(x, y) = 0\,.$$

Factoring $u(x, y)$ into $\varphi(x) \times \varphi(y)$, and integrating the system in (10.16) we find $\lambda_0 = 2\pi^2$.

10.3.2 EXAMPLE 1: $\chi(x, y) = Kxy(1 - x)(1 - y)$.

The calculations have been carried out for a value of ε equal to 5.10^{-3}. The results obtained are indicated in Graphs 9 and 9 *bis* in the form of those sections of the surfaces $\chi(x, y)$ and $\chi^*(x, y)$ by planes $y = constant$.

The Graph 9 *ter* provides results for $\varepsilon = 10^{-3}$.

10.3.3 EXAMPLE 2: *The function* $\chi(x, y)$ *piecewise linear*.

We took

$$
\begin{aligned}
\chi(x, y) &= x && \text{if } (x, y) \in D_1 \\
&= y && \text{if } (x, y) \in D_2 \\
&= 1 - x && \text{if } (x, y) \in D_3 \\
&= 1 - y && \text{if } (x, y) \in D_4 .
\end{aligned}
$$

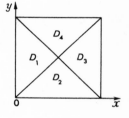

The surface $\chi(x, y)$ is a pyramid; the numerical results are given in Graphs 10, 10 *bis* and 10 *ter*.

Figure 2

11. AVERAGED FUNCTIONALS

11.1 Typical Example

Let us place ourselves in the framework of Sections 3 and 4. Let $u = u(t) = u(t; \xi)$ be the solution of (under the hypothesis of Theorem 3.1):

$$u' + A(t) u = 0 , \tag{11.1}$$

$$u(0) = \xi , \tag{11.2}$$

$$u \in L_2(0, T; V) , \qquad u' \in L_2(0, T; V') . \tag{11.3}$$

Let $\tau_0 > 0$ be given, with $\tau_0 < T$. We consider *the mean value*

$$\mathcal{M}u = \frac{1}{\tau_0} \int_{T-\tau_0}^{T} u(t)\, dt . \tag{11.4}$$

We then consider the *new functional*

$$J(\xi) = |\mathcal{M}u - \chi|^2 , \tag{11.5}$$

χ given in H.

This functional depends not only on the value of u at the time T, but also on values over the interval $(T - \tau_0, T)$; we say that $J(\xi)$ is a *thick functional*.[1] Our object is the study of $\underset{\xi \in H}{\mathrm{Inf}}\, J(\xi)$. ∎

Remark 11.1. Since $u \in L_2(0, T; V)$, we have $\mathcal{M}u \in V$. If then χ_1 is given *in* V, we can consider:

$$J_1(\xi) = \| \mathcal{M}u - \chi_1 \|^2 \qquad \text{[2]}. \quad ∎ \tag{11.6}$$

11.2 Another Example

Let $\chi = \chi(t)$ be a *function* given in $L_2(T - \tau_0, T; H)$. We can then consider the functional

$$\tilde{J}(\xi) = \int_{T-\tau_0}^{T} | u(t) - \chi(t) |^2 \, dt , \tag{11.7}$$

which is again a "thick functional," and again study $\underset{\xi \in H}{\mathrm{Inf}}\, \tilde{J}(\xi)$. ∎

Remark 11.2. For χ_1 given in $L_2(T - \tau_0, T; V)$, we could also consider

$$\tilde{J}_1(\xi) = \int_{T-\tau_0}^{T} \| u(t) - \chi_1(t) \|^2 \, dt . \tag{11.8}$$

---

[1]This type of functional arises often in a natural way in physical problems. It would be interesting, for example, to study the minimization of functionals of the form

$$\mathrm{Inf} \sum_{j=1}^{j_0} |\mathcal{M}_j u - \chi_j(x)|^2 ,$$

where $\mathcal{M}_j$ is a mean analogous to $\mathcal{M}$, taken with respect to the T_j, where the $\chi_j(x)$ are given.
[2]Recall that $\| \quad \|$ (or $| \quad |$) denotes the norm in V (or H).

Similarly we could consider functionals in which u' occurs:

$$\tilde{J}_2(\xi) = \int_{T-\tau_0}^{T} (\| u(t) - \chi_1(t) \|^2 + \| u'(t) - \chi_2(t) \|_*^2) \, dt, \qquad (^1)$$

with χ_2 given in $L_2(T - \tau_0, T; V')$ $(^2)$. ■

Following are some simple results on the limits inferior of thick functionals.

11.3 An Example Where Inf $J(\xi) = 0$.

Theorem 11.1: *Consider the hypotheses of Theorem 4.1. More precisely, $a(u, v) = a(v, u)$ and $a(v, v) \geqslant \alpha_0 \| v \|^2$, $v \in V$, $\alpha_0 > 0$. Then if $J(\xi)$ is as in (11.5), we have:*

$$\text{Inf}_{\xi} J(\xi) = 0. \qquad (11.9)$$

Proof. It reduces to showing that $\mathcal{M}u$ spans a sense space in H, or again (with the notation of the proof of Theorem 4.1) that:

$$\mathcal{U}\mathcal{M}u \text{ spans a space dense in } H. \qquad (11.10)$$

Since

$$\mathcal{U}\mathcal{M}u(\lambda) = \frac{1}{\tau_0} \int_{T-\tau_0}^{T} \hat{u}(\lambda, t) \, dt$$

and since

$$\hat{u}(\lambda, t) = e^{-\lambda t} \hat{\xi}(\lambda),$$

we have

$$\mathcal{U}\mathcal{M}u(\lambda) = \frac{2}{\tau_0} e^{-\lambda(T - \tau_0/2)} \frac{1}{\lambda} \text{sh}\left(\frac{\lambda\tau_0}{2}\right) \hat{\xi}(\lambda). \qquad (11.11)$$

$(^1)$In practice, only an expression under the integral sign of the form

$$\alpha \| u(t) - \chi(t) \|^2 + \beta \| u(t) - \chi_2(t) \|_*^2,$$

has a physical sense.
$(^2)$ $\| \quad \|_*$ = dual norm of $\| \quad \|$.

Let then $\eta \in \mathfrak{h}$, with

$$(\mathcal{U}\mathcal{M}u(\lambda), \eta(\lambda))_{\mathfrak{h}} = 0 \qquad \forall \xi \,;$$

using (11.11), we see that

$$e^{-\lambda(T-\tau_0/2)} \frac{1}{\lambda} \operatorname{sh}\left(\frac{\lambda\tau_0}{2}\right) \eta(\lambda) = 0$$

almost everywhere in λ for the measure μ, and therefore $\eta = 0$. ∎

Remark 11.3. The same reasoning shows under the same hypotheses that where J_1 is given by (11.6)

$$\operatorname{Inf} J_1(\xi) = 0 \,,$$

Remark 11.4. We conjecture that (11.9) is still valid under the hypotheses of Theorem 3.1. ∎

11.4   A Case Where There is No Density

Here is a *negative* result:

Theorem 11.2: *Consider the hypotheses of Theorem 11.1. Then, if $\tilde{J}(\xi)$ is given by (11.7), we do not in general have* $\operatorname*{Inf}_{\xi} \tilde{J}(\xi) = 0$.

Proof. To simplify a bit, suppose in addition that the identity mapping of V in H is compact. With the notation of Remark 4.1, we have:

$$u(t) = \sum_{j=1}^{\infty} (\xi, w_j) e^{-\lambda_j t} w_j \,. \tag{11.12}$$

Let j_0 be an arbitrary fixed index. Let φ_{j_0} be a function of $L_2(T - \tau_0, T)$ such that

$$\int_{T-\tau_0}^{T} e^{-\lambda_{j_0} t} \varphi_{j_0}(t) \, \mathrm{d}t = 0 \,. \tag{11.13}$$

Consider then

$$\varphi(t) = \varphi_{j_0}(t) w_{j_0} \,. \tag{11.14}$$

The function φ is in $L_2(T - \tau_0, T; H)$ and we are going to verify that it is orthogonal to $\forall \xi$ (which will demonstrate the result). Now

$$\int_{T-\tau_0}^{T} (u(t), \varphi(t))\, dt = (\hat{\xi}, w_{j_0}) \, | \, w_{j_0} |^2 \int_{T-\tau_0}^{T} e^{-\lambda_{j_0} t}\, \varphi_{j_0}(t)\, dt$$

and, from (11.13), this is zero.

11.5 Position of the Problems

The problem of the calculation of $\underset{\xi \in H}{\text{Inf}}\ \tilde{J}(\xi)$ is essentially open (cf., however, Remark 11.6). In the case where $\text{Inf}\ J(\xi) = 0$, we can pose a problem analogous to Problem 4.1:

Problem 11.1: For a given $\eta > 0$, find $\xi_\eta \in H$ such that

$$J(\xi_\eta) \leqslant \eta .\tag{11.15}$$

The "ideal" solution (compare Remark 4.1) would be to take u as the "solution" of

$$\left.\begin{array}{l} u'(t) + A(t)\, u(t) = 0 , \\[2mm] \dfrac{1}{\tau_0} \displaystyle\int_{T-\tau_0}^{T} u(t)\, dt = \chi , \end{array}\right\}\tag{11.16}$$

then to take $\xi = u(0)$. But in general, problem (11.16) is *improperly posed*. Let us consider in effect Theorem 11.1. If the ideal solution exists, it corresponds to

$$\frac{2}{\tau_0} e^{-\lambda(T - \tau_0/2)} \frac{1}{\lambda} \, \text{sh}\left(\frac{\lambda \tau_0}{2}\right) \hat{\xi}(\lambda) = \hat{\eta}(\lambda)\tag{11.17}$$

and for a given χ in H, *there does not exist* in general a ξ in H satisfying (11.17).

11.6 The QR Method

Since problem (11.16) is improperly posed, we are going to try—as in Section 5—to "approximate" this problem by properly posed problems! By analogy with Section 5, we are led to the following: we look for $u_\varepsilon(t)$ a solution of (in the notation of Section 5),

$$u_\varepsilon' + A(t)\, u_\varepsilon - \varepsilon A^*(t)\, A(t)\, u_\varepsilon(t) = 0 \,, \tag{11.17}$$

$$u_\varepsilon \in L_2(0, T; D(A(t)))\,, \qquad u_\varepsilon' \in L_2(0, T; D(A(t))')\,, \tag{11.18}$$

$$\frac{1}{\tau_0} \int_{T-\tau_0}^{T} u_\varepsilon(t)\, dt = \chi \tag{11.19}$$

then, *if this problem is properly posed*, we take

$$\xi = u_\varepsilon(0) \,. \ \blacksquare \tag{11.20}$$

Let us study this in the case of Theorem 11.1. We are going to demonstrate

Theorem 11.3: *Under the conditions of Theorem* 11.1, *problem* (11.7), (11.8), (11.9), *is properly posed. Let* U_ε *be the solution of*

$$U_\varepsilon' + A U_\varepsilon = 0 \,, \qquad U_\varepsilon(0) = u_\varepsilon(0) \,. \tag{11.21}$$

We have

$$\frac{1}{\tau_0} \int_{T-\tau_0}^{T} U_\varepsilon(t)\, dt \to \chi \quad in \quad H \quad when \quad \varepsilon \to 0 \,. \tag{11.22}$$

Proof: 1) We introduce $\hat{u}_\varepsilon(\lambda, t) = (\mathscr{U}u_\varepsilon(t))\,(\lambda) \ (\in \mathfrak{h}(\lambda))$, which satisfies

$$\frac{d\hat{u}_\varepsilon}{dt} + (\lambda - \varepsilon\lambda^2)\, \hat{u}_\varepsilon = 0$$

therefore

$$\hat{u}_\varepsilon(\lambda, t) = \theta_\varepsilon(\lambda)\, e^{-(\lambda - \varepsilon\lambda^2)t} \,,$$

where θ_ε is a field of vectors to be determined. Now (11.19) gives

$$\frac{1}{\tau_0} \int_{T-\tau_0}^{T} \hat{u}_\varepsilon(\lambda, t)\, dt = \hat{\chi}(\lambda)\ (= (\mathscr{U}\chi)\,(\lambda))$$

from which follows

$$\frac{1}{\tau_0}\,\theta_\varepsilon(\lambda)\,e^{-(\lambda-\varepsilon\lambda^2)(T-\tau_0/2)}\,\frac{2\,\mathrm{sh}\,((\lambda-\varepsilon\lambda^2)\,\tau_0/2)}{\lambda-\varepsilon\lambda^2} = \hat{\chi}(\lambda)\,. \tag{11.23}$$

We deduce from this

$$\left.\begin{array}{l} \hat{u}_\varepsilon(\lambda,\,t) = \xi(\lambda)\,e^{-(\lambda-\varepsilon\lambda^2)t}\,, \\[2mm] \theta_\varepsilon(\lambda) = \dfrac{\tau_0}{2}\,e^{(\lambda-\varepsilon\lambda^2)(T-\tau_0/2)}\,\dfrac{\lambda-\varepsilon\lambda^2}{\mathrm{sh}\,(\lambda-\varepsilon\lambda^2)\,\tau_0/2}\,\hat{\chi}(\lambda)\,, \end{array}\right\} \tag{11.24}$$

which demonstrates that, for *all* $\varepsilon > 0$, the problem (11.17), (11.18), (11.19) is properly posed.

2) Introducing $\hat{U}_\varepsilon(\lambda,\,t) = (\mathcal{U}U_\varepsilon(t))\,(\lambda)$, we have

$$\frac{d\hat{U}_\varepsilon}{dt} + \lambda\hat{U}_\varepsilon = 0\,, \qquad \hat{U}_\varepsilon(\lambda,\,0) = \hat{u}_\varepsilon(\lambda,\,0) = \theta_\varepsilon(\lambda)\,,$$

when

$$\hat{U}_\varepsilon(\lambda,\,t) = \theta_\varepsilon(\lambda)\,e^{-\lambda t}$$

and

$$\begin{aligned} \frac{1}{\tau_0}\int_{T-\tau_0}^{T}\hat{U}_\varepsilon(\lambda,\,t)\,dt &= \frac{1}{\tau_0}\,\frac{e^{-\lambda(T-\tau_0)}-e^{-\lambda T}}{\lambda}\,\theta_\varepsilon(\lambda) \\[2mm] &= \frac{1}{\tau_0}\,e^{-\lambda(T-\tau_0/2)}\frac{2\,\mathrm{sh}\,\lambda_0/2}{\lambda}\,\theta_\varepsilon(\lambda) \\[2mm] &= e^{-\varepsilon\lambda^2(T-\tau_0/2)}\,\frac{\lambda-\varepsilon\lambda^2}{\mathrm{sh}\,(\lambda-\varepsilon\lambda^2)\,\tau_0/2}\,\frac{\mathrm{sh}\,(\lambda\tau_0/2)}{\lambda}\,\hat{\chi}(\lambda)\,, \end{aligned}$$

from which (11.22) follows. ∎

Remark 11.5. We can also show that the problem (11.17), (11.18), (11.19) is properly posed *without* supposing that $A^* = A$, but supposing only that $A(t) = A$ does not depend on t. To simplify the statements a bit, consider the following problem (with t in the *increasing* sense):

$$u' + Au = 0\,, \tag{11.25}$$

$$\frac{1}{\tau_0}\int_0^{\tau_0}u(\sigma)\,d\sigma = u_0\,, \qquad u_0 \text{- given in } D(A)\,, \tag{11.26}$$

$$u \in L_2(0, T; V), \qquad u' \in L_2(0, T; V'). \tag{11.27}$$

We suppose that

$$a(v, v) \geqslant \alpha_0 \| v \|^2, \tag{11.28}$$

but we do not suppose that $a(u, v) = a(v, u)$, therefore $A^* \neq A$. We are going to show that this problem admits a unique solution.

Let us integrate (11.25) from 0 to τ_0,

$$u(\tau_0) - u(0) + \int_0^{\tau_0} Au(t)\, dt = 0,$$

which is to say

$$u(\tau_0) - u(0) = - A\left(\int_0^{\tau_0} u(t)\, dt \right) = - \tau_0\, Au_0 \in H,$$

using (11.26). Therefore, if u is a solution of the problem, then u is a solution of (11.25), (11.27) and

$$u(\tau_0) - u(0) = - \tau_0\, Au_0 \in H. \tag{11.29}$$

Reciprocally, let u be a solution of (11.25), (11.27), (11.29), then

$$A\left(\frac{1}{\tau_0} \int_0^{\tau_0} u(t)\, dt \right) = - \frac{1}{\tau_0}\, (u(\tau_0) - u(0)) = Au_0,$$

whence (11.26) follows.

Therefore the original problem is *equivalent* to (11.25), (11.27), (11.29) and *this problem admits a unique solution* ("periodic" solutions of the equations of evolution; cf. Lions [1], p. 50 et seq.).[1] ∎

Remark 11.6. We have seen (Theorem 11.2) that, in general, $\mathrm{Inf}\, \tilde{J}(\xi) > 0$. If, however, we are in a case where

$$\mathrm{Inf}\, \tilde{J}(\xi) = 0$$

___

[1] Once the solution determined in $(0, \tau_0)$, $u(\tau_0)$ is known and we then solve (i.e., for $t > \tau_0$) the usual problem

$$u' + Au = 0, \quad u(\tau_0)\ \text{known}, \quad t > \tau_0.$$

(and, evidently, such cases exist! The difficulty—a problem of "controllability"—is to know *a priori* if this is the case), then *the QR method can be applied,*[1] *at least heuristically.*

We introduce u_ε the solution of

$$u'_\varepsilon + A(t)\, u_\varepsilon - \varepsilon A^*(t)\, A(t)\, u_\varepsilon(t) - \varepsilon u_\varepsilon(t + 2\tau) = 0, \quad t < T - \tau, \quad \textbf{(11.30)}$$

$$u_\varepsilon \in L_2(0, T + \tau; D(A(t))), \quad u'_\varepsilon \in L_2(0, T; D(A(t))'), \quad \textbf{(11.31)}$$

$$u_\varepsilon(t) = \chi(t) \text{ on the interval } [T - \tau, T + \tau]. \quad (^2) \quad \textbf{(11.32)}$$

We then take, as usual,

$$u_\varepsilon(0) = \xi.$$

We have performed in this way some numerical trials which are presented in Section 11.8. ∎

11.7 Numerical Applications: (I) Functional *J*

11.7.1 THE PROBLEM

We must find $\xi(x)$ so that if $u(x, t; \xi) = u(x, t)$ is the solution of

$$\left. \begin{array}{l} \dfrac{\partial u}{\partial t} - \dfrac{\partial^2 u}{\partial x^2} = 0, \qquad t > 0, \qquad x \in [0, 1], \\[2mm] u(x, 0) = \xi(x), \\[2mm] u(0, t) = u(1, t) = 0, \end{array} \right\} \qquad \textbf{(11.33)}$$

we have

$$\int_0^1 \left[\frac{1}{2\tau} \int_{T-\tau}^{T+\tau} u(x, t)\, \mathrm{d}t - \chi(x) \right]^2 \mathrm{d}x \leqslant \eta. \qquad \textbf{(11.34)}$$

Using the results of 11.6, we are led to

$$\left. \begin{array}{l} \dfrac{\partial u_\varepsilon}{\partial t} - \dfrac{\partial^2 u_\varepsilon}{\partial x^2} - \varepsilon\, \dfrac{\partial^4 u_\varepsilon}{\partial x^4} = 0, \\[3mm] \dfrac{1}{2\tau} \displaystyle\int_{T-\tau}^{T+\tau} u_\varepsilon(x, t)\, \mathrm{d}t = \chi(x), \qquad x \in [0, 1], \qquad t \in\,]0, T + \tau], \\[3mm] u_\varepsilon(0, t) = u_\varepsilon(1,\ t) = \dfrac{\partial^2 u_\varepsilon}{\partial x^2}(0, t) = \dfrac{\partial^2 u_\varepsilon}{\partial x^2}(1, t) = 0\,; \end{array} \right\} \qquad \textbf{(11.35)}$$

---

[1] This remark is *general*, it applies to all functionals.
[2] Of course, this condition does not imply that u_ε satisfy (11.30) in this interval.

Then we take

$$\xi(x) = u_\varepsilon(x, 0).$$

Integrating in the sense of increasing time and setting

$$w(x, t) = u_\varepsilon(x, T - t),$$

we are led to find $\xi = w(x, T)$, where w is the solution of

$$\left.\begin{aligned}
&\frac{\partial w}{\partial t} + \frac{\partial^2 w}{\partial x^2} + \varepsilon \frac{\partial^4 w}{\partial x^4} = 0, \qquad x \in (0, 1), \qquad t \geqslant -\tau, \\
&\int_{-\tau}^{\tau} w(x, t)\, dt = \chi(x), \\
&w(0, t) = w(1, t) = \frac{\partial^2 w}{\partial x^2}(0, t) = \frac{\partial^2 w}{\partial x^2}(1, t) = 0.
\end{aligned}\right\} \tag{11.36}$$

Remark 11.7. We are thus led to a mixed problem where the initial condition is replaced by a "thick initial" condition [cf. the second equation (11.36)]. ■

The method followed will consist of estimating

$$w(x, \tau) = \varphi(x) \tag{11.37}$$

in order to lead (for $t \geqslant \tau$) to the integration of a standard mixed problem, where $\varphi(x)$ is the initial condition at time $t = \tau$.

11.7.2 NUMERICAL SOLUTION

11.7.2.1 *Notation*

Set

$$J \Delta x = 1,$$

$$N \Delta t = 1,$$

$$\tau = v \Delta t,$$

$$w_j^n \simeq w(j \Delta x, n \Delta t), \qquad j \in [0, J], \qquad n = -v, -v + 1, ..., 0, ..., N.$$

The function φ ("initial condition") will be given in discrete fashion by

$$\varphi_j(\simeq \varphi(j \Delta x)) = w_j^v, \qquad j \in [0, J]. \tag{11.38}$$

11.7.2.2 Calculation of φ_j.

We consider a discretization scheme of two levels for the first equation of (11.36). If w^n denotes the vector of dimension $J - 1$ whose components are the $w_j^n (j \in [1, J - 1])$, the scheme furnishes the algebraic relation:

$$A w^{n+1} = B w^n + C_n^{n+1}, \qquad n \geqslant -v + 1, \tag{11.39}$$

where C_n^{n+1} denotes the vector corresponding to the boundary conditions. In the case of (11.36), $C_n^{n+1} \equiv 0$ (all that follows is equally valid if $C_n^{n+1} \not\equiv 0$).

The relation (11.36) yields in particular

$$w^k = (A^{-1} B)^{v+k} w^{-v}, \qquad k \in -[v + 1, + v]. \tag{11.40}$$

It remains to take account of the integral relation of (11.36). To do this, we discretize with the aid of a quadrature formula in t (for $x = j \Delta x$) which allows us to write a linear relation of the type:

$$\sum_{k=-v}^{+v} \lambda_k w^k = X, \tag{11.41}$$

$$X = \{ \chi(j \Delta x) \} \quad \text{is known.}$$

The system (11.40), (11.41) yields:

$$\sum_{k=-v}^{+v} \lambda_k (A^{-1} B)^{v+k} w^{-v} = X. \tag{11.42}$$

Assuming the nonsingularity of the matrix

$$S = \sum_{k=-v}^{v} \lambda_k (A^{-1} B)^{v+k},$$

we deduce from (11.42):

$$w^{-v} = S^{-1} X.$$

Then, from (11.40)

$$\Phi = w^v = TX, \qquad T = (A^{-1} B)^{2v} S^{-1}. \tag{11.43}$$

The numerical integration of the scheme associated with (11.36) for $n \geqslant v + 1$ then gives

$$w^n = (A^{-1} B)^{n-v} TX.$$ (11.44)

Thus Ξ (the desired approximation for ξ) is given by

$$\Xi = w^N = (A^{-1} B)^{N-v} TX.$$ (11.45)

11.7.2.3 Details of Application

It remains to choose A, B (corresponding to the discretization) and the λ_k (corresponding to the quadrature formula).

The discretization scheme adapted is the following:

$$w^{n+1} = - \varepsilon \frac{\Delta t}{\Delta x^4} w^n_{j+2} + \left(4 \frac{\varepsilon \Delta t}{\Delta x^4} - \frac{\Delta t}{\Delta x^2}\right) w^n_{j+1} + \left(2 \frac{\Delta t}{\Delta x^2} - 6 \frac{\varepsilon \Delta t}{\Delta x^4} + 1\right) w^n_j +$$ (11.46)

$$+ \left(4 \frac{\varepsilon \Delta t}{\Delta x^4} - \frac{\Delta t}{\Delta x^2}\right) w^n_{j-1} - \varepsilon \frac{\Delta t}{\Delta x^4} w^n_{j-2}, \quad 2 \leqslant j \leqslant J - 2$$

which, taking account of the boundary conditions, determines B. With the explicit scheme used, the matrix A is the identity.

For quadrature formulas, we have simply taken the trapezoid formula.

11.7.2.4 Verification of the Solution Obtained

We integrate the initial system (11.33) with the aid of the explicit scheme

$$u^{n+1}_j = u^n_j + \frac{\Delta t}{\Delta x^2} (u^n_{j+1} + u^n_{j-1} - 2 u^n_j)$$

starting with

$$\{ u^0_j \} = \Xi \quad \text{[given by 11.7.2.3 and (11.45)].}$$

We approximate to $\frac{1}{2\tau} \int_{T-\tau}^{T+\tau} u(x, t) \, dt$ by the trapezoid formula, which gives an approximation for J.

11.7.3 NUMERICAL EXAMPLES

The numerical trials used the following functions $(x \in (0, 1))$:

$$\chi(x) = x(1 - x) \qquad \text{(Graphs 11 and 11 bis)} \qquad (1)$$

$$\chi(x) = e^{-\pi^2/10} \sin \pi x \qquad \text{(Graphs 12 and 12 bis)} \qquad (2)$$

$$\chi(x) = \begin{cases} x \text{ for } x \in (0,4/10) \\ 1 - x \text{ for } x \in (6/10, 1) \\ 0,4 + \dfrac{0,2}{\pi} \cos 5 \pi \left(x - \dfrac{1}{2}\right) \text{ for } x \in \left(\dfrac{4}{10}, \dfrac{6}{10}\right) \end{cases} \quad \text{(Graphs 13 and 13 bis)} \qquad (3)$$

$$\chi(x) = \begin{cases} 0 \text{ for } x \in \left(0, \dfrac{4}{10}\right) \\ 1 \text{ for } x \in \left(\dfrac{6}{10}, 1\right) \\ \dfrac{1}{2}\left(1 + \sin 5 \pi \left(x - \dfrac{1}{2}\right)\right) \text{ for } x \in \left(\dfrac{4}{10}, \dfrac{6}{10}\right). \end{cases} \quad \text{(Graphs 14 and 14 bis)} \qquad (4)$$

In all of the examples, we use

$$T = 10^{-1}, \qquad \tau = 2 . 10^{-3}$$

and we choose

$$\Delta x = 10^{-1}, \qquad \Delta t = 10^{-3}, \qquad v = 2 .$$

On Graphs 11 *ter* to 14 *ter* are shown the mean values which are here the objective of the problem:

$$\chi^{**} = \frac{1}{2\tau} \int_{T-\tau}^{T+\tau} \chi^* \, dt .$$

Remark 11.8. In these examples, we observe, *as was to be expected*, that the error

$$e_\varepsilon = \int_0^1 | \chi_\varepsilon - \chi |^2 \, dx ,$$

where

$$\chi_\varepsilon(x) = \frac{1}{2\tau} \int_{T-\tau}^{T+\tau} u_\varepsilon(x, t)\, dt\,,$$

(where these quantities are calculated using the discrete values obtained) *passes through a minimum as* $\varepsilon \to 0$. In all these cases the ε optimum is approximately at 5.10^{-3}. As an example we constructed e_ε for the fourth case (cf. Graph 15). ∎

11.8 Numerical Applications: (II) The Functional $\tilde{J}(\xi)$.

11.8.1 THE PROBLEM

It is a question (cf. Remark 11.6) of finding $\xi(x)$ such that if:

$u(x, t\,;\xi) = u(x, t)$ is the solution of

$$\frac{\partial u}{\partial t} - \Delta u = f(x, t)\,, \qquad t > 0\,, \qquad x \in [0, 1]\,, \qquad \textbf{(11.47)}$$

$$u(x, 0) = \xi(x)\,, \qquad\qquad \textbf{(11.48)} \quad \text{(I)}$$

$$u(x, t) = 0\,, \qquad \forall t,\, \text{at } \begin{cases} x = 0 \\ x = 1 \end{cases} \qquad \textbf{(11.49)}$$

we have

$$\int_{T-\tau}^{T+\tau} \int_0^1 [u(x, t) - \chi(x)]^2 \, dx \, dt \text{ minimum}\,, \qquad\qquad \textbf{(11.50)}$$

where T and τ are given, τ "small."

Remark 11.9. For $f \neq 0$, cf. Remark 5.8. ∎

The Quasi-Reversibility method leads to looking for $\xi(x) = u(x, 0)$, where $u(x, t)$ is the solution of:

$$\frac{\partial u}{\partial t} - \Delta u - \varepsilon \Delta^2 u - u(x, t + 2\,\tau) = f(x, t)\,, \qquad \textbf{(11.51)}$$

$$u(x, t) = \chi(x) \text{ on } t \in [T - \tau, T + \tau] \qquad\qquad \textbf{(11.52)} \quad \text{(II)}$$

$$\begin{cases} u(x, t) = 0 \text{ on } x = 0 \text{ and } x = 1 \\ \text{and} \\ \Delta u = 0 \text{ on } x = 0 \text{ and } x = 1\,. \end{cases} \qquad \textbf{(11.53)}$$

Integrating in the sense of increasing time and putting

$$w(x, t) = u(x, T - \tau - t),\tag{11.54}$$

we are led to look for $\xi = w(x, T)$, where $w(x, t)$ is the solution of:

$$\frac{\partial w}{\partial t} + \Delta w + \varepsilon \Delta^2 w + \varepsilon w(x, t - 2\tau) = -f(x, t), t \in [0, T - \tau],\tag{11.55}$$

$$w(x, t) = \chi(x), \quad t \in (-2\tau, 0),\tag{11.56}$$ (III)

$$\left.\begin{array}{l} w(x, t) = 0 \\[6pt] \Delta w = 0 \end{array}\right\} \text{ if } x = 0 \text{ or } 1 \text{ and } 0 < t < T - \tau.\tag{11.57}$$

Remark 11.10. We are led therefore to the integration of equations with partial derivatives with time-lag. The theory of the numerical integration of these equations is an extension, without difficulty, apart from the technique, of, for example, Raviart [1]. Since, however, practical examples do not abound in the literature, we have deemed it worthwhile to present two schemes for solution. ∎

11.8.2 NUMERICAL SOLUTION

11.8.2.1 *Notation and discretization*

Let

$$w_i^n = w(i \Delta x, n \Delta t), \text{ where } i = 0, ..., I, \quad \text{with} \quad I \Delta x = 1,$$

$$n = 0, ..., N, \quad \text{with} \quad N \Delta t = T;$$

$$n_0 = \frac{2\tau}{\Delta t} \quad \text{integer}$$

$$(\Delta w)_i \simeq \frac{w_{i+1} - 2 w_i + w_{i-1}}{\Delta x^2};$$

$$(\Delta^2 w)_i \simeq \frac{w_{i+2} - 4 w_{i+1} + 6 w_i - 4 w_{i-1} + w_{i-2}}{\Delta x^4},\tag{11.58}$$

$$\left(\frac{\partial w}{\partial t}\right)^n \simeq \frac{w^{n+1} - w^n}{\Delta t},$$

$$w(x, t - 2\tau) = w^{n - n_0}.$$

At the end of the tests, we choose successively two solution schemes: one explicit, the other of the Crank–Nicolson type.

11.8.2.2 Explicit Scheme

Resolution of system (III).—Taking account of what precedes, the system (III) can be written:

$$w_i^{n+1} = w_i^n \left[1 + \Delta t \left(\frac{2}{\Delta x^2} - \frac{6\,\varepsilon}{\Delta x^4} \right) \right] - (w_{i-1}^n + w_{i+1}^n) \left[\Delta t \left(\frac{1}{\Delta x^2} - \frac{4\,\varepsilon}{\Delta x^4} \right) \right]$$

$$- (w_{i+2}^n + w_{i+2}^n) \frac{\varepsilon\,\Delta t}{\Delta x^4} - (\varepsilon w_i^{n-n_0} + f_i^n)\,\Delta t , \tag{11.59}$$

$$w_i^{-n_0} = w_i^{1-n_0} = \cdots = w_i^0 = \chi_i \tag{11.60}$$

$$w_0^n = w_I^n = 0 \tag{11.61}$$

$$\left\{ \begin{array}{l} w_{-1}^n = 2\,w_0^n - w_1^n \\ w_{I+1}^n = 2\,w_I^n - w_{I-1}^n \end{array} \right\} \quad \forall t \in (0, T - \tau) \Bigg\}. \tag{11.62}$$

We wish to find $\xi(x) = u(x, T)$.

The relation (11.59) can be put in matrix form:

$$\left\{ \begin{array}{l} w^{n+1} = Aw^n + I(-\varepsilon w^{n-n_0} + f^n)\,\Delta t , \\ w^{n+1} = Aw^n + C , \end{array} \right. \tag{11.63}$$

where A is a five-diagonal symmetric matrix of dimension $(I - 1) = N_1$ and C is a vector of the same dimension; we have special formulas for $i = 1, 2, N_1 - 1, N_1$; to these values correspond the only non-zero components of C.

Verification of the solution obtained. Starting with $\xi(x)$ found from (11.63) we look for a solution χ^* of the system (I) and then we test:

$$\int_{T-\tau}^{T+\tau} \int_0^1 [\chi^*(x, t) - \chi(x)]^2 \, dx \, dt \leqslant \eta .$$

Discretization of the system (I). We choose an analogous explicit scheme; system (I) becomes

$$u_i^{n+1} = u_i^n \left[1 - 2\frac{\Delta t}{\Delta x^2} \right] + \frac{\Delta t}{\Delta x^2} [u_{i+1}^n + u_{i-1}^n] + f_i^n \Delta t \, ,$$

$$u_i^0 = \xi_i \, ,$$

$$u_i^n = 0 \qquad \forall n \ \text{at} \ i = 0 , \ i = I . \tag{11.64}$$

Let, in matrix form

$$u^{n+1} = Au^n + C$$

where A is tridiagonal, symmetric of dimension N_1 and C is a vector of the same dimension (with zero components except for $i = 1$ and $N_1 = I - 1$). It remains to evaluate the integral (11.50), which is to say, to verify:

$$\Delta x \, \Delta t \left\{ \sum_{n=N-\frac{n_0}{2}+1}^{n=N+\frac{n_0}{2}-1} \left[\sum_{i=1}^{I-1} (\chi_i^{*n} - \chi_i)^2 + \sum_{i=1}^{I-1} [(\chi_i^{*N-n_0/2} - \chi_i)^2 + (\chi_i^{*N+n_0/2} - \chi_i)^2] \right] \right\} \leqslant \eta ,$$

$$\tag{11.65}$$

where η is a given small quanity.

11.8.2.3 Implicit Scheme

Resolution of the system (III). We choose a scheme of Crank–Nicolson type with the same notation as for the explicit scheme: Equation (11.55) became:

$$\frac{w_i^{n+1} - w_i^n}{\Delta t} + \frac{w_{i+1}^{n+1} - 2 w_i^{n+1} + w_{i-1}^{n+1}}{\Delta x^2} + \frac{w_{i+1}^n - 2 w_i^n + w_{i+1}^n}{\Delta x^2} +$$

$$+ \varepsilon \left[\frac{w_{i+2}^{n+1} - 4 w_{i+1}^{n+1} + 6 w_i^{n+1} - 4 w_{i-1}^{n+1} + w_{i-2}^{n+1}}{\Delta x^4} + \frac{w_{i+2}^n - 4 w_{i+1}^n + 6 w_i^n - 4 w_{i-1}^n + w_{i-2}^n}{\Delta x^4} \right] +$$

$$+ \frac{\varepsilon}{2} [w_i^{n-n} {}^{+1_0} + w_i^{n-n_0}] = \frac{1}{2} [f_i^{n+1} + f_i^n] ,$$

$$\tag{11.66}$$

and we have, in addition, conditions at the boundary identical with those in the explicit case:

$$w_i^{-n_0} = w_i^{1-n_0} = \cdots = w_i^0 = \chi_i, \tag{11.60}$$

$$w_0^n = w_I^n = 0 \qquad \forall n, \tag{11.61}$$

$$w_{-1}^n = 2 w_0^n - w_1^n \qquad \forall n, \ t \in (0, T - \tau), \tag{11.62}$$

$$w_{I+1}^n = 2 w_I^n - w_{I-1}^n \qquad \forall n.$$

We can write (11.66) in matrix form

$$A w^{n+1} = B w^n + D[w^{n-n_0+1} + w^{n-n_0}] + [f_i^{n+1} + f_i^n], \tag{11.67}$$

where A and B, of dimension $N_1 = I - 1$, are two five-diagonal symmetric matrices whose coefficients along the non-principal diagonals are constant and where D is a diagonal matrix of constant term $(- \varepsilon \Delta t)$, of dimension N_1. There are still special formulas for $i = 1, 2, N_1 - 1, N_1$.

Verification of the solution obtained. We calculate, starting from ξ, the solution χ^* of the system (I) according to an implicit scheme of Crank–Nicolson type, with the same notation as above:

$$u_i^{n+1} \left[1 - \frac{\Delta t}{\Delta x^2} \right] + [u_{i-1}^{n+1} + u_{i-1}^{n+1}] \left(- \frac{\Delta t}{2 \Delta x^2} \right) =$$
$$= u_i^n \left[1 - \frac{\Delta t}{\Delta x^2} \right] + (u_{i-1}^n + u_{i+1}^n) \frac{\Delta t}{2 \Delta x^2} + (f_i^{n+1} + f_i^n) \frac{\Delta t}{2}, \tag{11.68}$$

with the conditions:

$$u_i^0 = \xi_i, \tag{11.69}$$

$$u_0^n = u_I^n = 0 \qquad \forall n, \tag{11.70}$$

a system which may be written in matrix form

$$A u^{n+1} = B u^n + C, \tag{11.71}$$

where C is a vector of dimension $N_1 = I - 1$, whose components are:

$$C(i) = (f_i^{n+1} + f_i^n) \Delta t,$$

and A and B are two tridiagonal symmetric matrices of the same dimension N_1.

Once the solution χ^* has been found, starting from (11.71), it remains to verify that

$$|\chi - \chi^*| \leqslant \eta,$$

This is done exactly the same way as for the explicit scheme.

11.8.3 NUMERICAL EXAMPLES

In all cases $f = 0$. The numerical examples were carried out using the following functions; both schemes gave practically equivalent results:

$\chi = x(1 - x)$ 　　　　　　　　　　　(Graphs 16, 16 *bis* and 16 *ter*)

$\chi = x$ on $[0 ; 0.4]$

　　$= 1 - x$ on $[0.6 ; 1]$ 　　　　　　(smoothing in 0.4 and 0.6)

　　　　　　　　　　　　　　　　　　　(Graphs 17, 17 *bis* and 17 *ter*)

$\chi = e^{-\pi^2 t} \sin \pi x \qquad t \in (T - \tau, T + \tau)$ 　　(Graphs 18 and 18 *bis*)

$\chi = e^{-\pi^2 T} \sin \pi x$ 　　　　　　　　(Graphs 19 and 19 *bis*)

$$\chi = \begin{cases} 0 & \text{on} \quad [0 ; 0.4] \\ 1 & \text{on} \quad [0.6 ; 1] \\ \dfrac{1}{2}\left(1 + \sin\left(\dfrac{x - 0.5}{0.1}\,\dfrac{\pi}{2}\right)\right) & \text{on} \quad [0.4 ; 0.6] \end{cases}$$

(Graphs 20 and 20 *bis*)

We applied the method of improvement of point 8.4.1 to the last example. Thanks to linearity, it is essentially equivalent to balance the ξ or the χ.

12. EQUATIONS WITH TIME-LAG

12.1 Equations of Evolution with Time-Lag

The notation is that of Section 3. In addition, let B be an operator with

$$B \in \mathscr{L}(V; V') \qquad (^1). \tag{12.1}$$

$(^1)$More generally we could consider a *family* $B(t) \in \mathscr{L}(V; V')$ such that $(B(t) v, w)$ is measurable $\forall v, w \in V$ and such that $\| B(t) \|_{\mathscr{L}(V;V')} \leqslant$ constant.

For $\tau_0 > 0$ fixed, consider the equation

$$u'(t) + A(t)\, u(t) + Bu(t - \tau_0) = f(t) , \qquad t > 0 \qquad (^1)\ (^2) \qquad (12.2)$$

with

$$u = 0 \qquad \text{for} \qquad t \in]-\tau_0, 0[, \tag{12.3}$$

$$u \in L_2(0, T; V) , \qquad u' \in L_2(0, T; V') , \tag{12.4}$$

$$u(0) = \xi , \qquad \xi \text{ given in } H . \tag{12.5}$$

This is an equation of evolution with a time-lag. The solution of the problem is immediate; under the hypothesis of Theorem 3.1, we have existence and uniqueness.

In effect, in $]0, \tau_0[$, (12.2) reduces to, taking account of (12.3),

$$u'(t) + A(t)\, u(t) = f(t) , \qquad t \in]0, \tau_0[, \tag{12.6}$$

which, combined with (12.5) defines u in a unique fashion in $(0, \tau_0)$. Write $u = u_1$, with

$$u_1 \in L_2(0, \tau_0 ; V) , \qquad u_1' \in L_2(0, \tau_0 ; V') . \tag{12.7}$$

In $(\tau_0, 2\,\tau_0)$, (12.2) may then be written

$$u'(t) + A(t)\, u(t) = -\, Bu_1(t - \tau_0) + f(t) \tag{12.8}$$

and, thanks to (1.27), $Bu_1(t - \tau_0) \in L_2(\tau_0, 2\,\tau_0 ; V')$, and thus (12.8) combined with

$$u(\tau_0) = u_1(\tau_0) , \tag{12.9}$$

defines u in a unique fashion in $(\tau_0, 2\,\tau_0)$. Write $u = u_2$ and so on. ∎

12.2 The Functional $J(\xi)$.

We are going to give a negative result which shows the essential difference between the cases with and without time-lag. We consider

---

$(^1)f$ is given in $L_2(0, T; V')$.
$(^2)$See the case of delay functions in t in Artola [1].

the "point" functional

$$J(\xi) = |u(T) - \chi|^2, \tag{12.10}$$

where χ is given in H, and $u = u(t; \xi)$, the solution of (12.2), . . . , (12.5) with $f = 0$.

Theorem 12.1: *For equations of evolution with time-lag, we do not in general have* Inf $J(\xi) = 0$.

Proof. 1) Let us suppose that $\underset{\xi \in H}{\text{Inf}}\ J(\xi) \neq 0$. Then there exists a $\psi \in H,\ \neq 0$, such that

$$(u(T), \psi) = 0 \qquad \forall \xi.$$

Let $v(t)$ be the solution of

$$\left.\begin{aligned}
& -\,v'(t) + A^*(t)\,v(t) + B^*\,v(t + \tau_0) = 0\,, \qquad t < T,\\
& v(t) = 0\,, \qquad t \in]T,\, T + \tau_0[\,,\\
& v(T) = \psi\,,
\end{aligned}\right\} \tag{12.12}$$

(a problem which admits a unique solution after 12.1). We have

$$\int_0^T [(u',v) + (u,v')]\,\mathrm{d}t = -\,(\xi, v(0))$$

[taking account of (12.11)] and, from (12.2) (where $f = 0$) and (12.12)

$$\int_0^T [(u',v) + u,v')]\,\mathrm{d}t =$$

$$= \int_0^T [(-A(t)u,v) - (Bu(t - \tau_0),v) + (u(t), A^*(t)v) + (u(t), B^*v(t + \tau_0))]\,\mathrm{d}t$$

$$= -\int_{-\tau_0}^0 (Bu(t), v(t + \tau_0))\,\mathrm{d}t - \int_0^{T-\tau_0} (Bu(t), v(t + \tau_0))\,\mathrm{d}t\ +$$

$$+ \int_0^{T-\tau_0} (Bu(t), v(t + \tau_0))\,\mathrm{d}t + \int_{T-\tau_0}^T (Bu(t), v(t + \tau_0))\,\mathrm{d}t\ ;$$

Using (12.3), the first integral is zero. From the second condition (12.2), the last integral is zero. Therefore

$$\int_0^T [(u', v) + (u, v')] \, dt = 0$$

and thus

$$(\xi, v(0)) = 0 \qquad \forall \xi \, ,$$

or

$$v(0) = 0 \, . \tag{12.13}$$

Reciprocally, if there exist T, τ_0, $A(t)$, B, $\psi \neq 0$, *such that the solution* v *of* (12.12) *satisfies* $v(0) = 0$, then $\forall \xi$ we have (12.11) and the theorem.

2) We are going to construct an example of this situation. We suppose

$$A^* = A, \qquad \text{independent of } T, \qquad V \to H \text{ compact.}$$

Let w_j, λ_j be characteristic functions and values of A:

$$A w_j = \lambda_j w_j \, , \qquad w_j \in V \, .$$

Let us fix j; write w and λ in place of w_j, λ_j.
Let τ_0 be chosen so that

$$\tau_0 = e^{-\lambda \tau_0} \tag{12.14}$$

and T such that

$$T = 2 \tau_0 \; ; \tag{12.15}$$

Let us take

$$B = \text{identity}, \tag{12.16}$$

$$\psi = w \, . \tag{12.17}$$

*We are going to show that the cor-
responding solution of* (12.12) *satisfies* $v(0) = 0$.

The solution v is of the form $v(t) = g(t) \, w$, $g(t)$ a scalar, with

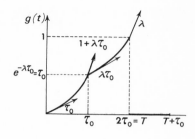

$$\left.\begin{array}{l} - g'(t) + \lambda g(t) + g(t + \tau_0) = 0 \,, \\[2mm] g(t) = 0 \qquad \text{in} \qquad]T, T + \tau_0[\,, \\[2mm] g(T) = 1 \,. \end{array}\right\}$$

Figure 3

We find then that

$$g(t) = e^{\lambda(t - T)} \quad \text{for} \quad t \in (T - \tau_0, T) \,,$$

then

$$g(t) = (t - (T - \tau_0) + e^{-\lambda \tau_0}) \, e^{\lambda(t + \tau_0 - T)} \quad \text{for} \quad t \in (T - 2 \tau_0, T - \tau_0) \,.$$

But, as a consequence of (12.15), we then have

$$g(0) = (- \tau_0 + e^{-\lambda \tau_0}) \, e^{-\lambda \tau_0}$$

and, from (12.14), $g(0) = 0$, whence $v(0) = 0$, and the result. ∎

Remark 12.1. The problem of the calculation of Inf $J(\xi)$ is open. Let us merely note here that the observations of Remark 11.6 may be applied to the present case.

13. OTHER TYPES OF EQUATIONS

13.1 Parabolic Equations in Noncylindrical Open Domains

The considerations of this chapter are valid for linear parabolic equations in noncylindrical domains (see Figure 4). We are given $\chi(x)$ in Ω_T and we wish to determine $\xi(x)$ (or Ω_0) so that the solution $u(x, t)$ of

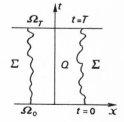

$$\frac{\partial u}{\partial t} - \Delta u = 0 \qquad\qquad (13.1)$$

in the noncylindrical domain, with

Figure 4

$$u(x, 0) = \xi(x), \tag{13.2}$$

$$u(x, t) = 0 \quad \text{if} \quad (x, t) \in \Sigma, \tag{13.3}$$

satisfies

$$\int_{\Omega_T} [u(x, T) - \chi(x)]^2 \, dx \leqslant \eta. \tag{13.4}$$

The QR methods previously studied can be adapted without difficulty to this case, subject to all of the variants already contemplated.

For the numerical integration of parabolic equations of evolution (or of other types) in noncylindrical domains, see A. Mignot [1] and Commincioli [1]. ■

13.2 Singular Equations

Analogous remarks hold for *singular* integro-differential evolution systems of parabolic type (arising in diverse applications) which have been studied in Y. Cherruault [1]. ■

13.3 Evolution Equations with Unbounded Coefficients

The theory reviewed in Section 3 may be extended to the case of operators $A(t) + \dfrac{d}{dt}$, where

$$(A(t)\, v, v) + \lambda(t)\,|\, v\,|^2 \geqslant \alpha\, \|\, v\, \|^2, \quad \alpha > 0, \tag{13.5}$$

where $t \to \lambda(t)$ is a *summable* function [and not a constant as in (3.7)]. In the same way the theory of *approximation* of solution may be extended (cf. Lions—Raviart [2]). The *QR method extends to this case* (at least *heuristically*), it would be interesting to study the *backward uniqueness* for operators $A(t) + \dfrac{d}{dt}$ satisfying (13.5).

Examples of this situation arise in the *linearization* of nonlinear evolution systems. ■

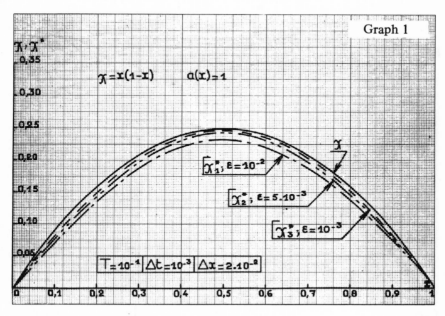

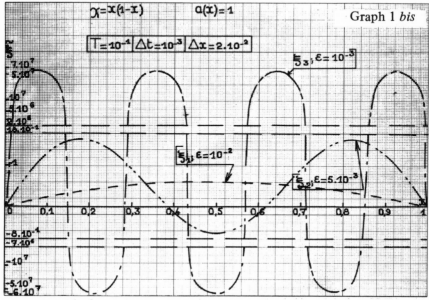

NOTE: In all graphs commas represent decimal points.

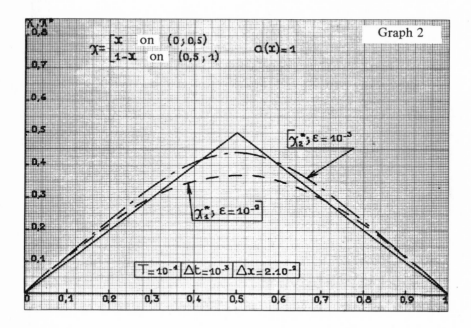

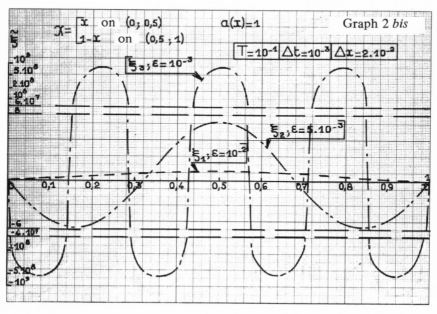

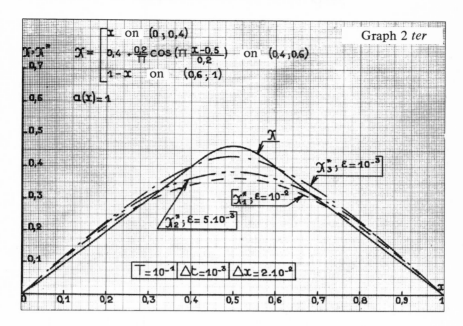

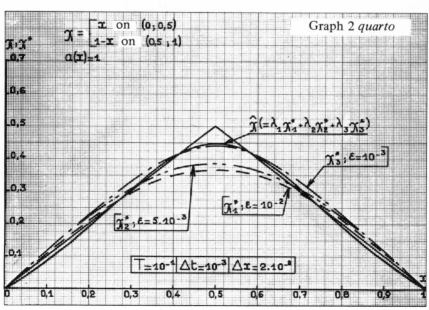

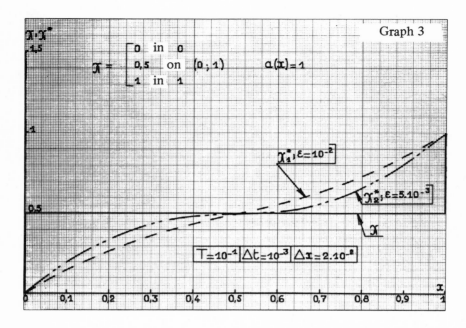

Graph 3

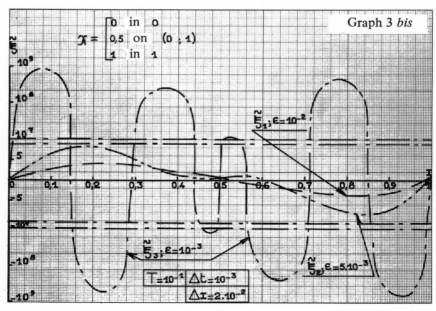

Graph 3 *bis*

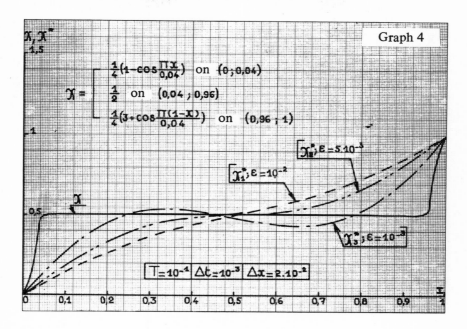

Graph 4

$$\mathfrak{X} = \begin{cases} \frac{1}{4}\left(1-\cos\frac{\pi x}{0,04}\right) & \text{on } (0;0,04) \\ \frac{1}{2} & \text{on } (0,04;0,96) \\ \frac{1}{4}\left(3+\cos\frac{\pi(1-x)}{0,04}\right) & \text{on } (0,96;1) \end{cases}$$

$\mathfrak{X}_{\mathfrak{n}}^{\ast}; \varepsilon = 5.10^{-3}$

$\mathfrak{X}_{1}^{\ast}; \varepsilon = 10^{-2}$

$\mathfrak{X}$

$\mathfrak{X}_{3}^{\ast}; \varepsilon = 10^{-3}$

$\boxed{T = 10^{-1}} \boxed{\Delta t = 10^{-3}} \boxed{\Delta x = 2.10^{-2}}$

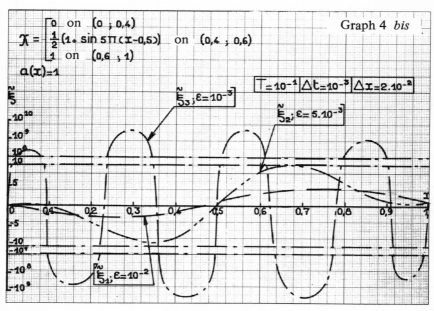

$$\mathfrak{X} = \begin{cases} 0 & \text{on } (0;0,4) \\ \frac{1}{2}(1+\sin 5\pi(x-0,5)) & \text{on } (0,4;0,6) \\ 1 & \text{on } (0,6;1) \end{cases}$$

$a(x)=1$

Graph 4 *bis*

$\tilde{\xi}_{3}; \varepsilon = 10^{-3}$

$\tilde{\xi}_{2}; \varepsilon = 5.10^{-3}$

$\boxed{T = 10^{-1}} \boxed{\Delta t = 10^{-3}} \boxed{\Delta x = 2.10^{-2}}$

$\tilde{\xi}_{1}; \varepsilon = 10^{-2}$

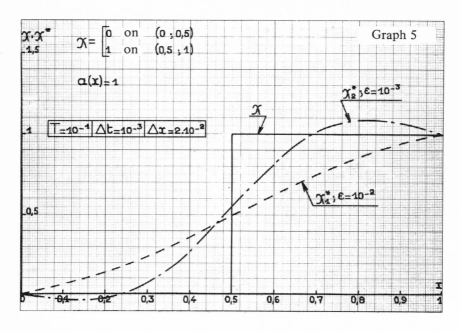

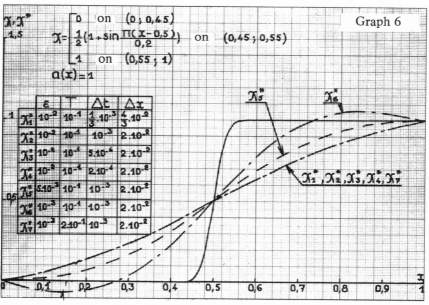

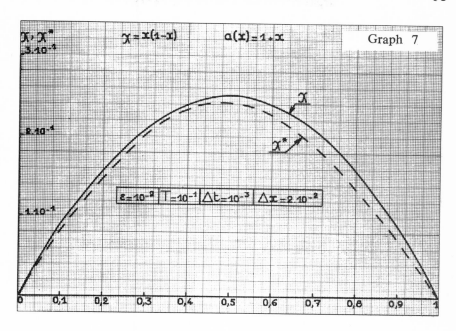

Graph 7

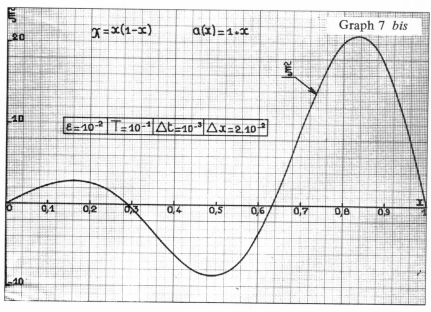

Graph 7 *bis*

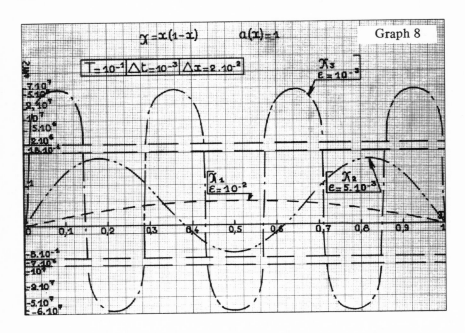

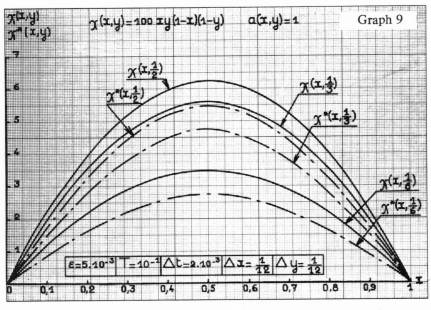

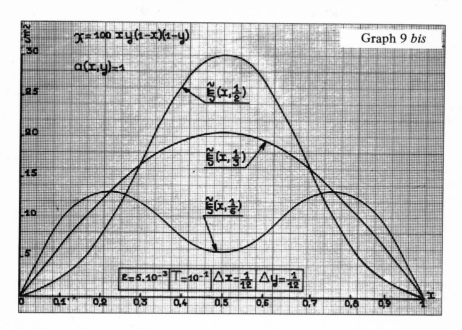

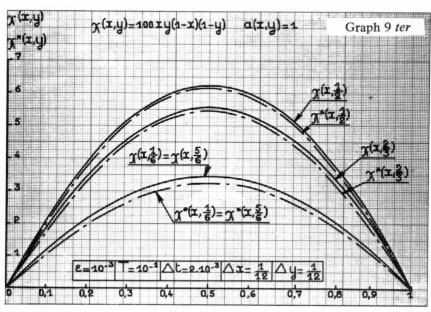

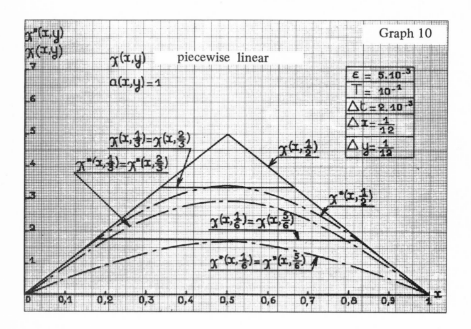

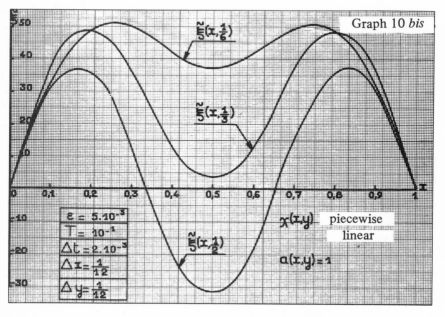

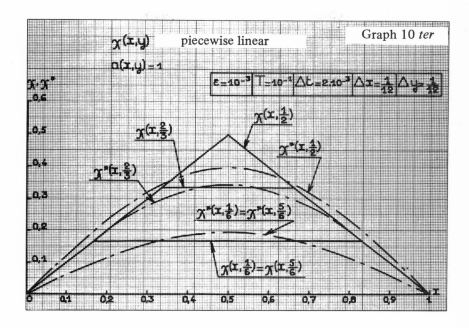

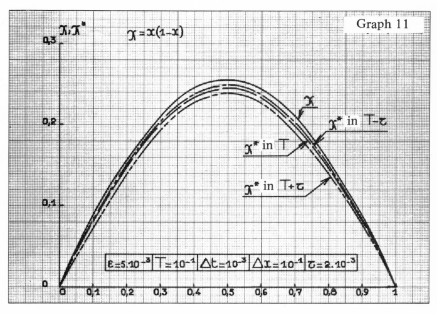

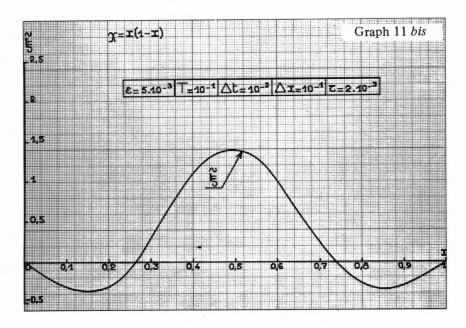

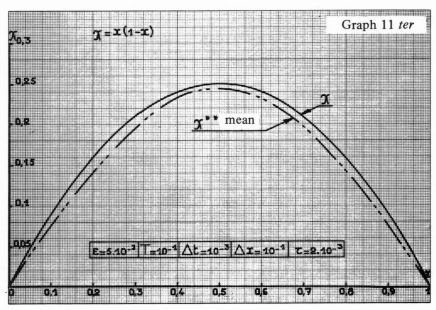

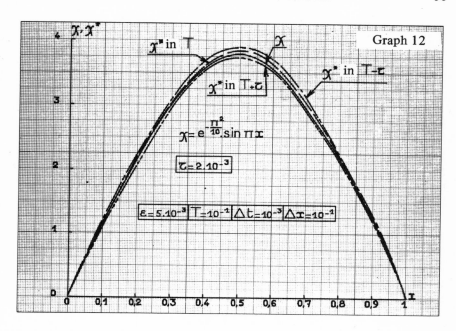

Graph 12

$\chi^{\ast}$ in T

χ

$\chi^{\ast}$ in $T+\tau$

$\chi^{\ast}$ in $T+\tau$

$\chi = e^{-\frac{\pi^2}{10}} \sin \pi x$

$\tau = 2 \cdot 10^{-3}$

$\varepsilon = 5 \cdot 10^{-3}$ $T = 10^{-1}$ $\Delta t = 10^{-3}$ $\Delta x = 10^{-1}$

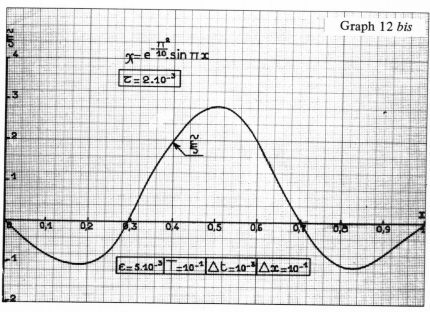

Graph 12 *bis*

$\chi = e^{-\frac{\pi^2}{10}} \sin \pi x$

$\tau = 2 \cdot 10^{-3}$

$\tilde{\xi}$

$\varepsilon = 5 \cdot 10^{-3}$ $T = 10^{-1}$ $\Delta t = 10^{-3}$ $\Delta x = 10^{-1}$

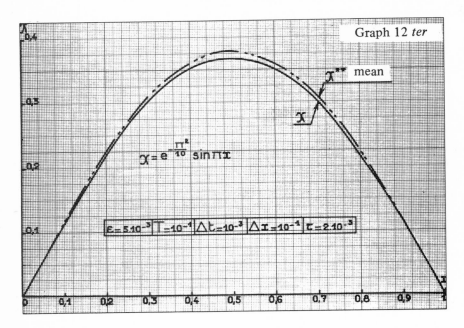

Graph 12 *ter*

$$\chi = e^{-\frac{\pi^2}{10}} \sin \pi x$$

$\varepsilon = 5 \cdot 10^{-3}$ | $T = 10^{-1}$ | $\triangle t = 10^{-3}$ | $\triangle x = 10^{-1}$ | $\tau = 2 \cdot 10^{-3}$

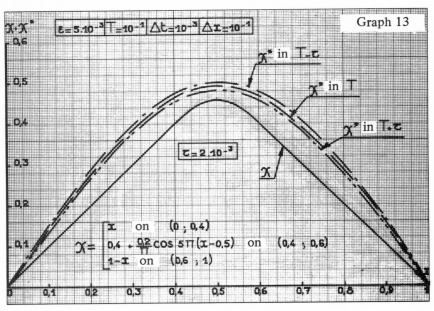

Graph 13

$\varepsilon = 5 \cdot 10^{-3}$ | $T = 10^{-1}$ | $\triangle t = 10^{-3}$ | $\triangle x = 10^{-1}$

χ^* in $T - \tau$

χ^* in T

χ^* in $T + \tau$

$\tau = 2 \cdot 10^{-3}$

$$\chi = \begin{bmatrix} x & \text{on} & (0 \, ; \, 0.4) \\ 0.4 + \frac{0.2}{\pi} \cos 5\pi(x-0.5) & \text{on} & (0.4 \, , \, 0.6) \\ 1-x & \text{on} & (0.6 \, ; \, 1) \end{bmatrix}$$

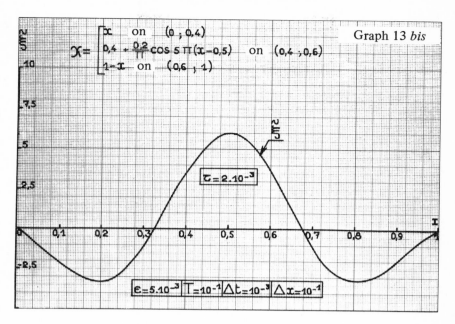

Graph 13 *bis*

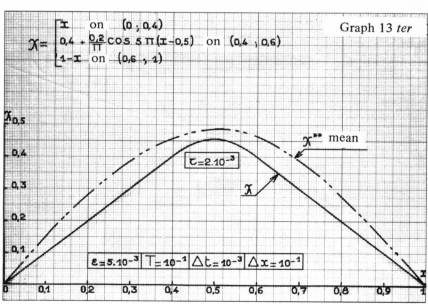

Graph 13 *ter*

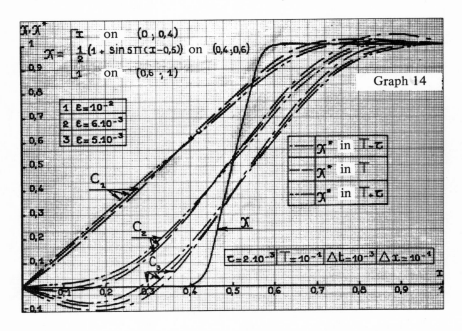

Graph 14

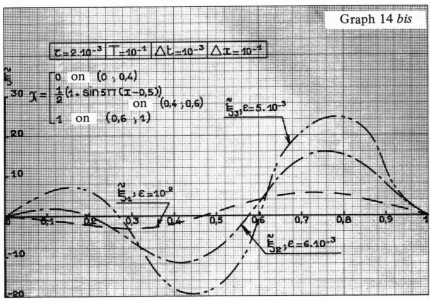

Graph 14 *bis*

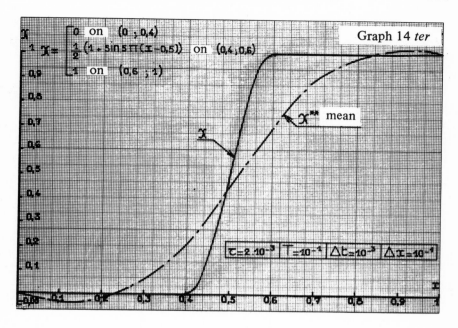

$$X = \begin{cases} 0 & \text{on} \quad (0 \; ; 0.4) \\ \frac{1}{2}(1 + \sin 5\pi(x - 0.5)) & \text{on} \quad (0.4 \; ; 0.6) \\ 1 & \text{on} \quad (0.6 \; ; 1) \end{cases}$$

Graph 14 *ter*

x^{**} mean

x

$$\boxed{\mathcal{C} = 2 \cdot 10^{-3}} \quad \boxed{T = 10^{-1}} \quad \boxed{\Delta t = 10^{-3}} \quad \boxed{\Delta x = 10^{-1}}$$

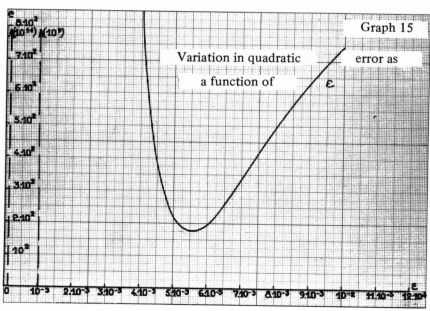

Graph 15

Variation in quadratic error as a function of ε

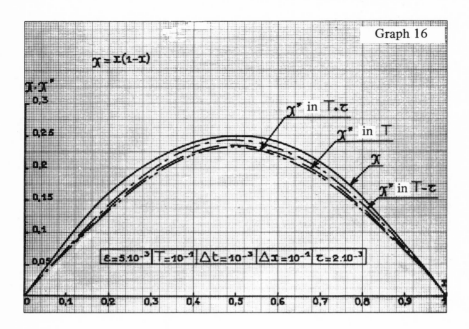

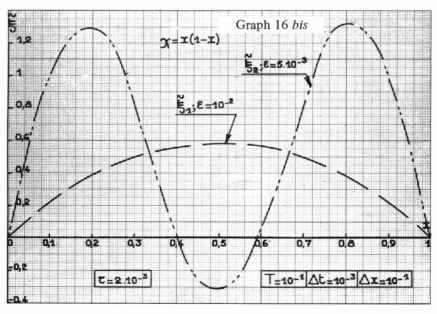

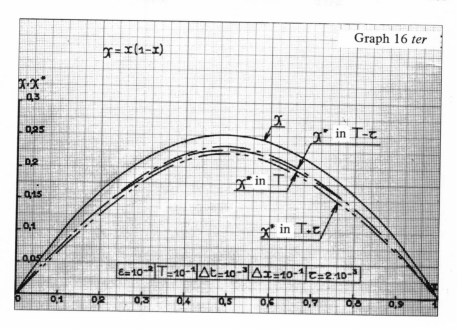

Graph 16 *ter*

$$\chi = x(1-x)$$

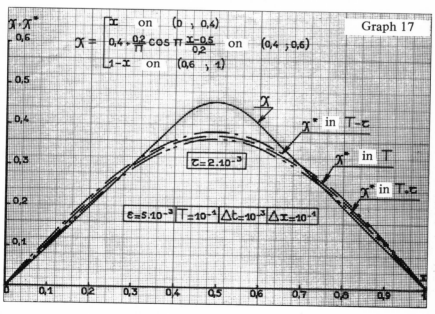

Graph 17

$$\chi = \begin{cases} x & \text{on} \quad (0 \ , \ 0,4) \\ 0,4 + \dfrac{0,2}{\pi}\cos\pi\,\dfrac{x-0,5}{0,2} & \text{on} \quad (0,4 \ , \ 0,6) \\ 1-x & \text{on} \quad (0,6 \ , \ 1) \end{cases}$$

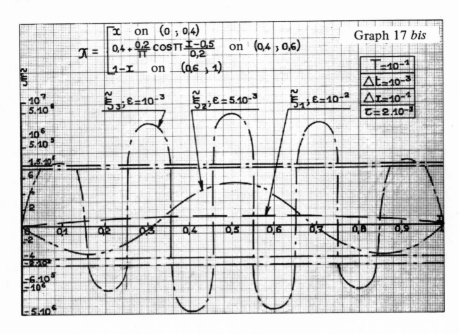

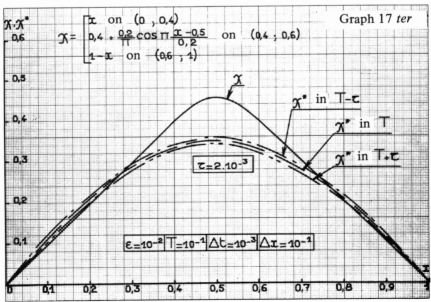

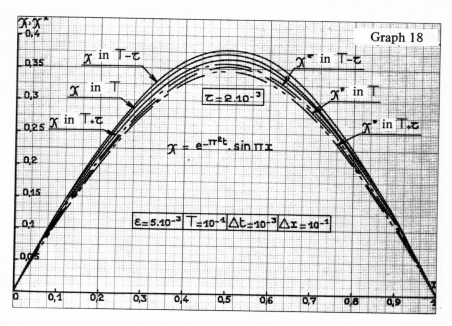

Graph 18

χ in $T-\tau$

χ^* in $T-\tau$

χ in T

χ^* in T

$\tau = 2 \cdot 10^{-3}$

χ in $T+\tau$

χ^* in $T+\tau$

$\chi = e^{-\pi^2 t} \cdot \sin \pi x$

$\varepsilon = 5 \cdot 10^{-3}$ $T = 10^{-1}$ $\Delta t = 10^{-3}$ $\Delta x = 10^{-1}$

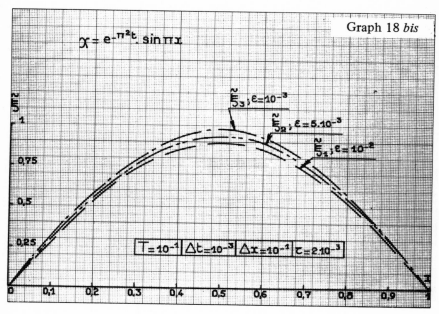

Graph 18 bis

$\chi = e^{-\pi^2 t} \cdot \sin \pi x$

$\tilde{\Xi}_3; \varepsilon = 10^{-3}$

$\tilde{\Xi}_2; \varepsilon = 5 \cdot 10^{-3}$

$\tilde{\Xi}_1; \varepsilon = 10^{-2}$

$T = 10^{-1}$ $\Delta t = 10^{-3}$ $\Delta x = 10^{-1}$ $\tau = 2 \cdot 10^{-3}$

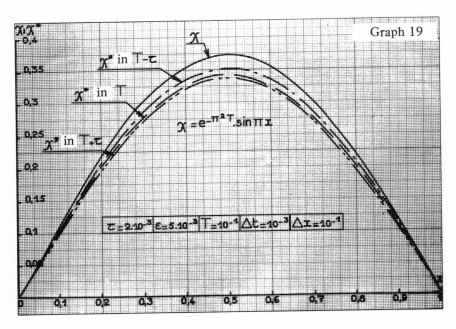

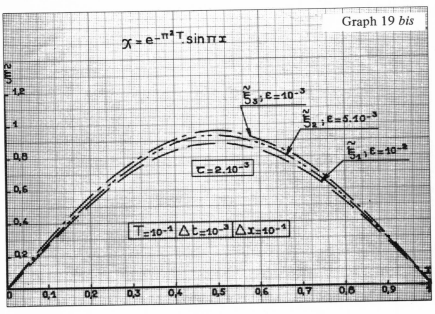

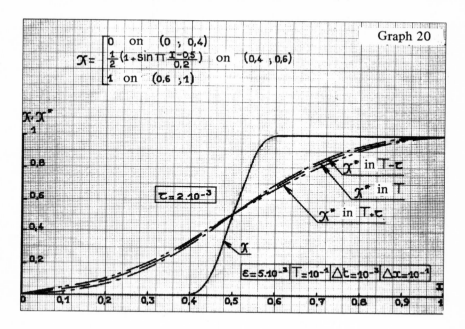

Graph 20

$$\chi = \begin{cases} 0 & \text{on } (0\ ;\ 0,4) \\ \frac{1}{2}\left(1+\sin\pi\frac{x-0,5}{0,2}\right) & \text{on } (0,4\ ;\ 0,6) \\ 1 & \text{on } (0,6\ ;\ 1) \end{cases}$$

$\chi\cdot\chi^*$

$\tau = 2\cdot10^{-3}$

χ^* in $T-\tau$

χ^* in T

χ^* in $T_+\tau$

χ

$\varepsilon = 5\cdot10^{-3}$ | $T = 10^{-1}$ | $\Delta t = 10^{-3}$ | $\Delta x = 10^{-1}$

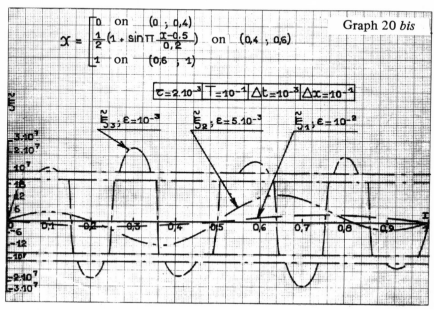

Graph 20 bis

$$\chi = \begin{cases} 0 & \text{on } (0\ ;\ 0,4) \\ \frac{1}{2}\left(1+\sin\pi\frac{x-0,5}{0,2}\right) & \text{on } (0,4\ ;\ 0,6) \\ 1 & \text{on } (0,6\ ;\ 1) \end{cases}$$

$\tau = 2\cdot10^{-3}$ | $T = 10^{-1}$ | $\Delta t = 10^{-3}$ | $\Delta x = 10^{-1}$

$\tilde{\xi}_3 ; \varepsilon = 10^{-3}$

$\tilde{\xi}_2 ; \varepsilon = 5\cdot10^{-3}$

$\tilde{\xi}_1 ; \varepsilon = 10^{-2}$

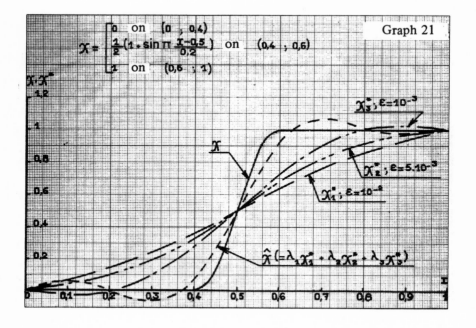

Chapter 2

NONPARABOLIC EQUATIONS
OF EVOLUTION

DETAILED OUTLINE

1. IRREVERSIBLE OPERATORS AND THE QR METHOD

1.1 Orientation

Up to now, we have applied the QR method to *parabolic* operators.

We will see in this chapter that the QR method is useful in analogous problems for evolution operators which are *nonparabolic where the sense of time cannot be reversed.*[1]

We will give three classes of such operators, in 1, 2, then 3, 4 and finally 5, 6 respectively.

1.2 First Example

Consider the spaces V, H, V' as in Chapter 1, Section 3.

We are given two families of operators $A(t)$, $B(t)$: the following hypotheses are made:

$$\left.\begin{array}{l} a(t\,;u,v) = (A(t)\,u,v) = a(t\,;v,u) \qquad \forall u, v \in V, \\ t \to a(t\,;u,v) \text{ is continuous with bounded measurable derivative in } [0, T] \\ a(t\,;v,v) + \lambda\,|\,v\,|^2 \geqslant \alpha\,\|\,v\,\|^2 \qquad \forall v \in V, \qquad \alpha > 0 \end{array}\right\} \text{(1.1)}$$

$$\left.\begin{array}{l} B(t) \in \mathscr{L}(V, V') \\ (B(t)\,u,v) \text{ is measurable and bounded } \forall u, \ v \in V, \\ (B(t)\,v,v) \geqslant 0 \qquad \forall v \in V. \end{array}\right\} \text{(1.2)}$$

It can be shown that (Lions [1]),

Theorem 1.1: *Under the hypotheses of* (1.1), (1.2), *there exists one and only one function u satisfying:*

$$u \in L_\infty(0, T\,;V), u' \in L_\infty(0, T\,;H), u'' \in L_2(0, T\,;V') \tag{1.3}$$

$$u'' + B(t)\,u' + A(t)\,u = f, \qquad f \text{ given in } L_2(0, T\,;H), \tag{1.4}$$

---

[1] We will see in Chapter 3, Section 2, that the QR method is useful when the unknown functions appear in the boundary conditions, even for operators where the time sense can be reversed.

$$u(0) = \xi_0, \qquad \xi_0 \quad \text{given} \quad V, \tag{1.5}$$

$$u'(0) = \xi_1, \qquad \xi_1 \quad \text{given} \quad H. \tag{1.6}$$

It can also be shown (Torelli [1], and Lions-Magenes [1], vol. 1) that u and u' are continuous from $[0, T] \rightarrow V$ and H respectively—after possible modification on a set of measure zero.

Remark 1.1. The hypotheses in (1.1), (1.2) are *not* the only ones under which the problem (1.4), (1.5), (1.6) is well posed. Another class of examples will be found in Section 3 and other cases in Lions [1]. ∎

Remark 1.2. Let us examine a particular case which will be useful in what follows. Suppose that

$$A(t) = A \text{ does not depend on } t$$

and that

$$B(t) = A.$$

Changing u into $e^{\lambda t} u$, we can suppose that

$$(Av, v) \geqslant \alpha_0 \| v \|^2, \qquad \alpha_0 > 0.$$

Let us make explicit the solution of (1.4), (1.5) and (1.6) with $f = 0$. This will be useful in what follows and demonstrates Theorem 1.1 in this particular case. ∎

We diagonolize A as in Chapter 1, Section 4, in the proof of Theorem 4.1. With the notation of this proof, set

$$\hat{u}(\lambda, t) = (\mathcal{U}u(t))(\lambda).$$

We have:

$$\frac{d^2\hat{u}}{dt^2} + \lambda \frac{d\hat{u}}{dt} + \lambda \hat{u} = 0, \tag{1.7}$$

with

$$\hat{u}(\lambda, 0) = \hat{\xi}_0(\lambda) \qquad (= (\mathcal{U}\xi_0)(\lambda)), \tag{1.8}$$

$$\frac{d\hat{u}}{dt}(\lambda, 0) = \hat{\xi}_1(\lambda) \qquad (= (\mathcal{U}\xi_1)(\lambda)).$$

(1.9)

Let r_1^0 and r_2^0 be the roots of the *characteristic equation*.[1]

$$r^2 + \lambda r + \lambda = 0.$$

(1.10)

Let us introduce the *fundamental solution* $E_i(\lambda, t)$, $i = 0, 1$, which satisfies

$$\frac{d^2}{dt^2} E_i(\lambda, t) + \lambda \frac{d}{dt} E_i(\lambda, t) + \lambda E_i(\lambda, t) = 0,$$

(1.11)

$$\left. \begin{array}{ll} E_0(\lambda, 0) = 1, & \frac{d}{dt} E_0(\lambda, 0) = 0, \\[2mm] E_1(\lambda, 0) = 0, & \frac{d}{dt} E_1(\lambda, 0) = 1. \end{array} \right\}$$

(1.12)

Explicitly:

$$E_0(\lambda, t) = \frac{1}{r_1^0 - r_2^0} \left[- r_2^0 \, e^{r_1^0 t} + r_1^0 \, e^{r_2^0 t} \right],$$

(1.13)

$$E_1(\lambda, t) = \frac{1}{r_1^0 - r_2^0} \left[e^{r_1^0 t} - e^{r_2^0 t} \right].$$

(1.14)

Then

$$\hat{u}(\lambda, t) = E_0(\lambda, t) \, \hat{\xi}_0(\lambda) + E_1(\lambda, t) \, \hat{\xi}_1(\lambda),$$

(1.15)

and using the values of r_1^0, r_2^0 (as $\lambda \to +\infty$), we deduce that the problem is properly-posed (in the sense of Therom 1.1). ∎

Remark 1.3. Let us take (using the notation of Chapter 1, Sections 1 and 6)

$$V = H_0^1(\Omega), \qquad H = L_2(\Omega),$$

$$A = B = -\Delta;$$

[1] $r_i^0 = r_i^0(\lambda)$ is a function of λ; to fix ideas we choose:

$$r_1^0(\lambda) = \frac{-\lambda + \sqrt{\lambda^2 - 4\lambda}}{2}, \qquad r_2^0(\lambda) = \frac{-\lambda - \sqrt{\lambda^2 - 4\lambda}}{2}$$

for $\lambda \to +\infty$.

the problem corresponding to (1.4), (1.5), (1.6) is then:

$$\frac{\partial^2 u}{\partial t^2} - \Delta \frac{\partial u}{\partial t} - \Delta u = 0 \qquad (f = 0), \tag{1.16}$$

$$\left. \begin{aligned} u(x, 0) &= \xi_0(x), \\ \frac{\partial u}{\partial t}(x, 0) &= \xi_1(x), \end{aligned} \right\} \tag{1.17}$$

with the boundary condition

$$u = 0 \quad \text{on} \quad \Sigma = \Gamma \times (0, T). \tag{1.18}$$

1.3   The Functional $J(\xi)$

Let us put ourselves in the context of Theorem 1.1 with $f = 0$. The solution $u(t)$ is a functional of

$$\xi = \{ \xi_0, \xi_1 \} \in V \times H \, ;$$

therefore

$$u(t) = u(t \, ; \xi).$$

Let χ_0, χ_1 be given with

$$\chi_0 \in V, \qquad \chi_1 \in H. \tag{1.19}$$

For a fixed $T > 0$, set

$$J(\xi) = \| u(T \, ; \xi) - \chi_0 \|^2 + | u'(T \, ; \xi) - \chi_1 |^2. \tag{1.20}$$

We *conjecture* that, under the hypotheses of Theorem 1.1, we have:

$$\mathop{\text{Inf}}_{\xi \in V \times H} J(\xi) = 0. \tag{1.21}$$

We will demonstrate a particular case:

Theorem 1.2: *We suppose that*

$$B(t) = A(t) = A, \qquad (Av, v) \geqslant \alpha_0 \| v \|^2 \quad (^1) \tag{1.22}$$

Then (1.21) *holds.*

$(^1)$ And thus the hypotheses of Remark 1.2 above holds.

Proof. We consider $\hat{u}(\lambda, t)$ (cf. Remark 1.2), given by (1.15).

It is a matter of showing that $\{\hat{u}(\lambda, T), \dfrac{d\hat{u}}{dt}(\lambda, T)\}$ spans a space dense in $\mathcal{U}V \times \mathcal{U}H$. But $\mathcal{U}H = \mathfrak{h}$ and $\mathcal{U}V = $ a field of vectors $g(\lambda) \in \mathfrak{h}$ such that $\lambda^{1/2} g \in \mathfrak{h}$.

Let therefore $\psi_0, \psi_1 \in \mathfrak{h}$, with

$$\int \left[\lambda(\hat{u}(\lambda, T), \psi_0(\lambda))_{\mathfrak{h}(\lambda)} + \left(\frac{d\hat{u}}{dt}(\lambda, T), \psi_1(\lambda) \right)_{\mathfrak{h}(\lambda)} \right] d\mu(\lambda) = 0 \qquad \forall \xi . \qquad (1.23)$$

We must show that ψ_0 and ψ_1 are zero.

Replacing $\hat{u}(\lambda, t)$ by its value (1.15), we see that (1.23) is equivalent to:

$$\left(\lambda E_0(\lambda, T) \, \hat{\xi}_0 + \lambda E_1(\lambda, T) \, \hat{\xi}_1, \psi_0 \right)_{\mathfrak{h}} +$$
$$+ (E_0'(\lambda, T) \, \hat{\xi}_0 + E_1'(\lambda, T) \, \hat{\xi}_1, \psi_1)_{\mathfrak{h}} = 0 \qquad \forall \xi \quad (^1)$$

which is to say

$$\begin{aligned}
E_0(\lambda, T) \, (\lambda\psi_0(\lambda)) + E_0'(\lambda, T) \, (\psi_1(\lambda)) &= 0 \\
E_1(\lambda, T) \, (\lambda\psi_0(\lambda)) + E_1'(\lambda, T) \, (\psi_1(\lambda)) &= 0
\end{aligned} \left. \right\} \qquad (1.24)$$

almost everywhere in λ for the measure μ.

But (Wronskian):

$$\begin{pmatrix} E_0(\lambda, T) & E_0'(\lambda, T) \\ E_1(\lambda, T) & E_1'(\lambda, T) \end{pmatrix} = e^{-\lambda T} \neq 0$$

therefore (1.24) is equivalent to:

$$\begin{cases} \lambda\psi_0(\lambda) = 0 \quad (\text{therefore } \psi_0(\lambda) = 0) \\ \lambda\psi_1(\lambda) = 0 \end{cases}$$

for μ—almost all λ, from whence the result. ∎

We can now pose the following problem (analogous to Problem 4.1):

$(^1)$ $E_i'(\lambda, T) = \dfrac{\partial}{\partial t} E_i(\lambda, t)|_{t=T}, \qquad i = 0, 1.$

Problem 1.1: Supposing that **(1.21)** is satisfied, with $\eta > 0$ given, it is desired to find $\xi_\eta \in V \times H$ such that

$$J(\xi_\eta) \leqslant \eta \cdot \blacksquare$$

Remark 1.4. This is analogous to Remark 1.2, Chapter 1, Section 1: there are an infinite number of ξ_η answering the question. Our object is to obtain a systematic method of construction of *certain* ξ_η answering the question. $\blacksquare$

The nontriviality of the problem results from the *irreversibility* of the problem, as we will now see.

1.4 Irreversibility of the Problem

The "ideal" solution corresponding to $J(\xi) = 0$ consists in the solution, u, of

$$\left. \begin{array}{l} u'' + B(t)\, u' + A(t)\, u = 0\,, \\[2mm] u(T) = \chi_0\,, \qquad u'(T) = \chi_1 \cdot \end{array} \right\} \tag{1.25}$$

But this problem is "in general"[1] improperly posed, as results from

Theorem 1.3: *Under the hypotheses of Theorem 1.2, the problem* **(1.25)** *is improperly posed.*

Proof. We use diagonalization, as in the proof of Theorem 1.2. The problem reduces to constructing $\hat{u}(\lambda, t)$, satisfying

$$\frac{d^2}{dt^2}\, \hat{u}(\lambda, t) + \lambda \frac{d}{dt}\, \hat{u}(\lambda, t) + \lambda \hat{u}(\lambda, t) = 0\,,$$

$$\hat{u}(\lambda, T) = \hat{\chi}_0(\lambda)\,, \qquad \frac{d\hat{u}}{dt}(\lambda, T) = \hat{\chi}_1(\lambda)\,.$$

Then $\hat{u}(\lambda, t)$ may be expressed as a linear combination of the $\chi_i(\lambda)$ with factors $\exp[r_i^0(\lambda)\,(t - T)]$, r_i^0 as in **(1.10)**.

---

[1]In all cases, under the hypotheses of the *type* of Theorem 1.1, and if *in addition B is "of the same force" as A;* it can also be properly posed, for example, if $B = 0$.

This does not define, in general, elements in $\mathfrak{h}$, because $t -$ $T < 0$. ∎

1.5 The QR Method

As already seen in Chapter 1, the QR method consists in "approximating" the system (1.25) by "nearby" systems which are properly posed. We are thus led to: denote by u_ε for $\varepsilon > 0$ "the" solution of

$$
\left.
\begin{aligned}
& u_\varepsilon'' + B(t)\, u_\varepsilon' + A(t)\, u_\varepsilon - \varepsilon B^*(t)\, B(t)\, u_\varepsilon'(t) = 0 \\
& u_\varepsilon(T) = \chi_0\,, \qquad u_\varepsilon'(T) = \chi_1\,.
\end{aligned}
\right\}
\qquad (1.26)
$$

We are going to prove

Theorem 1.4: *Under the hypotheses of Theorem 1.1, problem* (1.26) *is properly posed: there exists a unique* u_ε *satisfying* (1.26) *and*

$$
u_\varepsilon \in L_\infty(0,\,T\,;\,V)\,, \qquad u_\varepsilon' \in L_\infty(0,\,T\,;\,H)\,, \qquad (1.27)
$$

$$
u_\varepsilon' \in T_2(0,\,T\,;\,D(B(t))\ (^1)\,. \qquad (1.28)
$$

Proof. Everything reduces, with the aid of standard techniques (cf., for example, Lions [1]), to proving suitable a priori inequalities.

Let us suppose that u_ε is a "regular" solution of (1.26); performing a scalar multiplication by u' (suppressing the index ε in what follows), we have

$$
(u''(t), u'(t)) + a(t\,;\,u(t), u'(t)) + (B(t)\,u'(t), u'(t)) - \varepsilon\,|\,B(t)\,u'(t)\,|^2 = 0\,;
$$

setting

$$
a'(t\,;\,u,\,v) = \frac{\mathrm{d}}{\mathrm{d}t}\,a(t\,;\,u,\,v) \qquad \forall u,\,v \in V,
$$

we have

$$
\frac{1}{2}\,\frac{\mathrm{d}}{\mathrm{d}t}\,\big[\,|\,u'(t)\,|^2 + a(t\,;\,u(t), u(t))\big] - a'(t\,;\,u(t), u(t))
$$
$$
+ (B(t)\,u'(t), u'(t)) - \varepsilon\,|\,B(t)\,u'(t)\,|^2 = 0\,.
$$

(^{1}) We suppose that the unbounded operators $B(t)$ are closed in H, with the space $D(B(t))$ possessing a norm with a graph of measurable sum.

Integrating from t to T, we then have:

$$| u'(t) |^2 + a(t \; ; u(t), u(t)) + 2 \, \varepsilon \int_t^T | B(\sigma) \, u'(\sigma) |^2 \, d\sigma +$$

$$+ 2 \int_t^T a'(\sigma \; ; u(\sigma), u(\sigma)) \, d\sigma - 2 \int_t^T (B(\sigma) \, u'(\sigma), u'(\sigma)) \, d\sigma =$$

$$= | \chi_1 |^2 + a(T \; ; \chi_0, \chi_0)$$

whence[1] (noting as a consequence of Cauchy–Schwarz)

$$2 \int_t^T (B(\sigma) \, u'(\sigma), u'(\sigma)) \, d\sigma \leqslant \varepsilon \int_t^T | B(\sigma) \, u'(\sigma) |^2 \, d\sigma + \frac{1}{\varepsilon} \int_t^T | u'(\sigma) | \, d\sigma^2) :$$

$$| u'(t) |^2 + \alpha \parallel u(t) \parallel^2 + 2 \, \varepsilon \int_t^T | B(\sigma) \, u'(\sigma) |^2 \, d\sigma \leqslant C \int_t^T \parallel u(\sigma) \parallel^2 \, d\sigma +$$

$$+ \varepsilon \int_t^T | B(\sigma) \, u'(\sigma) |^2 \, d\sigma + \frac{1}{\varepsilon} \int_t^T | u'(\sigma) |^2 \, d\sigma + C \, | \chi_1 |^2 + C \parallel \chi_0 \parallel^2 \quad (2)$$

whence

$$| u'(t) |^2 + \parallel u(t) \parallel^2 + \varepsilon \int_t^T | B(\sigma) \, u'(\sigma) |^2 \, d\sigma \leqslant C \int_t^T \parallel u(\sigma) \parallel^2 \, d\sigma + \left. \begin{array}{c} \\ \\ \\ \\ \end{array} \right\} \quad (1.29)$$

$$+ \frac{C}{\varepsilon} \int_t^T | u'(\sigma) |^2 \, d\sigma + C$$

whence the desired result follows with the aid of the Bellman–Gronwall inequality. ∎

The QR method is then the following:

1) solve (1.26);
2) take $\xi_0 = u_\varepsilon(0)$, $\xi_1 = u'_\varepsilon(0)$. ∎

Exactly as in Chapter 1 the problems are now the following:

(i) if $\xi_\varepsilon = \{ u_\varepsilon(0), u'_\varepsilon(0) \}$, does $J(\xi_\varepsilon) \to 0$ when $\varepsilon \to 0$?
(ii) how do we choose ε so that $J(\xi_\varepsilon) \leqslant \eta$?
(iii) numerical solution of (1.26).

We examine (i) in the following section. ∎

[1] We can suppose that $a(t \; ; v, v) \geqslant \alpha \parallel v \parallel^2$.
[2] The C denotes various constants.

1.6 A Result Concerning Convergence

Theorem 1.5: *Assume the hypotheses of Theorem 1.2. Let U_ε be the solution of*

$$U_\varepsilon'' + B(t) U_\varepsilon' + A(t) U_\varepsilon = 0, \left.\vphantom{\begin{matrix}a\\b\end{matrix}}\right\} \qquad (1.30)$$
$$U_\varepsilon(0) = u_\varepsilon(0), \qquad U_\varepsilon'(0) = u_\varepsilon'(0),$$

with u_ε the solution of (1.26). *Then, when $\varepsilon \to 0$, we have:*

$$U_\varepsilon(T) \to \chi_0 \quad in \quad V, \qquad U_\varepsilon'(T) \to \chi_1 \quad in \quad H.$$

Proof. As in the demonstration of Theorem 1.2, we use diagonalization. Let

$$\hat{U}_\varepsilon(\lambda, t) = (\mathscr{U} U_\varepsilon(t) (\lambda)).$$

Let $F_{i\varepsilon}(\lambda, t)$ be the "fundamental solutions":

$$\frac{d^2}{dt^2} F_{i\varepsilon}(\lambda, t) + (\lambda - \varepsilon\lambda^2) \frac{d}{dt} F_{i\varepsilon}(\lambda, t) + \lambda F_{i\varepsilon}(\lambda, t) = 0,$$

$$F_{0\varepsilon}(\lambda, T) = 1, \qquad F_{0\varepsilon}'(\lambda, T) = 0 \quad (^1),$$

$$F_{1\varepsilon}(\lambda, T) = 0, \qquad F_{1\varepsilon}'(\lambda, T) = 1;$$

then

$$\hat{u}_\varepsilon(\lambda, t) = F_{0\varepsilon}(\lambda, t) \hat{\chi}_0(\lambda) + F_{1\varepsilon}(\lambda, t) \hat{\chi}_1(\lambda)$$

and we deduce [suitably using (1.15)]:

$$\hat{U}_\varepsilon(\lambda, t) = E_0(\lambda, t) [F_{0\varepsilon}(\lambda, 0) \hat{\chi}_0 + F_{1\varepsilon}(\lambda, 0) \hat{\chi}_1] +$$
$$+ E_1(\lambda, t) [F_{0\varepsilon}'(\lambda, 0) \hat{\chi}_0 + F_{1\varepsilon}'(\lambda, 0) \hat{\chi}_1],$$

whence

$$\hat{U}_\varepsilon(\lambda, T) = [E_0(\lambda, T) F_{0\varepsilon}(\lambda, 0) + E_1(\lambda, T) F_{0\varepsilon}'(\lambda, 0)] \hat{\chi}_0 + \left.\vphantom{\begin{matrix}a\\b\end{matrix}}\right\} \qquad (1.32)$$
$$+ [E_0(\lambda, T) F_{1\varepsilon}(\lambda, 0) + E_1(\lambda, T) F_{1\varepsilon}'(\lambda, 0)] \hat{\chi}_1$$

$(^1)$ $F_{i\varepsilon}'(\lambda, t) = \frac{\partial}{\partial t} F_{i\varepsilon}(\lambda, t)$.

and

$$\frac{d}{dt}\,\hat{U}_\varepsilon(\lambda,t)\bigg|_{t=T} = \left.\begin{array}{l} [E'_0(\lambda,T)\,F_{0\varepsilon}(\lambda,0) + E'_1(\lambda,T)\,F'_{0\varepsilon}(\lambda,0)]\,\hat{\chi}_0 + \\[2mm] + [E'_0(\lambda,T)\,F_{1\varepsilon}(\lambda,0) + E'_1(\lambda,T)\,F'_{1\varepsilon}(\lambda,0)]\,\hat{\chi}_1\,. \end{array}\right\} \quad (1.33)$$

But we can find $F_{i\varepsilon}(\lambda,t)$ explicitly. Let $r_i^0(\lambda;\varepsilon)$ be the roots of the characteristic equation

$$r^2 + (\lambda - \varepsilon\lambda^2)\,r + \lambda = 0\,,$$

indexed in such a way that

$$r_i(\lambda;\varepsilon) \to r_i^0(\lambda) \quad \text{when} \quad \varepsilon \to 0\,, \qquad (\varepsilon > 0)\,.$$

Then, writing r_i in place of $r_i(\lambda;\varepsilon)$:

$$F_{0\varepsilon}(\lambda,t) = \left[\frac{1}{r_1 - r_2} - r_2\,e^{r_1(t-T)} + r_1\,e^{r_2(t-T)}\right],$$

$$F_{1\varepsilon}(\lambda,t) = \frac{1}{r_1 - r_2}\left[e^{r_1(t-T)} - e^{r_2(t-T)}\right].$$

The verification of the result can then be carried out using (1.32), (1.33) and an elementary calculation, using the expressions $E_i(\lambda,T)$ and $F_{i\varepsilon}(\lambda,0)$. ∎

2. NUMERICAL APPLICATIONS: AN EXAMPLE DERIVED FROM VISCOELASTICITY

2.1 The Problem

Consider the system

$$\frac{\partial^2 u}{\partial t^2} - \frac{\partial}{\partial t}\frac{\partial^2 u}{\partial x^2} - \frac{\partial^2 u}{\partial x^2} = 0 \qquad \begin{array}{l} 0 \leqslant x \leqslant 1, \\ t > 0 \end{array} \tag{2.1}$$

$$u(x,0) = \xi_0(x), \tag{2.2}$$

$$\frac{\partial u}{\partial t}(x,0) = \xi_1(x), \tag{2.3}$$

$$u(0,t) = u(1,t) = 0\,. \tag{2.4}$$

Let χ_0 and χ_1 be two arbitrary elements of $L_2(0, 1)$.

We set (cf. the note at the bottom of the page for Remark 11.2, Chapter 1 for the normalization):

$$J(\xi_0, \xi_1) = \frac{\| u(x, T) - \chi_0 \|}{\| \chi_0 \|} + \frac{\left| \frac{\partial u}{\partial t}(x, T) - \chi_1(x) \right|}{| \chi_1 |}. \tag{2.5}$$

The *problem* is to determine ξ_0 and ξ_1, so that the solution $u(x, t)$ of the system (2.1) to (2.4) satisfies

$$J(\xi_0, \xi_1) \leqslant \eta \tag{2.6}$$

where η is a given number.

This problem falls within the domain of Section 1; cf. Remark 1.3.

2.2 The Associated QR Problem

Let $\varepsilon > 0$ be a "small" number. We must solve:

$$\frac{\partial^2 u_\varepsilon}{\partial t^2} - \frac{\partial}{\partial t} \frac{\partial^2 u_\varepsilon}{\partial x^2} - \varepsilon \frac{\partial}{\partial t} \frac{\partial^4 u_\varepsilon}{\partial x^4} - \frac{\partial^2 u_\varepsilon}{\partial x^2} = 0, \quad t < T, \quad 0 < x < 1 \tag{2.7}$$

$$u_\varepsilon(x, T) = \chi_0(x), \tag{2.8}$$

$$\frac{\partial u_\varepsilon}{\partial t}(x, T) = \chi_1(x), \tag{2.9}$$

$$u_\varepsilon(0, t) = u_\varepsilon(1, t) = 0, \tag{2.10}$$

$$\frac{\partial^2 u_\varepsilon}{\partial x^2}(0, t) = \frac{\partial^2 u_\varepsilon}{\partial x^2}(1, t) = 0, \tag{2.11}$$

then we take

$$\xi_0(x) = u_\varepsilon(x, 0), \tag{2.12}$$

$$\xi_1(x) = \frac{\partial u_\varepsilon}{\partial t}(x, 0). \tag{2.13}$$

Remark 2.1. The number ε is to be chosen suitably as a function of η. ∎

If we change t into $T - t$, problem (2.7) to (2.11) may be written, setting

$$u_\varepsilon(x, T - t) = V(x, t),$$

$$\frac{\partial^2 V}{\partial t^2} + \frac{\partial}{\partial t}\frac{\partial^2 V}{\partial x^2} + \varepsilon\frac{\partial}{\partial t}\frac{\partial^4 V}{\partial x^4} - \frac{\partial^2 V}{\partial x^2} = 0 \qquad\qquad (2.14)$$

$$0 < t < T$$

$$V(x, 0) = \chi_0(x) \qquad\qquad (2.15)$$

$$\frac{\partial V}{\partial t}(x, 0) = -\chi_1(x) \qquad\qquad (2.16)$$

$$V(0, t) = V(1, t) = 0 \qquad\qquad (2.17)$$

$$\frac{\partial^2 V}{\partial x^2}(0, t) = \frac{\partial^2 V}{\partial x^2}(1, t) = 0 \qquad\qquad (2.18)$$

and we then take

$$\xi_0(x) = V(x, T), \qquad\qquad (2.19)$$

$$\xi_1(x) = -\frac{\partial V}{\partial t}(x, T). \qquad\qquad (2.20)$$

2.3 Numerical Solution

2.3.1 NOTATION

We divide the interval $(0, 1)$ into J equal intervals and the interval $(0, T)$ into N equal intervals. Then:

$$1 = J\,\Delta x, \qquad T = N\,\Delta t \qquad\qquad (2.21)$$

and we set:

$$V_j^n = V(j\,\Delta x, n\,\Delta t).$$

Take for the derivatives t at time $n\,\Delta t$

$$\left(\frac{\partial V}{\partial t}\right)^n \cong \frac{V^{n+1} - V^{n-1}}{2\,\Delta t} \qquad\qquad (2.22)$$

$$\left(\frac{\partial^2 V}{\partial t^2}\right)^n \cong \frac{V^{n+1} - 2V^n + V^{n-1}}{\Delta t^2} \qquad\qquad (2.23)$$

and for derivatives in space at the point $j \, \Delta x$

$$\left(\frac{\partial V}{\partial x}\right)_j \simeq \frac{V_{j+1} - V_{j-1}}{2 \, \Delta x} \tag{2.24}$$

$$\left(\frac{\partial^2 V}{\partial x^2}\right)_j \simeq \frac{V_{j+1} - 2 \, V_j + V_{j-1}}{\Delta x^2} \tag{2.25}$$

$$\left(\frac{\partial^4 V}{\partial x^4}\right)_j \simeq \frac{V_{j+2} - 4 \, V_{j+1} + 6 \, V_j - 4 \, V_{j-1} + V_{j-2}}{\Delta x^4} . \tag{2.26}$$

2.3.2 DISCRETIZATION

The equation (2.14) with partial derivatives is then approximated by

$$\frac{V_j^{n+1} - 2 V_j^n + V_j^{n-1}}{\Delta t^2} + \frac{1}{2\Delta t}\left[\frac{V_{j+1}^{n+1} - 2 V_j^{n+1} + V_{j-1}^{n+1}}{\Delta x^2} - \frac{V_{j+1}^{n-1} - 2 V_j^{n-1} + V_{j-1}^{n-1}}{\Delta x^2}\right] +$$

$$+ \frac{\varepsilon}{2\,\Delta t}\left[\frac{V_{j+2}^{n+1} - 4 \, V_{j+1}^{n+1} + 6 \, V_j^{n+1} - 4 \, V_{j-1}^{n+1} + V_{j+2}^{n+1}}{\Delta x^4} - \right.$$

$$\left. - \frac{V_{j+2}^{n-1} - 4 V_{j+1}^{n-1} + 6 V_j^{n-1} - 4 V_{j-1}^{n-1} + V_{j-2}^{n-1}}{\Delta x^4}\right] - \frac{1}{\Delta x^2}[V_{j+1} - 2 V_j + V_{j-1}] = 0$$

which gives, after multiplication by Δt^2 and rearrangement of terms

$$V_j^{n+1}\left(1 - \frac{\Delta t}{\Delta x^2} + \frac{3\,\varepsilon\,\Delta t}{\Delta x^4}\right) + (V_{j+1}^{n+1} + V_{j-1}^{n+1})\left(\frac{\Delta t}{2\,\Delta x^2} - 2\frac{\varepsilon\,\Delta t}{\Delta x^4}\right) +$$

$$+ (V_{j+2}^{n+1} + V_{j-2}^{n+1})\left(\frac{\varepsilon\Delta t}{2\Delta x^4}\right) = V_j^n\left(2 - \frac{2\Delta t^2}{\Delta x^2}\right) + (V_{j+1} + V_{j-1})\left(\frac{\Delta t^2}{\Delta x^2}\right) +$$

$$+ V_j^{n-1}\left(-1 - \frac{\Delta t}{\Delta x^2} + \frac{3\,\varepsilon\,\Delta t}{\Delta x^4}\right) + (V_{j+1}^{n-1} + V_{j-1}^{n-1})\left(\frac{\Delta t}{2\,\Delta x^2} - 2\frac{\varepsilon\,\Delta t}{\Delta x^4}\right) +$$

$$+ (V_{j+2}^{n-1} + V_{j-2}^{n-1})\left(\frac{\varepsilon\,\Delta t}{2\,\Delta x^4}\right). \tag{2.28}$$

The initial conditions become in discrete form $(j = 0, ..., J)$

$$V_j^0 = \chi_0(j \, \Delta x), \tag{2.29}$$

$$V_j^1 - V_j^{-1} = - 2 \, \Delta t \, \chi_1(j \, \Delta x), \tag{2.30}$$

and the boundary conditions $(n = 0, ..., N)$

$$V_0^n = F_0(n \, \Delta t) \, ,$$

$$V_J^n = F_1(n \, \Delta t) \, ,$$

$$V_1^n - 2 \, V_0^n + V_{-1}^n = 0 \, ,$$

$$V_{J+1}^n - 2 \, V_J^n + V_{J-1}^n = 0 \, .$$

The system of equations (2.28), $j = 1, ..., J - 1$ and the relations (2.29) to (2.34) permit us to calculate the solution at the time $(n + 1) \, \Delta t$ as a function of the solution at time $n \, \Delta t$ and $(n - 1) \, \Delta t$. To start the calculation, the system (2.28) holds; however, (2.29) and (2.30) permit us to express the solution at time Δt as a function of the given quantities $\chi_0(x)$ and $\chi_1(x)$.

The system (2.28) can be written in matrix form

$$AV^{n+1} = BV^n + CV^{n-1} + D \qquad (2.35)$$

where V^{n+1}, V^n, V^{n-1} represent vectors of dimension $J - 1$ whose components are V_j^{n+1}, V_j^n and V_j^{n-1}.

A, B, C are square matrices of order $J - 1$, independent of n and D is a vector formed using the boundary conditions (2.31) to (2.34). The solution of (2.35) furnishes at the end of $(N + 1)$ steps in time:

$$\xi_0(j \, \Delta x) = V_j^N \qquad (2.36)$$

and

$$\xi_1(j \, \Delta x) \doteq \frac{V_j^{N-1} - V_j^{N+1}}{2 \, \Delta t} \, . \qquad (2.37)$$

2.4 Direct Integration of the System

At the end of the verification, we integrate the system (2.1) to (2.4) with the functions found at time T in the resolution of the retrograde problem (2.7) to (2.11) as initial functions ξ_0 and ξ_1.

Adapting the notation given in 2.3.1, the equation (2.1) may be written in discrete form:

$$\frac{u_j^{n+1} - 2\,u_j^n + u_j^{n-1}}{\Delta t^2} - \frac{1}{2\,\Delta t}\,\frac{u_{j+1}^{n+1} - 2\,u_j^{n+1} + u_{j-1}^{n+1}}{\Delta x^2} -$$

$$-\frac{u_{j+1}^{n-1} - 2\,u_j^{n-1} + u_{j-1}^{n-1}}{\Delta x^2} - \frac{u_{j+1}^n - 2\,u_j^n + u_{j-1}^n}{\Delta x^2} = 0 \qquad (2.38)$$

$$j = 1, ..., J - 1, \qquad n = 0, ..., N - 1.$$

The conditions (2.2) to (2.4) may be written:

$$u_j^0 = \xi_0(j\,\Delta x), \qquad (2.39)$$

$$u_j^1 - u_j^{-1} = 2\,\Delta t\xi_1(j\,\Delta x), \qquad (2.40)$$

$$u_0^n = u_J^n = 0. \qquad (2.41)$$

Denote by χ_0^* and χ_1^* the functions obtained at time T.

2.5 Numerical Results

2.5.1 EXAMPLE 1

We took:

$$\begin{cases} \chi_0(x) = \sin \pi x, \\ \chi_1(x) = 0, \end{cases}$$

and the boundary conditions

$$u(0, t) = u(1, t) = 0.$$

The value of T has been taken equal to $1/2$.

Graphs 1 and 2 represent the functions χ_0 and χ_0^*, χ_1 and χ_1^* for the values of ε equal to 10^{-2} and $1/2$.

Graphs 3 and 4 furnish the values $\tilde{\xi}_0$ and $\tilde{\xi}_1$ calculated by the QR method.

2.5.2 EXAMPLE 2

We took

$$\begin{cases} \chi_0(x) = \cos \pi x, \\ \chi_1(x) = \sin \pi x, \end{cases}$$

and the boundary conditions

$$\begin{cases} u(0, t) = + 1, \\ u(1, t) = - 1. \end{cases}$$

The value of T was taken equal to 10^{-1}.

Graphs 5 and 6 represent χ_0 and χ_0^*, χ_1 and χ_1^*; Graphs 7 and 8 give $\tilde{\xi}_0$ and $\tilde{\xi}_1$.

3. COUPLED PROBLEMS OF EVOLUTION AND THE QR METHOD

3.1 Coupled Problems

(We follow in 3.1 the note J. L. Lions–P. A. Raviart [1], no. 1). Let V_j, $j = 1, 2$ be two Hilbert spaces with

$$V_j \subset H, \qquad j = 1, 2 ;$$

let

$$V = V_1 \cap V_2 ,$$

$$\| \quad \|_j = \quad \text{norm in} \quad V_j ,$$

$$\| v \| = (\| v \|_1^2 + \| v \|_2^2)^{1/2} = \quad \text{norm in} \quad V,$$

$$| \quad | = \quad \text{norm in} \quad H .$$

We have:

$$V \subset V_j \subset H , \qquad j = 1, 2 ,$$

and we suppose that each space is dense in the following. Then, with X' denoting the dual of X in general fashion, and identifying H with its dual—as already done in Chapter 1—we have:

$$V \subset V_j \subset H \subset V_j' \subset V' , \qquad j = 1, 2 . \tag{3.1}$$

We will denote by $(\ , \ )$ the scalar product between V and V' (or V_j and V_j').

We then assume:

$$A_j \in \mathscr{L}(V ; V') , \qquad j = 1, 2 ,$$

with

$$(A_1 v, v) + \lambda |v|^2 \geqslant \alpha_1 \|v\|_1^2, \qquad \forall v \in V_1, \alpha_1 > 0, \lambda \quad \text{suitable} \quad (^1) \qquad (3.3)$$

$$A_2^* = A_2, \qquad (A_2 v, v) + \lambda |v|^2 \leqslant \alpha_2 \|v\|_2^2 \qquad \forall v \in V_2, \alpha_2 > 0. \qquad (3.4)$$

and finally

$$B \in \mathscr{L}(V_1 ; V_1') . \blacksquare \qquad (3.5)$$

Remark 3.1. We can more generally consider $A_j = A_j(t), B = B(t)$. Cf. Remark 1.1 of Lions–Raviart [1]. $\blacksquare$

We show

Theorem 3.1: *Under the hypotheses* (3.2), ... , (3.5), *there exists a function u satisfying*

$$u \in L_\infty(0, T; V), \qquad (3.6)$$

$$u' \in L_\infty(0, T; H) \cap L_2(0, T; V_1), \qquad (3.7)$$

$$u'' \in L_2(V_1') + L_\infty(V_2') \qquad (3.8)$$

$$u'' + A_1 u' + (A_2 + B) u = f, \qquad f \quad \text{given in} \quad L_2(0, T; V_1'), \qquad (3.9)$$

$$u(0) = \xi_0, \qquad \xi_0 \quad \text{given in} \quad V, \qquad (3.10)$$

$$u'(0) = \xi_1, \qquad \xi_1 \quad \text{given in} \quad H. \qquad (3.11)$$

We can show in addition that, with possible modification on a set of measure zero, $u(u')$ is continuous from $[0, T] \to V$(resp. H). $\blacksquare$

3.2 An Example

(Cf. R. D. Richtmyer [1], p. 171.)
Consider the coupled sound-heat system:

$$\left. \begin{array}{l} \dfrac{\partial w}{\partial t} - c \dfrac{\partial \tilde{u}}{\partial x} = 0, \\[2mm] \dfrac{\partial \tilde{u}}{\partial t} - c \dfrac{\partial w}{\partial x} + c(\gamma - 1) \dfrac{\partial e}{\partial x} = 0, \\[2mm] \dfrac{\partial e}{\partial t} + c \dfrac{\partial \tilde{u}}{\partial x} - \sigma \dfrac{\partial^2 e}{\partial x^2} = 0 \qquad (^2) \end{array} \right\} \qquad (3.12)$$

(¹)But we do not necessarily have $A_1^* = A_1$.
(²)We write $\tilde{u}$ in place of u, as in Richtmyer [1], so that there is no confusion with the notation in the general case, 3.1.

where c, γ, σ are positive constants $> 0, \gamma > 1$, and where $x \in \mathbf{R}$ or $x \in {]}0, \infty{[}$; in the second case, we have the boundary condition $e(0, t) = 0$, $\tilde{u}(0, t) = 0$.

Eliminating w we can write (differentiating the last equation in 3.12 with respect to t for reasons of symmetry)

$$\left.\begin{aligned}
\frac{\partial^2 e}{\partial t^2} - \sigma \frac{\partial^2}{\partial x^2} \frac{\partial e}{\partial t} + c \frac{\partial^2 \tilde{u}}{\partial x\, \partial t} &= 0, \\[2mm]
\frac{\partial^2 \tilde{u}}{\partial t^2} + c(\gamma - 1) \frac{\partial^2 e}{\partial x\, \partial t} - c^2 \frac{\partial^2 \tilde{u}}{\partial x^2} &= 0.
\end{aligned}\right\} \tag{3.13}$$

Set

$$u = \{ e, \tilde{u} \}.$$

Then (3.13) may be written in the form (3.9) with:

$$B = 0;$$

$$\left.A_1 = \begin{pmatrix} -\sigma \dfrac{\partial^2}{\partial x^2} & c \dfrac{\partial}{\partial x} \\[4mm] c(\gamma - 1) \dfrac{\partial}{\partial x} & 0 \end{pmatrix}, \quad A_2 = \begin{pmatrix} 0 & 0 \\[4mm] 0 & -c^2 \dfrac{\partial^2}{\partial x^2} \end{pmatrix}. \right\} \tag{3.14}$$

Let us take the case where $x \in \Omega = {]}0, \infty{[}$. Set

$$H = L_2(\Omega) \times L_2(\Omega),$$
$$V_1 = H_0^1(\Omega) \times L_2(\Omega), \qquad V_2 = L_2(\Omega) \times H_0^1(\Omega).$$

We have, for $v \in V_1$, if $v = \{v_1, v_2\}$

$$(A_1 v, v) = \left\langle -\sigma \frac{d^2 v_1}{dx^2}, v_1 \right\rangle + \left\langle c \frac{dv_2}{dx}, v_1 \right\rangle + \left\langle c(\gamma - 1) \frac{dv_1}{dx}, v_2 \right\rangle$$

where the first and second brackets denote the duality between $H^{-1}(\Omega)$[1] and $H_0^1(\Omega)$ and the third bracket denotes the duality between $L_2(\Omega)$ and $L_2(\Omega)$. Then

[1]Dual of $H_0^1(\Omega)$.

$$(A_1 v, v) = \int_0^\infty \left(\frac{dv_1}{dx}\right)^2 dx + c(\gamma - 2)\left\langle \frac{dv_1}{dx}, v_2 \right\rangle,$$

therefore

$$(A_1 v, v) \geqslant \sigma \int_0^\infty \left(\left(\frac{dv_1}{dx}\right)^2 + \frac{v_1^2}{2}\right)dx - \frac{\sigma}{2}\int_0^\infty v_1^2 \, dx -$$

$$- c(\gamma - 2)\left(\int_0^\infty \left(\frac{dv_1}{dx}\right)^2 dx\right)^{1/2}\left(\int_0^\infty v_2^2 \, dx\right)^{1/2}.$$

Using the inequality

$$ab \leqslant \alpha \frac{a^2}{2} + \frac{b^2}{2\alpha},$$

we have (with α taking the value σ):

$$c(\gamma - 2)\left[\int_0^\infty \left(\frac{dv_1}{dx}\right)^2 dx\right]^{1/2}\left[\int_0^\infty v_2^2 \, dx\right]^{1/2} \leqslant$$

$$\leqslant \frac{\sigma}{2}\int_0^\infty \left(\frac{dv_1}{dx}\right)^2 dx + \frac{c^2(\gamma - 2)^2}{2}\int_0^\infty v_2^2 \, dx$$

Hence

$$(A_1 v, v) \geqslant \frac{\sigma}{2}\left[\int_0^\infty \left(\left(\frac{dv_1}{dx}\right)^2 + v_1^2\right)dx + \int_0^\infty v_2^2 \, dx\right] - \frac{\sigma}{2}\int_0^\infty v_1^2 \, dx$$

$$- \frac{c^2(\gamma - 2)^2}{2\sigma}\int_0^\infty v_2^2 \, dx - \frac{\sigma}{2}\int_0^\infty v_2^2 \, dx .$$

Finally:

$$(A_1 v, v) \geqslant \frac{\sigma}{2}\left[\int_0^\infty \left(\left(\frac{dv_1}{dx}\right)^2 + v_1^2\right)dx + \int_0^\infty v_2^2 \, dx\right] - C\int_0^\infty (v_1^2 + v_2^2) \, dx ,$$

which implies

$$(A_1 v, v) = \frac{\sigma}{2} \|v\|_1^2 - C |v|^2 , \quad \text{with} \quad C = \frac{c^2(\gamma - 2)^2}{2\sigma} + \frac{\sigma}{2},$$

whence (3.3).

Let us verify (3.4):

$$(A_2 v, v) = c^2 \int_0^\infty \left(\frac{dv_2}{dx}\right)^2 dx ,$$

whence immediately the result.

Therefore the system (3.13)—or (3.14)—falls within the province of Theorem 3.1.([1])

3.3 The Functional $J(\xi)$

Let us return to the general case of 3.1. Let χ_0, χ_1 be given functions respectively in V and H As in Section 1, we consider the functional

$$J(\xi) = \| u(T) - \chi_0 \|^2 + | u'(T) - \chi_1 |^2 \qquad (3.15)$$

where

$$u = u(t) = u(t ; \xi) , \qquad \xi = \{ \xi_0, \xi_1 \}$$

is *the* solution of (3.6), . . . , (3.11), *with* $f = 0$.

As in the previous sections, our aim is the study of

$$\text{Inf } J(\xi) , \qquad \xi \in V \times H . \blacksquare$$

We can only conjecture, here again, that under the hypotheses of Theorem 3.1, we have:

$$\underset{\xi \in V \times H}{\text{Inf }} J(\xi) = 0 .$$

We can demonstrate the result in a particular case:

Theorem 3.2: *Assume the hypotheses of Theorem 3.1, with in addition*

$$B = 0 , \qquad A_1^* = A_1 , \; (^2) \qquad (3.17)$$

and A_1 and A_2 commutative. Then (3.16) holds.

([1])If $\Omega = \mathbf{R}$, we have: $V_1 = H^1(\mathbf{R}) \times L_2(\mathbf{R})$, $V_2 = L_2(\mathbf{R}) \times H^1(\mathbf{R})$.
([2]) In the sense: the resolvents $(A_1 + \lambda_1)^{-1}$ and $(A_2 + \lambda_2)^{-1}$, when defined, commute.

Proof. We can find (cf. Dixmier [1]), using the hypothesis of commutativity a measurable hilbertian sum

$$\mathfrak{h} = \int \mathfrak{h}(\lambda_1, \lambda_2) \, d\mu(\lambda_1, \lambda_2), \qquad \lambda_j \geqslant \lambda_j^0 \, ;$$

where $\mathfrak{h}(\lambda_1, \lambda_2) = $ a measurable field of hilbert spaces, and an isometry $\mathscr{U}$ from H to $\mathfrak{h}$, such that the image of A_j under this isometry is the *operator of multiplication* by λ_j.

Let then

$$\hat{u}(\lambda, t) = \hat{u}(\lambda_1, \lambda_2, t) = (\mathscr{U}u(t))(\lambda), \qquad \lambda = \{ \lambda_1, \lambda_2 \} \, ;$$

equation (3.9) with $f = 0$ becomes:

$$\frac{d^2\hat{u}}{dt^2} + \lambda_1 \frac{d\hat{u}}{dt} + \lambda_2 \, \hat{u} = 0 \tag{3.18}$$

and (3.10), (3.11) become:

$$\hat{u}(\lambda, 0) = \hat{\xi}_0(\lambda), \qquad \frac{d\hat{u}}{dt}(\lambda, 0) = \hat{\xi}_1(\lambda).$$

As in Section 1, Remark 1.2, we introduce $E_i(\lambda, t)$, which satisfies

$$\left.\begin{aligned}
&\frac{d^2}{dt^2} E_i(\lambda, t) + \lambda_1 \frac{d}{dt} E_i(\lambda, t) + \lambda_2 E_i(\lambda, t) = 0, \qquad i = 0, 1 \\[2mm]
&E_0(\lambda, 0) = 1, \qquad \frac{d}{dt} E_0(\lambda, 0) = 0, \\[2mm]
&E_1(\lambda, 0) = 0, \qquad \frac{d}{dt} E_1(\lambda, 0) = 1.
\end{aligned}\right\} \tag{3.20}$$

Then

$$\hat{u}(\lambda, t) = E_0(\lambda, t) \, \hat{\xi}_0(\lambda) + E_1(\lambda, t) \, \hat{\xi}_1(\lambda). \tag{3.21}$$

We wish to demonstrate that the space spanned in $\mathscr{U}V \times \mathfrak{h}$ by

$$\left\{ \hat{u}(\lambda, T), \quad -\frac{d\hat{u}}{dt}(\lambda, T) \right\}$$

is dense. Let $\{\psi_0, \psi_1\} \in \mathscr{U}V \times \mathfrak{h}$ be orthogonal to this space. Then[1]

$$\iint \left[(1 + \lambda_1 + \lambda_2) \, (\hat{u}(\lambda, T), \psi_0(\lambda))_{\mathfrak{h}(\lambda_1, \lambda_2)} + \right.$$

$$\left. + \left(\frac{d\hat{u}}{dt}(\lambda, T), \psi_1(\lambda) \right)_{\mathfrak{h}(\lambda_1, \lambda_2)} \right] d\mu(\lambda_1, \lambda_2) = 0 \qquad \forall \xi_0, \xi_1 \,. \tag{3.22}$$

Using (3.21) and setting

$$\tilde{\psi}_0(\lambda) = (1 + \lambda_1 + \lambda_2) \, \psi_0(\lambda) \,,$$

(3.22) is equivalent to

$$\left. \begin{aligned} &E_0(\lambda, T) \, \tilde{\psi}_0(\lambda) + \frac{d}{dt} \, E_0(\lambda, T) \, \psi_1(\lambda) = 0 \\[2mm] &E_1(\lambda, T) \, \tilde{\psi}_0(\lambda) + \frac{d}{dt} \, E_1(\lambda, T) \, \psi_1(\lambda) = 0 \end{aligned} \quad \right\} \quad d\mu - \text{almost everywhere} \qquad (3.23)$$

and since

$$\det \begin{bmatrix} E_0(\lambda, T) & \dfrac{d}{dt} E_0(\lambda, T) \\[3mm] E_1(\lambda, T) & \dfrac{d}{dt} E_1(\lambda, T) \end{bmatrix} = e^{-\lambda_1 T} \,,$$

we have from (3.23) that

$$\tilde{\psi}_0(\lambda) \ (\text{therefore } \psi_0(\lambda)) = 0 \,, \qquad d\mu - \text{almost everywhere},$$

$$\psi_1(\lambda) = 0 \,, \qquad d\mu - \text{almost everywhere},$$

which establishes the theorem. ∎

Remark 3.2. The preceding demonstration is evidently essentially *the same* as that of Theorem 1.2. The same result holds whenever we can *simultaneously* diagonalize the operators appearing in the system under study. ∎

[1] We take as scalar product in $\mathscr{U}V$:

$$\iint (1 + \lambda_1 + \lambda_2) \, (\hat{v}(\lambda), \hat{w}(\lambda))_{\mathfrak{h}(\lambda_1, \lambda_2)} \, d\mu(\lambda_1, \lambda_2) \,.$$

Remark 3.3. Here is an *example* of an application of the preceding remark. Let us take the example of 3.2, with $\Omega = \mathbf{R}$. Diagonalize the differentiation with respect to x operator by means of the *Fourier transformation. We deduce that for the system* 3.13, *the property* 3.16 *is valid.* ∎

We can now pose

Problem 3.1: (Analogous to Problem 1.1.)

For a given $\eta > 0$, find $\xi_\eta = \{\xi_{0\eta}, \xi_{1\eta}\} \in V \times H$ such that

$$J(\xi_\eta) \leqslant \eta . \blacksquare \tag{3.24}$$

Exactly as in Section 1, 1.4 (demonstration of Theorem 1.3) we show

Theorem 3.3: (*Irreversibility*). *Under the hypotheses of Theorem 3.2, for* $\{\chi_0, \chi_1\}$ *given in* $V \times H$, *the problem*

$$\left.\begin{array}{c} u'' + A_1 u' + A_2 u = 0, \quad t < T, \\ u(T) = \chi_0, \qquad u'(T) = \chi_1 \end{array}\right\} \tag{3.25}$$

is improperly posed.[1]

This justifies the introduction of the QR method, formally analogous to that introduced in Section 1.

3.4 The QR Method

Let us first demonstrate

Theorem 3.4: *Assume the hypotheses of Theorem 3.1, with* $B \in \mathscr{L}$ $(H; V_1')$. *For given* $\varepsilon > 0$, *there exists one and only one function* u_ε *satisfying*

$$u_\varepsilon'' + A_1 u_\varepsilon' - \varepsilon A_1^* A_1 u_\varepsilon' + (A_2 + B) u_\varepsilon = 0, \quad t < T, \tag{3.26}$$

$$u_\varepsilon(T) = \chi_0, \qquad u_\varepsilon'(T) = \chi_1, \qquad (\chi_0 \in V, \ \chi_1 \in H), \tag{3.27}$$

---

[1] For u in the space $L_\infty(0, T; V)$, with $u' \in L_\infty(0, T; H) \cap L_2(0, T; V_1)$.

$$u_\varepsilon \in L_\infty(0, T; V),$$

$$u'_\varepsilon \in L_\infty(0, T; H) \cap L_2(0, T; D(A_1)). \qquad (^1)$$

$$\left.\begin{array}{r}\\ \\\end{array}\right\} \qquad (3.28)$$

Proof. As already mentioned, it is necessary to obtain suitable a priori inequalities. Let us write u in place of u_ε. Multiply (3.26) by u'_ε and integrate from t to T; writing

$$a_2(v) = a_2(v, v),$$

we have:

$$\tfrac{1}{2}\left(|u'(T)|^2 + a_2(u(T))\right) - \tfrac{1}{2}|u'(t)|^2 - \tfrac{1}{2}a_2(u(t)) -$$

$$- \varepsilon \int_t^T |A_1 u'(\sigma)|^2 \, d\sigma + \int_t^T (A_1 u'(\sigma), u'(\sigma)) \, d\sigma +$$

$$+ \int_t^T (Bu(\sigma), u'(\sigma)) \, d\sigma = 0,$$

whence

$$\left.\begin{array}{l} |u'(t)|^2 + a_2(u(t)) + 2\,\varepsilon \displaystyle\int_t^T |A_1 u'(\sigma)|^2 \, d\sigma \leqslant \\[3mm] \leqslant 2\displaystyle\int_t^T |A_1 u'(\sigma)||u'(\sigma)| \, d\sigma + 2\displaystyle\int_t^T |(Bu(\sigma), u'(\sigma))| \, d\sigma + C\ (^2). \end{array}\right\} \qquad (3.29)$$

The second term of (3.29) is majorized by

$$\frac{\varepsilon}{2}\int_t^T (A_1 u'(\sigma))^2 \, d\sigma + \frac{2}{\varepsilon}\int_t^T |u'(\sigma)|^2 \, d\sigma + 2\,C_1 \int_t^T |u(\sigma)|\,\|u'(\sigma)\|_1 \, d\sigma + C.$$

Since

$$\|v\|_1^2 \leqslant C_1 |A_1 v|^2 + C_2 |v|^2,$$

this can be *majorized* by

$$\varepsilon \int_t^T |A_1 u'(\sigma)|^2 \, d\sigma + \frac{C_3}{\varepsilon}\int_t^T \left(|u(\sigma)|^2 + |u'(\sigma)|^2\right) d\sigma + C$$

$(^1)$ $D(A_1)$ = domain of A_1 with the norm of the graph:

$$(|v|^2 + |A_1 v|^2)^{1/2}.$$

$(^2)$ $C, C_1, \ldots,$ denote constants.

and (3.29) yields

$$
\left.
\begin{aligned}
|\, u'(t)\,|^2 + \alpha_2 \,\|\, u(t)\,\|_2^2 + \varepsilon \int_t^T |\, A_1\, u'(\sigma)|^2 \, d\sigma \leqslant \\
\leqslant \frac{C_3}{\varepsilon} \int_t^T (|\, u(\sigma)\,|^2 + |\, u'(\sigma)\,|^2)\, d\sigma + C + C_4 \,|\, u(t)\,|^2 .
\end{aligned}
\right\}
\tag{3.30}
$$

But

$$
u(t) = - \int_t^T u'(\sigma)\, d\sigma + \chi_0
$$

gives

$$
|\, u(t)\,|^2 \leqslant C_5 \int_t^T |\, u'(\sigma)\,|^2\, d\sigma + |\, \chi_0\,|^2
$$

and (3.30) therefore yields

$$
\left.
\begin{aligned}
|\, u'(t)\,|^2 + \alpha_2 \,\|\, u(t)\,\|_2^2 + \varepsilon \int_t^T |\, A_1\, u'(\sigma)|^2 \, d\sigma \leqslant \\
\leqslant \frac{C_6}{\varepsilon} \int_t^T (|\, u(\sigma)\,|^2 + |\, u'(\sigma)\,|^2)\, d\sigma + C_7
\end{aligned}
\right\}
\tag{3.31}
$$

whence the desired inequalities follow as a consequence of the Bellman–Gronwall inequality. ∎

We can now make the QR method precise:

1) *We solve* (3.26), (3.27), (3.28);
2) *We take*

$$
\xi_0 = u_\varepsilon(0) , \qquad \xi_1 = u'_\varepsilon(0) . \quad \blacksquare
$$

For the *convergence* of this method, we can prove (as in Section 1, proof of Theorem 1.5)

Theorem 3.5: *Let the hypotheses of 3.2 be satisfied. Let U_ε be the solution of:*

$$
\left.
\begin{aligned}
& U''_\varepsilon + A_1\, U'_\varepsilon + A_2\, U_\varepsilon = 0 , \\
& U_\varepsilon(0) = u_\varepsilon(0) , \qquad U'_\varepsilon(0) = u'_\varepsilon(0) , \\
& U_\varepsilon \in L_\infty(0, T; V) , \qquad U'_\varepsilon \in L_\infty(0, T; H) \cap L_2(0, T; V_1) .
\end{aligned}
\right\}
\tag{3.32}
$$

Then, when $\varepsilon \to 0$

$$U_\varepsilon(T) \to \chi_0 \quad \text{in} \quad V, \qquad U'_\varepsilon(T) \to \chi_1 \quad \text{in} \quad H.$$

4. NUMERICAL APPLICATIONS: COUPLED SOUND-HEAT EQUATION

4.1 The Problem

Consider the system (cf. Richtmyer [1] and 3.2):

$$\frac{\partial \tilde{u}}{\partial t} = c \frac{\partial}{\partial x} (w - (\gamma - 1) e) \qquad x \in [0, 1], \qquad t > 0 \tag{4.1}$$

$$\frac{\partial w}{\partial t} = c \frac{\partial \tilde{u}}{\partial x} \tag{4.2}$$

$$\frac{\partial e}{\partial t} = \sigma \frac{\partial^2 e}{\partial x^2} - c \frac{\partial \tilde{u}}{\partial x} \tag{4.3}$$

where c and σ are positive constants, and $\gamma > 1$.
Take $u = \{e, \tilde{u}\}$, and eliminate w. We have [cf. (3.13)]:

$$\left. \begin{array}{l} \dfrac{\partial^2 \tilde{u}}{\partial t^2} - c^2 \dfrac{\partial^2 \tilde{u}}{\partial x^2} - c(\gamma - 1) \dfrac{\partial^2 e}{\partial x \, \partial t} = 0 \\[4mm] \dfrac{\partial^2 e}{\partial t^2} - \sigma \dfrac{\partial}{\partial t} \dfrac{\partial^2 e}{\partial x^2} + c \dfrac{\partial^2 \tilde{u}}{\partial x \, \partial t} = 0 \end{array} \right\} \qquad \begin{array}{l} (4.4) \\[8mm] (4.5) \end{array}$$

or, in matrix form:

$$\frac{d^2 u}{dt^2} + A_1 \frac{du}{dt} + A_2 u = 0, \qquad t > 0 \tag{4.6}$$

where

$$A_1 = \begin{bmatrix} -\sigma \dfrac{\partial^2}{\partial x^2} & c \dfrac{\partial}{\partial x} \\[4mm] c(\gamma - 1) \dfrac{\partial}{\partial x} & 0 \end{bmatrix} \tag{4.7}$$

and

$$
A_2 = \begin{bmatrix} 0 & 0 \\ 0 & -c^2 \dfrac{\partial^2}{\partial x^2} \end{bmatrix} \tag{4.8}
$$

We wish to determine functions e and $\tilde{u}$ components of the vector u, satisfying (4.6) and the initial conditions:

$$
u(x, 0) = \xi_0(x) = \{\, \xi_{01}(x), \xi_{02}(x) \,\} \tag{4.9}
$$

$$
\frac{\partial u}{\partial t}(x, 0) = \xi_1(x) = \{\, \xi_{11}(x), \xi_{12}(x) \,\} \tag{4.10}
$$

and the boundary conditions:

$$
u(0, t) = u(1, t) = 0. \tag{4.11}
$$

Remark 4.1. The calculations can be carried out with a condition (4.11′) more general than that of (4.11), namely:

$$
\left.\begin{aligned}
u(0, t) &= \{\, \Omega_{01}(t), \Omega_{02}(t) \,\} = \Omega_0(t) \\
u(1, t) &= \{\, \Omega_{11}(t), \Omega_{12}(t) \,\} = \Omega_1(t)
\end{aligned}\right\} \tag{4.11′}
$$

where $\Omega_{01}(t)$, $\Omega_{02}(t)$, $\Omega_{11}(t)$, $\Omega_{12}(t)$ are known functions. ∎

Denote by $\chi_{01}^*(x)$, $\chi_{11}^*(x)$, $\chi_{02}^*(x)$, $\chi_{12}^*(x)$ respectively the solutions obtained for $e(x, T), \dfrac{\partial e}{\partial t}(x, T), \tilde{u}(x, T), \dfrac{\partial \tilde{u}}{\partial t}(x, T)$ at the given time T.

The problem defined by (4.6), (4.9), (4.10) and (4.11′) will be called the "direct problem." ∎

Problem 3.1 consists of determining functions $\xi_0(x)$ and $\xi_1(x)$ such that $u(x, t)$, the solution of (4.6), (4.9), (4.10) and (4.11′) (or direct problem) satisfies at time $t = T$:

$$
\int_0^1 \left(\frac{|u(x, T) - \chi_0(x)|^2}{|\chi_0|^2} + \frac{\left|\dfrac{\partial u}{\partial t}(x, T) - \chi_1(x)\right|^2}{|\chi_1|^2} \right) dx \leqslant \eta \tag{4.12}
$$

where η is a "small" number given a priori, and where:

$$
\chi_0(x) = \{\, \chi_{01}(x), \chi_{02}(x) \,\} \quad \text{and} \quad \chi_1(x) = \{\, \chi_{11}(x), \chi_{12}(x) \,\}
$$

are given arbitrary functions.

4.2 The Associated QR Problem

The method of Quasi-Reversibility of Section 3 leads to resolving the following equation

$$\frac{d^2u_\varepsilon}{dt^2} + (A_1 - \varepsilon A_1^0 \, A_1^0) \frac{du_\varepsilon}{dt} + \acute{A}_2 \, u_\varepsilon = 0 \,, \qquad t < T \tag{4.13}$$

where

$$A_1^0 = \begin{bmatrix} -\sigma \dfrac{\partial^2}{\partial x^2} & 0 \\[2mm] 0 & 0 \end{bmatrix}$$

is the "principal part" of A_1 and ε is a "small" number, a function of η.

Put $u_\varepsilon = \{e_\varepsilon, \tilde{u}_\varepsilon\}$, with the initial conditions:

$$u_\varepsilon(x, T) = \chi_0(x) \,, \tag{4.14}$$

$$\frac{\partial u_\varepsilon}{\partial t}(x, T) = \chi_1(x) \,, \tag{4.15}$$

and with the boundary conditions:

$$u_\varepsilon(0, t) = \Omega_0(t) \,, \tag{4.16}$$

$$u_\varepsilon(1, t) = \Omega_1(t) \,, \tag{4.17}$$

$$\frac{\partial^2}{\partial x^2} e_\varepsilon(0, t) = \frac{\partial^2}{\partial x^2} e_\varepsilon(1, t) = 0 \,. \tag{4.18}$$

We then take:

$$\xi_0(x) = u_\varepsilon(x, 0) \tag{4.19}$$

$$\xi_1(x) = \frac{\partial u_\varepsilon}{\partial t}(x, 0) \,. \tag{4.20}$$

For convenience, make the change of notation in (4.13) to (4.18)

$$v(x, t) = u_\varepsilon(x, T - t)$$

and set:

$$v(x, t) = \{ g(x, t), \tilde{v}(x, t) \} \,; \tag{4.21}$$

(4.13) becomes:

$$\frac{d^2v}{dt^2} - (A_1 - \varepsilon A_1^0 A_1^0)\frac{dv}{dt} + A_2 v = 0 \tag{4.22}$$

where the initial and boundary conditions may be written:

$$v(x, 0) = \chi_0(x) \tag{4.23}$$

$$\frac{\partial v}{\partial t}(x, 0) = -\chi_1(x) \tag{4.24}$$

$$v(0, t) = \{\,\Omega'_{01}(t),\, \Omega'_{02}(t)\,\} = \Omega'_0(t) = \Omega_0(T - t) \tag{4.25}$$

$$v(1, t) = \{\,\Omega'_{11}(t),\, \Omega'_{12}(t)\,\} = \Omega'_1(t) = \Omega_1(T - t) \tag{4.26}$$

$$\frac{\partial^2}{\partial x^2} g(0, t) = \frac{\partial^2}{\partial x^2} g(1, t) = 0 \,. \tag{4.27}$$

Denote respectively by $\xi^*_{01}(x)$, $\xi^*_{11}(x)$, $\xi^*_{02}(x)$ and $\xi^*_{12}(x)$ the solutions obtained for $g(x, T)$, $-\frac{\partial g}{\partial t}(x, T)$, $\tilde{v}(x, T)$ and $-\frac{\partial \tilde{v}}{\partial t}(x, T)$. The problem (4.22) to (4.27) will be denoted in what follows as the "retrograde" problem, by contrast with the direct problem.

4.3 Numerical Solution

4.3.1 PRELIMINARY REMARK

The problems, direct and backward, can be considered as two particular cases of the following general problem: Find a pair of functions $V(x, T) = \{X(x, t), Y(x, t)\}$ satisfying the system of partial differential equations:

$$\frac{d^2V}{dt^2} + B_1 \frac{dV}{dt} + B_2 V = 0 \tag{4.28}$$

where

$$B_1 = \begin{bmatrix} \alpha \dfrac{\partial^2}{\partial x^2} + \beta \dfrac{\partial^4}{\partial x^4} & \varphi \dfrac{\partial}{\partial x} \\[3mm] \psi \dfrac{\partial}{\partial x} & 0 \end{bmatrix}$$

and

$$B_2 = \begin{bmatrix} 0 & 0 \\ 0 & -c^2 \dfrac{\partial^2}{\partial x^2} \end{bmatrix}$$

with the initial conditions:

$$V(x, 0) = H_0(x) = \{ H_{01}(x), H_{02}(x) \} \tag{4.29}$$

$$\frac{\partial V}{\partial t}(x, 0) = H_1(x) = \{ H_{11}(x), H_{12}(x) \} \tag{4.30}$$

and the boundary conditions:

$$V(0, t) = F_0(t) = \{ F_{01}(t), F_{02}(t) \}, \tag{4.31}$$

$$V(0, t) = F_0(t) = \{ F_{11}(t), F_{12}(t) \}, \tag{4.32}$$

$$\frac{\partial^2 X}{\partial x^2}(0, t) = \frac{\partial^2 X}{\partial x^2}(1, t) = 0. \tag{4.33}$$

4.3.1.1 The *direct problem* corresponds to:

$$\alpha = -\sigma, \quad \beta = 0, \quad \varphi = c, \quad \psi = c(\gamma - 1),$$

$$H_0(x) = \xi_0(x), \quad H_1(x) = \xi_1(x),$$

$$F_0(t) = \Omega_0(t), \quad F_1(t) = \Omega_1(t).$$

4.3.1.2 The *backward problem* corresponds to:

$$\alpha = \sigma, \quad \beta = \varepsilon\sigma^2, \quad \varphi = -c, \quad \psi = -c(\gamma - 1),$$

$$H_0(x) = \chi_0(x), \quad H_1(x) = -\chi_1(x),$$

$$F_0(t) = \Omega'_0(t), \quad \dot{F}_1(t) = \Omega'_1(t).$$

4.3.2 THE DISCRETIZATION SCHEME FOR THE SYSTEM (4.28) TO (4.33)

4.3.2.1 Notation

Set:

$$N \Delta t = T,$$

$$J \Delta x = 1.$$

The value V_j^n of the function $V(x, t)$ at the point of discretization defined by j and n is calculated in the following fashion (see Figure 1):
if:

$$0 < j \, \Delta x < 1 \, ,$$

equivalently:

$$0 < j < J \, ,$$
$$V_j^n = X(j \, \Delta x, n \, \Delta t) \, ;$$

if:

$$1 \leqslant j \, \Delta x < 1 + (J - 1) \, \Delta x \, ,$$

equivalently:

$$J \leqslant j < 2 \, J - 1 \, ,$$
$$V_j^n = Y([j - (J - 1)] \, \Delta x, n \, \Delta t) \, .$$

Thus at a given discretization point, we go from $X(x_0, n \, \Delta t)$ to $Y(x_0, n \, \Delta t)$ by going from V_j^n to V_{j+1}^n, which is to say by increasing the j index of the fixed quantity $\boxed{i = J - 1}$.

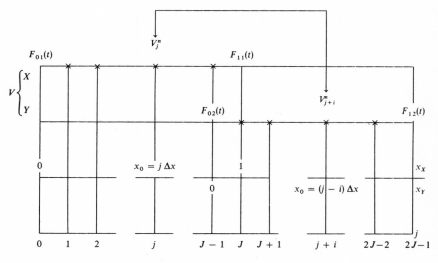

Figure 1

4.3.2.2 The derivatives are discretized in the following way:

$$\left(\frac{\partial V}{\partial t}\right)^n \simeq \frac{V^{n+1} - V^{n-1}}{2\,\Delta t}\,, \qquad \left(\frac{\partial^2 V}{\partial t^2}\right)^n \simeq \frac{V^{n+1} - 2\,V^n + V^{n-1}}{\Delta t^2}\,,$$

$$\left(\frac{\partial V}{\partial x}\right)_j \simeq \frac{V_{j+1} - V_{j-1}}{\Delta x}\,, \quad \text{etc.}$$

Remark 4.1. These formulas do not hold for:

$$j = 1\,, \quad j = 2\,; \quad j = J - 2 = i - 1\,, \quad j = J - 1 = i\,,$$

$$j = 1 + i\,; \qquad\qquad j = J - 1 + i = 2\,i\,,$$

$$n = 0\,.$$

4.3.2.3 Discretization of (4.28) at the time $n\,\Delta t$

We have:

$$\frac{\partial^2}{\partial t^2}\begin{pmatrix} X \\ Y \end{pmatrix} + \begin{pmatrix} \alpha\dfrac{\partial^2}{\partial x^2} + \beta\dfrac{\partial^4}{\partial x^4} & \varphi\dfrac{\partial}{\partial x} \\[2mm] \psi\dfrac{\partial}{\partial x} & 0 \end{pmatrix}\frac{\partial}{\partial t}\begin{pmatrix} X \\ Y \end{pmatrix} + \begin{pmatrix} 0 & 0 \\[2mm] 0 & -c^2\dfrac{\partial^2}{\partial x^2} \end{pmatrix}\begin{pmatrix} X \\ Y \end{pmatrix} = \begin{pmatrix} 0 \\ 0 \end{pmatrix}.$$

(i) *Formulation for* $1 \leqslant j \leqslant J - 1, i.e., V \equiv X$.
    General case: $3 \leqslant j \leqslant J - 3$.

We can then write:

$$\left(\frac{\partial^2 X}{\partial t^2}\right)_j^n \simeq \frac{V_j^{n+1} - 2\,V_j^n + V_j^{n-1}}{\Delta t^2}\,,$$

$$\left(\alpha\frac{\partial^2}{\partial x^2}\frac{\partial X}{\partial t}\right)_j^n \simeq \frac{\alpha}{2\,\Delta x^2\,\Delta t}(V_{j+1}^{n+1} - 2\,V_j^{n+1} + V_{j-1}^{n+1} - V_{j+1}^{n-1} + 2\,V_j^{n-1} - V_{j-1}^{n-1})\,,$$

$$\left(\beta\frac{\partial^4}{\partial x^4}\frac{\partial X}{\partial t}\right)_j^n \simeq \frac{\beta}{2\,\Delta x^4\,\Delta t}\,(V_{j+2}^{n+1} - 4\,V_{j+1}^{n+1} + 6\,V_j^{n+1} - 4\,V_{j-1}^{n+1} +$$

$$+ V_{j-2}^{n+1} - V_{j+2}^{n-1} + 4\,V_{j+1}^{n-1} - 6\,V_j^{n-1} + 4\,V_{j-1}^{n-1} - V_{j-2}^{n-1})\,,$$

$$\left(\varphi\frac{\partial}{\partial x}\frac{\partial Y}{\partial t}\right)_j^n \simeq \frac{\varphi}{4\,\Delta x\,\Delta t}\,(V_{j+1+i}^{n+1} - V_{j-1+i}^{n+1} - V_{j+1+i}^{n-1} + V_{j-1+i}^{n-1})\,;$$

whence the equation in finite differences

$$V_j^{n+1}\left(\frac{1}{\Delta t^2} - \frac{\alpha}{\Delta x^2\,\Delta t} + \frac{3\,\beta}{\Delta x^4\,\Delta t}\right) + (V_{j+1}^{n+1} + V_{j-1}^{n+1})\left(\frac{\alpha}{2\,\Delta x^2\,\Delta t} - \frac{2\,\beta}{\Delta x^4\,\Delta t}\right) +$$

$$+ (V_{j+2}^{n+1} + V_{j-2}^{n+1})\left(\frac{\beta}{2\,\Delta x^4\,\Delta t}\right) + (V_{j+1+i}^{n+1} - V_{j-1+i}^{n+1})\left(\frac{\varphi}{4\,\Delta x\,\Delta t}\right) =$$

$$= 2\,V_j^n\left(\frac{1}{\Delta t^2}\right) + V_j^{n-1}\left(-\frac{\alpha}{\Delta x^2\,\Delta t} + \frac{3\,\beta}{\Delta x^4\,\Delta t} - \frac{1}{\Delta t^2}\right) +$$

$$+ (V_{j+1}^{n-1} + V_{j-1}^{n-1})\left(\frac{\alpha}{2\,\Delta x^2\,\Delta t} - \frac{2\,\beta}{\Delta x^4\,\Delta t}\right) + (V_{j+2}^{n-1} + V_{j-2}^{n-1})\left(\frac{\beta}{2\,\Delta x^4\,\Delta t}\right) +$$

$$+ (V_{j+1+i}^{n-1} - V_{j-1+i}^{n-1})\left(\frac{\varphi}{4\,\Delta x\,\Delta t}\right).$$

Let us multiply by Δt^2 and set:

$$S_1 = \frac{\alpha\,\Delta t}{\Delta x^2},$$

$$S_2 = \frac{\beta\,\Delta t}{2\,\Delta x^4},$$

$$S_3 = 1 - 2\,S_1 + 6\,S_2,$$

$$S_4 = \frac{\varphi\,\Delta t}{4\,\Delta x},$$

$$S_5 = S_1 - 4\,S_2\,;$$

the finite difference equation may then be written:

$$S_3\,V_j^{n+1} + S_5(V_{j+1}^{n+1} + V_{j-1}^{n+1}) + S_2(V_{j+2}^{n+1} + V_{j-2}^{n+1}) +$$

$$+ S_4(V_{j+1+i}^{n+1} - V_{j-1+i}^{n+1}) = 2\,V_j^n + (S_3 - 2)\,V_j^{n-1} + S_5(V_{j+1}^{n-1} + V_{j-1}^{n-1}) + \qquad \text{(4.34)}$$

$$+ S_2(V_{j+2}^{n-1} + V_{j-2}^{n-1}) + S_4(V_{j+1+i}^{n-1} - V_{j-1+i}^{n-1}).$$

PARTICULAR CASES: *Introduction of boundary conditions.*

These conditions may be written:

—for X:

$$V_0^n = F_{01}(n\,\Delta t) \qquad\qquad \text{from (4.31)},$$

$$V_{-1}^n = 2\,V_0^n - V_1^n \qquad\qquad \text{from (4.33)},$$

$$V_J = V_{i+1} = F_{11}(n\,\Delta t) \qquad \text{from (4.32),}$$

$$V_{J+1}^n = V_{i+2}^n = 2\,V_{i+1}^n - V_i^n \qquad \text{from (4.33);}$$

—for Y:

$$V_i^n = F_{02}(n\,\Delta t) \qquad \text{from (4.31),}$$

$$V_{j+i}^n = V_{2i+1}^n = F_{12}(n\,\Delta t) \qquad \text{from (4.32).}$$

We thus deduce the equations which correspond to particular values of j:

$\underline{j = 1}$

$$S_3\,V_1^{n+1} + S_5\,V_2^{n+1} + S_2\,V_3^{n+1} + S_4\,V_{i+2}^{n+1} + S_5\,V_0^{n+1} + 2\,S_2\,V_0^{n+1} -$$
$$- S_2\,V_1^{n+1} - S_4\,V_i^{n+1} = 2\,V_1^n + (S_3 - 2)\,V_1^{n-1} + S_5\,V_2^{n-1} +$$
$$+ S_4\,V_{i+2}^{n-1} + S_5\,V_0^{n-1} + 2\,S_2\,V_0^{n-1} - S_2\,V_1^{n-1} - S_4\,V_i^{n-1}$$

let:

$$(S_3 - S_2)\,V_1^{n+1} + S_5\,V_2^{n+1} + S_2\,V_3^{n+1} + S_4\,V_{i+2}^{n+1} =$$
$$= 2\,V_1^n + (S_3 - S_2 - 2)\,V_1^{n-1} + S_5\,V_2^{n-1} + S_2\,V_3^{n-1} + S_4\,V_{i+2}^{n-1} -$$
$$- (S_5 + 2\,S_2)\,[F_{01}[(n + 1)\,\Delta t] - F_{01}[(n - 1)\,\Delta t]] +$$
$$+ S_4[F_{02}[(n + 1)\,\Delta t] - F_{02}[(n - 1)\,\Delta t]] \tag{4.35}$$

$\underline{j = 2}.$

$$S_3\,V_2^{n+1} + S_5(V_3^{n+1} + V_1^{n+1}) + S_2\,V_4^{n+1} + S_4(V_{i+3}^{n+1} - V_{i+1}^{n+1}) =$$
$$= 2\,V_2^n + (S_3 - 2)\,V_2^{n-1} + S_5(V_3^{n-1} + V_1^{n-1}) + S_2\,V_4^{n-1} + \tag{4.36}$$
$$+ S_4(V_{i+3}^{n-1} - V_{i+1}^{n-1}) - S_2[F_{01}[(n + 1)\,\Delta t] - F_{01}[(n - 1)\,\Delta t]]$$

$\underline{j = i - 1}.$

$$S_3\,V_{i-1}^{n+1} + S_5(V_i^{n+1} + V_{i-2}^{n+1}) + S_2\,V_{i-3}^{n+1} + S_4(V_{2i}^{n+1} - V_{2i-2}^{n+1}) =$$
$$= 2\,V_{i-1}^n + (S_3 - 2)\,V_{i-1}^{n-1} + S_5(V_i^{n-1} + V_{i-2}^{n-1}) + S_2\,V_{i-3}^{n-1} + \tag{4.37}$$
$$+ S_4(V_{2i}^{n-1} - V_{2i-2}^{n-1}) - S_2[F_{11}[(n + 1)\,\Delta t] - F_{11}[(n - 1)\,\Delta t]]$$

$j = i$.

$$(S_3 - S_2) V_i^{n+1} + S_5 V_{i-2}^{n+1} + S_2 V_{i-2}^{n+1} - S_4 V_{2i-1}^{n+1} =$$
$$= 2 V_i^n + (S_3 - S_2 - 2) V_i^{n-1} + S_5 V_{i-1}^{n-1} + S_2 V_{i-2}^{n-1} - S_4 V_{2i-1}^{n-1} -$$
$$- (S_5 + 2 S_2) F_{11}[[(n + 1) \Delta t] - F_{11}[(n - 1) \Delta t]] \qquad (4.38)$$
$$- S_4[F_{12}[(n + 1) \Delta t] - F_{12}[(n - 1) \Delta t]] .$$

PARTICULAR CASES: *Introduction of initial conditions.*

These conditions may be written [from (4.29) and (4.30)]:

$$V_j^0 = H_{01}(j \Delta x),$$
$$V_j^{-1} = V_j^1 - 2 \Delta t H_{11}(j \Delta x),$$
$$V_{-1}^{-1} = 2 V_0^{-1} - V_1^{-1} = 2 V_0^{-1} - V_1^1 + 2 \Delta t H_{11}(\Delta x),$$
$$= 2 V_0^1 - 4 \Delta t H_{11}(0) - V_1^1 + 2 \Delta t H_{11}(\Delta x),$$
$$V_{i+2}^{-1} = 2 V_{i+1}^{-1} - V_i^{-1} = 2 V_{i+1}^{-1} - V_i^{-1} + 2 \Delta t H_{11}(1 - \Delta x),$$
$$= 2 V_{i+1}^1 - 4 \Delta t H_{11}(1) - V_i^1 + 2 \Delta t H_{11}(1 - \Delta x).$$

We deduce the corresponding equations for $n = 0$:

$n = 0,\quad 1 < j < i$

$$S_3 V^1 + S_5 (V_{j+1}^1 + V_{j-1}^1) + S_2 (V_{j+2}^1 + V_{j-2}^1) + S_4 (V_{j+1+i}^1 - V_{j-1+i}^1) =$$
$$= 2 V_j^0 + (S_3 - 2) [V_j^1 - 2 \Delta t . H_{11}(j \Delta x)] +$$
$$+ S_5 [V_{j+1}^1 - 2 \Delta t H_{11}[(j+1) \Delta x] + V_{j-1}^1 - 2 \Delta t H_{11}[(j-1) \Delta x]] \qquad (4.39)$$
$$+ S_2 [V_{j+2}^1 - 2 \Delta t H_{11}[(j+2) \Delta x] + V_{j-2}^1 - 2 \Delta t H_{11}[(j-2) \Delta x]]$$
$$+ S_4 [V_{j+1+i}^1 - 2 \Delta t H_{12}[(j+1) \Delta x] - V_{j-1+i}^1 + 2 \Delta t H_{12}[(j-1) \Delta x]] ,$$

let:

$$V_j^1 = H_{01}(j \Delta x) - \Delta t[(S_3 - 2) H_{11}(j \Delta x) + S_5[H_{11}(j + 1) \Delta x] +$$
$$+ H_{11}[(j - 1) \Delta x]] + S_2[H_{11}[(j + 2) \Delta x] + H_{11}[(j - 2) \Delta x]] + \qquad (4.40)$$
$$+ S_4[H_{12}[(j + 1) \Delta x] - H_{12}[(j - 1) \Delta x]]$$

$n = 0,\quad j = 1,$

$$V_1^1 = H_{01}(j \Delta x) - \Delta t[(S_3 - 2) H_{11}(\Delta x) +$$
$$+ S_5[H_{11}(2 \Delta x) + H_{11}(0)] + S_2[H_{11}(3 \Delta x) - H_{11}(\Delta x) + \qquad (4.41)$$
$$+ 2 H_{11}(0)] + S_4[H_{12}(2 \Delta x) - H_{12}(0)]]$$

$n = 0, \quad j = i,$

$$V^1 = H_{01}(1 - \Delta x) - \Delta t[(S_3 - 2) H_{11}(1 - \Delta x) +$$
$$+ S_5[H_{11}(1 - 2 \Delta x) + H_{11}(1)] + S_2[H_{11}(1 - 3 \Delta x) - \qquad (4.42)$$
$$- H_{11}(1 - \Delta x) + 2 H_{11}(1)] - S_4[H_{12}(1 - 2 \Delta x) - H_{12}(1)]].$$

(ii) *Formulation for* $i + 1 \leqslant j \leqslant 2 i$, *i.e.*, $V \equiv Y$.
General case: $i + 2 \leqslant j \leqslant 2 i - 1$.

We can then write:

$$\left(\frac{\partial^2 Y}{\partial t^2}\right)^n_j \simeq \frac{V^{n+1}_j - 2 V^n_j + V^{n-1}_j}{\Delta t^2}$$

$$\left(\psi \frac{\partial}{\partial x} \frac{\partial X}{\partial t}\right)^n_j \simeq \frac{\psi}{4 \Delta x \Delta t} (V^{n+1}_{j+1-i} - V^{n+1}_{j-1-i} - V^{n-1}_{j+1-i} + V^{n-1}_{j-1-i})$$

$$\left(- c^2 \frac{\partial^2 Y}{\partial x^2}\right)^n_j \simeq \frac{- c^2}{\Delta x^2} (V^n_{j+1} - 2 V^n_j + V^n_{j-1}) ,$$

whence the finite difference equation:

$$V^{n+1}_j \left(\frac{1}{\Delta t^2}\right) + (V^{n+1}_{j+1-i} - V^{n+1}_{j-1-i})\left(\frac{\psi}{4 \Delta x \Delta t}\right) =$$

$$= V^n_j \left(\frac{2}{\Delta t^2} - \frac{2 c^2}{\Delta x^2}\right) - V^{n-1}_j \left(\frac{1}{\Delta t^2}\right) - (V^n_{j+1} + V^n_{j-1})\left(\frac{- c^2}{\Delta x^2}\right) +$$

$$+ (V^{n-1}_{j+1-i} - V^{n-1}_{j-1-i})\left(\frac{\psi}{4 \Delta x \Delta t}\right).$$

Multiply by Δt^2 and set:

$$T_1 = \frac{\psi \Delta t}{4 \Delta x},$$

$$T_2 = 2 - \frac{2 c^2 \Delta t^2}{\Delta x^2},$$

$$T_3 = \frac{c^2 \Delta t^2}{\Delta x^2} ;$$

the finite difference equation may be written:

$$V_j^{n+1} + T_1(V_{j+1-i}^{n+1} - V_{j-1-i}^{n+1}) =$$
$$= T_2 V_j^n + T_3(V_{j+1}^n + V_{j-1}^n) + T_1(V_{j+1-i}^{n-1} - V_{j-1-i}^{n-1}) - V_j^{n-1}. \tag{4.43}$$

PARTICULAR CASES: *Introduction of boundary conditions.*

From the conditions, which have been formulated in (i), we deduce the following equations, corresponding to particular values of j:

$\underline{j = i + 1}$

$$V_{i+1}^{n+1} + T_1 V_2^{n+1} = T_2 V_{i+1}^n + T_3 V_{i+2}^n + T_1 V_2^{n-1} - V_{i+1}^{n-1} +$$
$$+ T_1[F_{01}[(n + 1) \Delta t] - F_{01}[(n - 1) \Delta t]] + T_3 F_{02}(n \Delta t); \tag{4.44}$$

$\underline{j = 2i}$

$$V_{2i}^{n+1} - T_1 V_{i-1}^{n+1} = T_2 V_{2i}^n + T_3 V_{2i-1}^n - T_1 V_{i-1}^{n-1} - V_{2i}^{n-1} -$$
$$- T_1[F_{11}[(n + 1) \Delta t] - F_{11}[(n - 1) \Delta t]] + T_3 F_{12}(n \Delta t). \tag{4.45}$$

PARTICULAR CASES: *Introduction of initial conditions.*

These conditions may be written:

$$V_j^0 = H_{02}(j \Delta x),$$
$$V_j^{-1} = V_j^1 - 2 \Delta t H_{12}(j \Delta x).$$

We deduce the equation corresponding to $n = 0$, valid for $i + \leqslant 1j \leqslant 2i$

$$V_j^1 + T_1(V_{j+1-i}^1 - V_{j-1-i}^1) = T_2 V_j^0 + T_3(V_{j+1}^0 + V_{j-1}^0) +$$
$$+ T_1[V_{j+1-i}^1 - 2 \Delta t H_{11}[(j + 1) \Delta x] - V_{j-1-i}^1 + 2 \Delta t H_{11}[(j - 1) \Delta x]] -$$
$$- V_j^1 + 2 \Delta t H_{12}(j \Delta x),$$

let:

$$V_j^1 = \tfrac{1}{2}[T_2 H_{02}(j \Delta x) + T_3[H_{02}[(j + 1) \Delta x] + H_{02}[(j - 1) \Delta x]]] -$$
$$- \Delta t[T_1[H_{11}[(j + 1) \Delta x] - H_{11}[(j - 1) \Delta x]] - H_{12}(j \Delta x)]. \tag{4.46}$$

4.3.3 MATRIX FORM OF DISCRETIZED SYSTEM

The system (4.28) and conditions (4.29) to (4.33), after discretization, can therefore be written in matrix form:

$$AV^{n+1} = B_n,\qquad(4.47)$$

where A is a matrix of $2i$ rows and $2i$ columns whose coefficients are independent of n; V^{n+1} is the column-vector with components V_j^{n+1} (j = 1 to $2i$), and B_n is a column vector with $2i$ components, obtained from V^n, V^{n-1} and the boundary conditions, using (4.34) to (4.46). The form of the matrix is given in Figure 2.

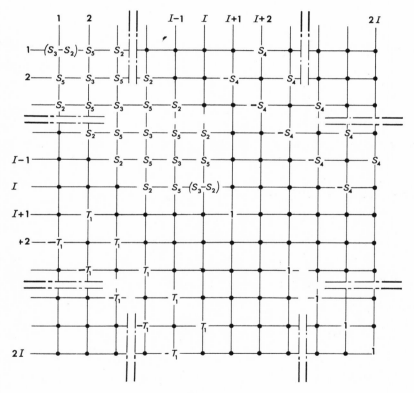

Figure 2

We can then calculate, from V^0 and V^1 (known from the initial conditions), V^2, ..., V^n succesively, and thus $V^N = V(x, T)$.

4.3.4 CALCULATION OF THE ERROR

From $\chi_0(x)$ and $\chi_1(x)$, the approximate solution of the retrograde problem yields $\xi_0^*(x)$ and $\xi_1^*(x)$.

The exact solution of the direct problem with $\xi_0^*(x)$ and $\xi_1^*(x)$ as initial conditions is the function $u^*(x, T)$ and its derivative $\frac{\partial u^*}{\partial t}(x, T)$.

Condition (4.12) is then:

$$E = \int_0^1 \left(\frac{|u^*(x, T) - \chi_0(x)|^2}{|\chi_0|^2} + \frac{\left|\frac{\partial u^*}{\partial t}(x, T) - \chi_1(x)\right|^2}{|\chi_1|^2} \right) dx \leqslant \eta.$$

Generally we only know $u^*(x, T)$ and $\frac{\partial u^*}{\partial t}(x, T)$ in their approximate forms $\chi_0^*(x)$ and $\chi_1^*(x)$, solutions of the direct problem with initial conditions $\xi_0^*(x)$ and $\xi_1^*(x)$. We thus calculate not E, but

$$E^* = \int_0^1 \left(\frac{|\chi_0^*(x) - \chi_0(x)|^2}{|\chi_0|^2} + \frac{|\chi_1^*(x) - \chi_1(x)|^2}{|\chi_1|^2} \right) dx$$

or more precisely an approximation $\tilde{E}$ to E^*, obtained, for example, by the trapezoid method:

$$\tilde{E} = \frac{\sum_{j=1}^{2i} [(\chi_0^*)_j - (\chi_0)_j]^2}{\sum_{j=1}^{2i} (\chi_0)_j^2} + \frac{\sum_{j=1}^{2i} [(\chi_1^*)_j - (\chi_1)_j]^2}{\sum_{j=1}^{2i} (\chi_1)_j^2}. \tag{4.48}$$

In each numerical application, the value of $\tilde{E}$ yields an indication of the precision of the results.

4.4 Numerical Results

4.4.1 DATA

In all of the trials, we used the following numerical values:

$$\sigma = 1, \quad c = 3, \quad \gamma = 1.4$$

$$\Delta x = 10^{-1} \quad \text{or} \quad 5.10^{-2},$$

$$\Delta t \leqslant 5.10^{-3}.$$

4.4.2 THE OPTIONS

The program was written so as to treat the following sequences:

—*Option 1*: solution of only the direct problem; we ask for $\chi_0^*(x)$ and $\chi_1^*(x)$, starting with $\xi_0(x)$ and $\xi_1(x)$.

—*Option 2*: solution of the backward problem, i.e., determination of $\xi_0^*(x)$ and $\xi_1^*(x)$ starting with $\chi_0(x)$ and $\chi_1(x)$, then at the end of the verification, solution of the direct problem starting with $\xi_0^*(x)$ and $\xi_1^*(x)$, yielding χ_0^* and χ_1^* (we then calculate the error $\tilde{E}$).

—*Option 3*: we begin with the solution of the direct problem starting with $\xi_0(x)$ and $\xi_1(x)$, and starting from the solutions obtained we finish as in Option 2; the solutions are denoted successively by ξ_0^*, ξ_1^* and χ_0^{**}, χ_1^{**}.

4.4.3 INITIAL AND BOUNDARY CONDITIONS

The initial and boundary conditions imposed are not independent. In effect, they must satisfy:

—*for the direct problem* [from (4.9), (4.10) to (4.11)]:

$$u(0, 0) = \xi_0(0) = \Omega_0(0) ,$$

$$u(1, 0) = \xi_0(1) = \Omega_1(0) ,$$

$$\frac{\partial u}{\partial t}(0, 0) = \xi_1(0) = \frac{\partial}{\partial t}\Omega_0(0) ,$$

$$\frac{\partial u}{\partial t}(1, 0) = \xi_1(1) = \frac{\partial}{\partial t}\Omega_1(0) ;$$

—*for the backward problem* [from (4.34) to (4.21)]:

$$v(0, 0) = \chi_0(0) = \Omega_0'(0) ,$$

$$v(1, 0) = \chi_0(1) = \Omega_1'(0) ,$$

$$\frac{\partial v}{\partial t}(0, 0) = - \chi_1(0) = \frac{\partial \Omega_0'}{\partial t}(0) ,$$

$$\frac{\partial v}{\partial t}(1, 0) = - \chi_1(1) = \frac{\partial \Omega_1'}{\partial t}(0) ,$$

$$\frac{\partial^2}{\partial x^2} g(0, 0) = \frac{\partial^2}{\partial x^2} \chi_{01}(0) = 0 \,,$$

$$\frac{\partial^2}{\partial x^2} g(1, 0) = \frac{\partial^2}{\partial x^2} \chi_{01}(1) = 0 \,.$$

4.4.4 FIRST SERIES OF TRAILS

These correspond to Option 3 (4.42).

Example 1: Initial and boundary conditions

$$\xi_0(x) = \{ 0, 0 \} \,,$$
$$\xi_1(x) = \{ -3x + 2, 0 \} \,,$$
$$\Omega_0(t) = \{ 2t, 0 \} \,,$$
$$\Omega_1(t) = \{ -t, 0 \} \,.$$

We take:

$$T = 4.10^{-3} \,, \qquad \Delta t = 4.10^{-4} \,, \qquad \Delta x = 10^{-1} \,;$$

for

$$\varepsilon = 5.10^{-3} \,.$$

we obtain, with $\tilde{E}$ given by (4.48),

$$\tilde{E} = 1.6.10^{-10} \,.$$

Example 2: Initial and boundary conditions

$$\xi_0(x) = \{ 0, 0 \} \,,$$
$$\xi_1(x) = \left\{ \sin\left(-\frac{\pi}{2} x\right), 0 \right\} \,,$$
$$\Omega_0(t) = \{ 2t^3, 0 \} \,,$$
$$\Omega_1(t) = \{ -t, 0 \} \,.$$

We take

$$T = 10^{-1} \,, \qquad \Delta t = 2 \; 10^{-3} \,, \qquad \Delta x = 10^{-1} \,;$$

for

$$\varepsilon = 5.10^{-3},$$

we obtain

$$\tilde{E} = 1.4.10^{-2}$$

The curves representative of the functions obtained have been traced in Graphs 9 and 10.

Example 3. Let us repeat the previous trial, with the same initial and boundary conditions, and the same value of Δx; the only quantity modified is:

$$T = 3.10^{-1}.$$

This time for $\varepsilon = 5.10^{-3}$, we obtain:

$$\tilde{E} = 1.6.10^{6}.$$

If we choose for ε the value which gives the best results in Example 5 (cf. below), i.e.:

$$\varepsilon = 4.10^{-2},$$

we obtain

$$\tilde{E} = 1.2.10^{-2}.$$

Thus, for the same trial, the act of taking $\varepsilon = 4.10^{-2}$ in place of 5.10^{-3} permits us to reestablish convergence and to obtain good results for $T = 3.10^{-1}$. The corresponding curves are shown in Graphs 11 and 12.

4.4.5 SECOND SERIES OF TRIALS

These correspond to Option 2 (4.4.2).

Examples 4 and 5: Initial and boundary conditions

$$\chi_0(x) = \{ -0.175\,x + 0.075,\ 10^{-2}\sin \pi x \},$$

$$\chi_1(x) = \{ 2.5\,x - 1.5,\ 10^{-1}\sin \pi x \},$$

$$\Omega'_0(t) = \{ 7.5(T - t)^2, 0 \},$$

$$\Omega'_1(t) = \{ -(T - t), 0 \}.$$

Example 4: We take

$$T = 10^{-1}, \qquad \Delta l = 2.10^{-3}, \qquad \Delta x = 10^{-1}$$

We obtain the following results

ε	$\tilde{E}$
10^{-3}	$3.2 \ .10^{10}$
8.10^{-3}	$3.3 \ .10^{-1}$
5.10^{-2}	$1.21.10^{-1}$
10^{-1}	$2.1 \ .10^{-1}$

ε	$\tilde{E}$
5.10^{-3}	1.8
10^{-2}	$2.2 \ .10^{-1}$
8.10^{-2}	$1.68.10^{-1}$
5.10^{-1}	$6.5 \ .10^{-1}$

Example 5: We take

$$T = 10^{-1}. \qquad \Delta t = 2.10^{-3}, \qquad \Delta x = 5.10^{-2}$$

We obtain

ε	$\tilde{E}$
10^{-2}	$1.82.10^{-1}$
3.10^{-2}	$1.04.10^{-1}$
5.10^{-2}	$1.02.10^{-1}$

ε	$\tilde{E}$
2.10^{-2}	$1.24.10^{-1}$
4.10^{-2}	$9.8 \ .10^{-2}$

Note that for these two values of Δx, there exists a value of ε for which $\tilde{E}$ is a minimum (cf. Remark 5.9, Chapter 1).

We have shown some of the corresponding curves on Graphs 13 to 16 *bis*, and the curve $\tilde{E}(\varepsilon)$ on Graph 17.

Example 6: Initial and boundary conditions

$$\chi_0(x) = \{ -0.975 \, x + 0.675, \ 10^{-2} \sin \pi x \},$$
$$\chi_1(x) = \{ 5.5 \, x - 4.5, \ 10^{-1} \sin \pi x \},$$
$$\Omega'_0(t) = \{ 7.5(T - t)^2, 0 \},$$
$$\Omega'_1(t) = \{ -(T - t), 0 \}.$$

We take:

$$T = 3.10^{-1}, \qquad \Delta t = 2.10^{-3}, \qquad \Delta x = 5.10^{-2}.$$

For ε we took the value which gave the best results in Example 5, that is:

$$\varepsilon = 4.10^{-2};$$

we obtain

$$\tilde{E} = 3,4.10^{-1}.$$

Thus, contrary to Example 3, the scheme is still convergent for $T = 3.10^{-1}$.

The corresponding curves are shown on Graphs 18 and 19.

Example 7: We reran Example 2, changing only the value of ε; we take:

$$\varepsilon = 4.10^{-2},$$

we obtain

$$\tilde{E} = 1,8.10^{-3},$$

which shows that the results are much better than before. The representative curves have been drawn on Graphs 20 and 21, which should be compared with Graphs 9 and 10.

5. TRANSPORT EQUATIONS AND THE QR METHOD

5.1 General Remarks

In a Hilbert space $H(^1)$ on **R**, let A be an unbounded operator, where

$$\left. \begin{array}{l} A \text{ is the infinitesimal generator of a continuous semi-group} \\ G(t) \text{ (i.e., } \forall f \in H,\ t \to G(t)\,f \text{ is continuous for } t \geqslant 0 \to H). \end{array} \right\} \quad (5.1)$$

$(^1)$Where the scalar product, as previously, is denoted by (,), and the corresponding norm by| |.

(For the theory of semi-groups, cf. Hille–Phillips [1], K. Yosida [1]). We suppose that:

$$G(t) \ cannot \ \text{be extended into a group.} \qquad (5.2)$$

Let $G^*(t)$ be the *adjoint semi-group*:

$$(G^*(t) f, g) = (f, G(t) g) \qquad \forall f, g \in H \, .$$

Let $- A^*$ be the infinitesimal generator of $G^*(t)$; A^* is then the adjoint (in the sense of unbounded operators) of A.

We now make the following hypothesis:

$$G^*(T), \ (T > 0 \text{ fixed}), \text{ is } one\text{-}to\text{-}one \text{ in } H. (^1) \quad \blacksquare \qquad (5.3)$$

The functional $J(\xi)$ (compare Section 4, Chapter 1).

For ξ given in H, let $u = u(t) = u(t ; \xi)$ be *the* solution of:

$$u' + Au = 0 \qquad \left(u' = \frac{du}{dt} \right), \qquad (5.4)$$

$$u(0) = \xi \, . \qquad (5.5)$$

This solution may be expressed as:

$$u(t) = G(t).\xi \, ;$$

the properties of u are as follows:

1) if $\xi \in D(A)$ (domain of A), then $t \to u(t)$ is continuous for $t \geqslant 0 \to D(A)$, with the derivative $u'(t)$ continuous for $t \geqslant 0 \to H$;
2) if $\xi \in H$, then $t \to u(t)$ is only continuous for $t \geqslant 0 \to H$ and u is a *generalized solution* of (5.4) in the following sense:

the operator A^* is linear and continuous from $D(A^*) \to H$ (²) therefore by passage to the adjoint: $(A^*)^* \in \mathscr{L}(H ; D(A^*)')$, and $(A^*)^*$ coincides with A in $D(A)$; since $D(A)$ is dense in H, we see that A extends by continuity into an operator, still denoted by A, with

$$A \in \mathscr{L}(H ; D(A^*)') \, ;$$

(¹) If this is not the case, what follows can be adapted by supposing that (5.8) holds.
(²) $D(A^*)$ has the norm of the graph: $(|u|^2 + |A^*u|^2)^{1/2}$.

then **(5.4)** is an equality in $D(A^*)'$ $(\forall t \geqslant 0)$; the function u' is continuous in $t \geqslant 0 \to D(A^*)'$. ∎

For χ given in H, we then set:

$$J(\xi) = |u(T) - \chi|^2 = |G(T)\xi - \chi|^2 . \blacksquare \tag{5.7}$$

We will demonstrate.

Theorem 5.1: *Under the hypotheses* **(5.1), (5.2), (5.3),** *we have*:

$$\operatorname*{Inf}_{\xi \in D(A)} J(\xi) = 0 . \tag{5.8}$$

Proof. It reduces to showing that $u(T) = G(T)\xi$ spans a space dense in H when ξ spans $D(A)$.

Suppose that $\psi \in H$ with

$$(G(T)\,\xi, \psi) = 0 \qquad \forall \xi ;$$

then this is equivalent to

$$(\xi, G^*(T)\,\psi) = 0 \qquad \forall \xi \in D(A)$$

or

$$G^*(T)\,\psi = 0$$

and, from **(5.3)**, $\psi = 0$, whence the result. ∎

Remark 5.1. We will have **(5.8)** when ξ varies over any space dense in H. ∎

Remark 5.2. The preceding remark extends, under analogous hypotheses, and with the same demonstration, to the case of semi-groups in reflexive Banach spaces. ∎

Remark 5.3. Hypothesis **(5.3)** is a property of retrograde uniqueness. In effect, *the* solution of

$$v' + A^*\, v = 0 , \qquad v(0) = \psi ,$$

is

$$v(t) = G^*(t)\,\psi$$

(a generalized solution if $\psi \in H$) and (5.3) means: if $v(T) = 0$ then $v(t) = 0$ for $t < T$. ∎

We can now pose

Problem 5.1:   For a given $\eta > 0$, find ξ_η such that

$$J(\xi_\eta) \leqslant \eta . \; ∎ \tag{5.9}$$

Remark 5.4. The "ideal" solution would be to find u satisfying

$$u' + Au = 0 ,$$

$$u(T) = \chi .$$

But this solution *does not exist* (in the space H), since $G(t)$ is not extendable to a *group*. ∎

5.2   The QR Method

The QR method we are going to give is formally analogous to that of Section 5, Chapter 1.

We define $u_\varepsilon(t)$ as *the* solution of

$$u'_\varepsilon + Au_\varepsilon - \varepsilon A^* Au_\varepsilon = 0 , \tag{5.10}$$

$$u_\varepsilon(T) = \chi . \tag{5.11}$$

We have

Theorem 5.2: *Under the hypothesis* (5.1),([1]) *problem* (5.10), (5.11) *admits a solution satisfying*

$$u_\varepsilon \in L_2(0, T; D(A)) , \qquad u'_\varepsilon \in L_2(0, T; D(A)') .$$

Proof. This is a particular case of Theorem 3.1, Chapter 1 (change t into $T - t$ to convert it into the case of "increasing t"). ∎

The QR method then consists in

1) *solving* (5.10), (5.11);
2) *taking*

$$\xi = u_\varepsilon(0) . \; ∎$$

([1]) In reality, it suffices that A be a *closed* operator.

Remark 5.5. Let A_0 be the "principal part" of A, where A_0 is an operator "simpler" than A and "equivalent" to A.[1] We then replace (5.10) by:

$$u'_\varepsilon + A u_\varepsilon - \varepsilon A_0^* A_0 u_\varepsilon = 0 \qquad\qquad (5.13)$$

and still take $\xi = u_\varepsilon(0)$, where u_ε is the solution of (5.13) and (5.11). ∎

Remark 5.6. For a *verification* of the results, we take U_ε the solution of

$$U'_\varepsilon + A U_\varepsilon = 0, \qquad U_\varepsilon(0) = u_\varepsilon(0),$$

or

$$U_\varepsilon(t) = G(t)\, u_\varepsilon(0),$$

then we compare $U_\varepsilon(T)$ and χ.

We do not know if

$$G(T)\, u_\varepsilon(0) \to \chi \quad\text{as}\quad \varepsilon \to 0$$

(we verified this result in Chapter 1 when A is self-adjoint and > 0). ∎

5.3 Transport Equations

Let:

Ω be an open region in $\mathbf{R}^N$, with generic point x;

ω be a locally compact space in $\mathbf{R}^N$, with a postive measure $d\mu(\omega)$, ω a generic point of $\underline{\omega}$;

$H = L_2(\Omega \times \underline{\omega})$ = space of (classes of) functions f, measurable on $\Omega \times \underline{\omega}$ for the measure $dx.d\mu(\omega)$ and such that

$$\int_{\Omega \times \omega} |f(x, \omega)|^2 \, dx \, d\mu(\omega) < \infty . \quad ∎$$

---

[1] This becomes precise when we consider examples.

The Operator A_0

We consider first the space of functions v such that

$$
\left.
\begin{array}{l}
v \in H, \\[2mm]
\displaystyle\sum_{i=1}^{n} \omega_i \frac{\partial v}{\partial x_i} \in H \quad (^1).
\end{array}
\right\}
\tag{5.14}
$$

For such functions v, we set

$$
A_0 v = \sum_{i=1}^{n} \omega_i \frac{\partial v}{\partial x_i},
\tag{5.15}
$$

and we define the *domain $D(A_0)$* by

$$
D(A_0) = \left\{ v \mid v \text{ satisfying (5.14) and} \right.
$$
$$
\left. v = 0 \quad \text{if} \quad \sum_{i=1}^{n} \omega_i \cos (v, x_i) < 0, \quad x \in \Gamma \right\}
\tag{5.16}
$$

(where Γ is the boundary of Ω, supposed a variety C^1; $v = $ normal to Γ, exterior of Ω). ∎

The Operator A

The operator A is defined by:

$$
\left.
\begin{array}{l}
D(A) = D(A_0), \\[2mm]
Av = A_0 v + Kv,
\end{array}
\right\}
\tag{5.17}
$$

where K is an *(integral) operator continuous from H to H*. To fix ideas (and this occurs in applications):

$$
Kv(x, \omega) = \int_\omega K(x, \omega, \omega') \, v(x, \omega') \, d\mu(\omega') . \; ∎
\tag{5.18}
$$

Under the preceding conditions, it can be shown (K. Jorgens [1]) that (5.1) holds.

$(^1)$ The derivatives $\partial u/\partial x_i$ are taken in the sense of distributions.

5.4 The QR Method and Transport Equations

We apply Remark (5.5).
We are thus led to solve:

$$\frac{\partial u_\varepsilon}{\partial t} + A u_\varepsilon - \varepsilon A_0^* A_0 u_\varepsilon = 0 , \tag{5.19}$$

where

$$A_0^* A_0 = -\left[\sum_{i=1}^{n} \omega \frac{\partial}{\partial x_i} \right]^2 , \tag{5.20}$$

with

$$u_\varepsilon(x, \omega, T) = \chi(x, \omega) , \tag{5.21}$$

and the *boundary conditions*:

$$\begin{cases} u_\varepsilon(x, \omega, t) = 0 & \text{if} \quad x \in \Gamma , \quad \Sigma \omega_i \cos(v, x_i) < 0 , \\ A_0 u_\varepsilon(x, \omega, t) = 0 & \text{if} \quad x \in \Gamma , \quad \Sigma \omega_i \cos(v, x_i) < 0 . \end{cases}$$

6. TRANSPORT EQUATIONS: NUMERICAL APPLICATIONS

6.1 The Problem

We study here the equation of transport for neutrons in a one-dimensional geometry. We consider[1]

$$\frac{1}{v} \frac{\partial N}{\partial t} + \mu \frac{\partial N}{\partial x} + \sigma N - \frac{c}{2} \int_{-1}^{1} N(x, t, \mu) \, d\mu = 0 , \tag{6.1}$$

with the boundary conditions:

$$N = 0 \quad \text{if} \quad v.n < 0 \quad \text{where} \quad \begin{cases} v \text{ is the neutron velocity} \\ n \text{ is the exterior normal} \end{cases}$$

implying the entering neutron flux is zero.

---

[1] In Section 5, we used the notation of the general theory; here it seems preferable to employ the usual physical notation.

We have therefore

$$N = 0 \quad \text{for} \quad \left.\begin{array}{l} x = 0, \quad 0 < \mu < 1 \\ x = a, -1 < \mu < 0. \end{array}\right\} \quad (6.2)$$

Therefore, let $\chi(x, \mu)$ be given in $L_2[(0, a) \times (-1, +1)]$, and

η and T be given positive quantities;

it is a matter of finding

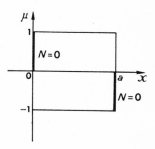

Figure 3

$$\xi = \xi(x, \mu) = \xi_\eta(x, \mu)$$

so that the solution of:

$$\frac{1}{v} \frac{\partial N}{\partial t} + \mu \frac{\partial N}{\partial x} + \sigma N - \frac{c}{2} \int_{-1}^{1} N(x, t, \mu) \, d\mu = 0, \quad (6.1)$$

$$N(x, \mu, 0) = \xi(x, \mu), \quad (6.2) \quad \Big\} \quad \text{(I)}$$

$$N = 0 \quad \text{if} \quad \begin{cases} \mu < 0, \quad x = a \quad \mu \in [-1, 1], \\ \mu > 0, \quad x = 0 \end{cases} \quad (6.3)$$

satisfies

$$\left[\int_0^a \int_{-1}^1 [N(x, \mu, T) - \chi(x, \mu)]^2 \, d\mu \, dx \right]^{1/2} \leqslant \eta. \quad (6.4)$$

We set:

$$\mathscr{A}N = \mu \frac{\partial N}{\partial x}$$

$$\mathscr{B}N = \sigma N - \frac{c}{2} \int_{-1}^{1} N \, d\mu.$$

We are thus led to determine $\xi = N(x, \mu, T)$ from the solution of the backward system

$$-\frac{1}{v}\frac{\partial N}{\partial t} + \mathscr{A}N + \mathscr{B}N - \varepsilon\mathscr{A}\mathscr{A}^* N = 0\,; \qquad (6.5)$$

$$N(x, \mu, 0) = \chi(x, \mu)\,; \qquad (6.6)$$

$$\mu\frac{\partial N}{\partial x} = 0 \qquad \begin{cases} x = 0 & \mu < 0\,, \\ x = a & \mu > 0\,; \end{cases} \qquad (6.7)$$

$$N = 0 \qquad \begin{cases} x = 0 & \mu > 0\,, \\ x = a & \mu < 0\,. \end{cases} \qquad (6.8)$$

(II)

Note that (6.5) may be written:

$$-\frac{1}{v}\frac{\partial N}{\partial t} + \mathscr{A}N + \mathscr{B}N + \varepsilon\mu^2\frac{\partial^2 N}{\partial x^2} = 0\,. \qquad (6.9)$$

6.2 Solution of Problem (II)

We integrate the system (II) numerically with the aid of an explicit scheme of the Wick–Chandrasekhar type (cf., for example, Richtmyer [1]).

6.2.1 APPROXIMATE QUADRATURE

The general idea of the method consists in replacing the integral apparing in (6.5) by an approximate quadrature.

For that, we consider the set of n_μ discrete values μ_j of the variable μ and we write:

$$\int_{-1}^{+1} N(x, \mu, t)\,\mathrm{d}\mu = \sum_{(k)} \omega_k N_{,k}\,, \qquad (6.10)$$

where we have set:

$$N_{,k} = N(x, \mu_k, t)\,.$$

In the case where we use (as we have) a Gauss quadrature, the μ_j are the zeros of the Legendre polynomial of order n_μ and the ω_k

are the associated Christoffel coefficients which we denote by Q_k. Replacing, in (6.5), μ by μ_j, we obtain a system of n_μ equations in the $N(x, \mu_j, t)$. This system may be written:

$$-\frac{1}{v}\frac{\partial N_{,j}}{\partial t} + \mu_j \frac{\partial N_{,j}}{\partial x} + \sigma N_{,j} - \frac{c}{2} \sum_{(k)} Q_k N_{,k} + \varepsilon\mu_j^2 \frac{\partial^2 N_{,j}}{\partial x^2} = 0,$$

$$j = 1, ..., n_\mu.$$

(6.11)

The initial conditions are:

$$N_{,j}(x, 0) = \chi(x, \mu_j)$$

(6.12)

and the boundary conditions (supposing $\mu_j \neq 0$)

$$\frac{\partial N_{,j}}{\partial x} = 0 \quad \begin{cases} x = 0 & \mu_j < 0, \\ x = a & \mu_j > 0; \end{cases}$$

(6.13)

$$N_{,j} = 0 \quad \begin{cases} x = 0 & \mu_j > 0, \\ x = a & \mu_j < 0. \end{cases}$$

(6.14)

Taking $n_\mu = 2L$, the μ_j are the zeros of the Legendre polynomial of order $2L$, an even polynomial whose zeros are $\pm \mu_j$. The Christoffel coefficients corresponding to μ_j and $-\mu_j$ are equal.

6.2.2 DISCRETIZATION OF (6.11)

We divide the interval $(0, 1)$ of the x-axis into i equal intervals and $(0, T)$ into J_T intervals. We then set:

$$N_{i,j}^n = N(i\,\Delta x, \mu_j, n\,\Delta t)$$

(6.15)

$$i = 0, ..., I, \qquad j = 1, ..., n_\mu, \qquad n = 0, ..., J_T;$$

and we take for the derivatives:

$$\left(\frac{\partial N}{\partial t}\right)^n \simeq \frac{N^{n+1} - N^n}{\Delta t},$$

$$\left(\frac{\partial N}{\partial x}\right)_i \simeq \frac{N_{i+1} - N_{i-1}}{2\,\Delta x}, \quad \text{etc.}$$

The system (6.11) may then be written, introducing a parameter θ:

$$- \frac{1}{v}\left(\frac{\partial N_{,j}}{\partial t}\right)^n_i + \mu_j \left(\frac{\partial N_{,j}}{\partial x}\right)^n_i + \sigma[\theta N^{n+1}_{i,j} + (1 - \theta) N^n_{i,j}] -$$

$$- \frac{c}{2} \sum_{(k)} Q_k N^n_{i,k} + \varepsilon\mu_j^2 \left(\frac{\partial^2 N_{,j}}{\partial x^2}\right)^n_i = 0, \qquad 0 \leqslant \theta \leqslant 1. \tag{6.15}$$

Remark 6.1. The discretization scheme used is explicit: to evaluate the term σN we take a weighted mean (with θ) of the values of N at time n and $n + 1$, which conserves the explicit nature of the scheme, while perceptibly improving the numerical results (cf. Graphs 23 and 25). ∎

6.2.3 MATRIX REPRESENTATION

System (6.15) may be written in its full form:

$$- \frac{N^{n+1}_{i,j} - N^n_{i,j}}{v\,\Delta t} + \mu_j \frac{N^n_{i+1,j} - N^n_{i-1,j}}{2\,\Delta x} + \sigma[\theta N^{n+1}_{i,j} + (1 - \theta) N^n_{i,j}] - $$

$$- \frac{c}{2} \sum_{l=1}^{n_\mu} Q_l N^n_{i,l} + \varepsilon\mu_j^2 \frac{N^n_{i+1,j} - 2 N^n_{i,j} + N^n_{i-1,j}}{\Delta x^2} = 0; \tag{6.16}$$

$$N^0_{i,j} = \chi_{i,j} \tag{6.17}$$

$$N^n_{i,j} = 0 \qquad \forall n \begin{cases} i = 0 & 0 < \mu < 1 \quad L + 1 \leqslant j \leqslant n, \\ i = I & -1 < \mu < 0 \qquad 1 \leqslant j \leqslant L, \end{cases} \tag{6.18}$$

$$N^n_{i+1,j} = N^n_{i-1,j} \;\; \forall n \begin{cases} i = 0 & 1 \leqslant j \leqslant L, \\ i = I & L + 1 \leqslant j \leqslant n. \end{cases} \tag{6.19}$$

In the general case, we write (6.16) in the form

$$N^{n+1}_{i,j} = \frac{1}{1 - v\sigma\theta\,\Delta t}\left[N^n_{i,j}\left[1 + v\,\Delta t\left(\sigma(1 - \theta) - 2\,\mu_j^2 \frac{\varepsilon}{\Delta x^2}\right)\right] + \right.$$

$$\left. + N^n_{i+1,j}\,v\,\Delta t\left[\frac{\mu_j}{2\,\Delta x} + \frac{\varepsilon\mu_j^2}{\Delta x^2}\right] + N_{i-1,j}\,v\,\Delta t\left[-\frac{\mu_j}{2\,\Delta x} + \frac{\varepsilon\mu_j^2}{\Delta x^2}\right] - \frac{c}{2}\left(\sum_{l=1}^{n_\mu} Q_l N^n_{i,l}\right)v\,\Delta t\right]. \tag{6.20}$$

6.2.4 PARTICULAR CASES OF THE REPRESENTATION OF (6.20)

Since $\mu_j < 0$, $j \leqslant L$, Eq. (6.16) is considered for i varying from 0 to $I - 1$, since $N_{I,j}^n = 0$; for $\mu_j > 0$, i varies from 1 to I because $N_{0,j}^n = 0$.

6.2.4.1 $\mu_j < 0$ (or $0 \leqslant j \leqslant L$)

At $x = 0$, $(i = 0)$, (6.19) gives: $N_{-1,j}^n = N_{1,j}^n$ $\forall n$ and in this case

$$
N_{0,j}^{n+1} = \frac{1}{1 - v\sigma\theta\,\Delta t}\left[N_{0,j}^n \left[1 + v\,\Delta t \left(\sigma(1 - \theta) - 2\mu_j^2\,\frac{\varepsilon}{\Delta x^2} \right) \right] + \right.
$$
$$
\left. + 2N_{1,j}^n \frac{\varepsilon\mu_j^2}{\Delta x^2} v\,\Delta t\cdots - \frac{cv\,\Delta t}{2} \sum_{l=1}^{n_\mu} (Q_l\,N_{0,l}^n) \right]. \tag{6.21}
$$

At $x = a$, $(i = I)$, $N_{I,j}^n = 0$ $\forall n$ from (6.18), and thus:

$$
N_{I-1,j} = \frac{1}{1 - v\sigma\theta\,\Delta t}\left[N_{I-1,j}^n \left[1 + v\,\Delta t \left(\sigma(1 - \theta) - 2\mu_j^2\,\frac{\varepsilon}{\Delta x^2} \right) \right] + \right.
$$
$$
\left. + N_{I-2,j} v\,\Delta t \left[-\frac{\mu_j}{2\,\Delta x} + \frac{\varepsilon\mu_j^2}{\Delta x^2} \right] - \frac{c}{2} v\,\Delta t \sum_{l=1}^{n_\mu} (Q_l\,N_{I-1,l}^n) \right]. \tag{6.22}
$$

6.2.4.2 $\mu_j > 0$, $L + 1 \leqslant j < n_\mu$.

At $x = 0$, $N_{0,j}^n = 0$ $\forall n$ from (6.18),

$$
N^{n+1} = \frac{1}{1 - v\sigma\theta\,\Delta t}\left[N_{1,j}^n \left[1 + v\,\Delta t \left(\sigma(1 - \theta) - 2\mu_j^2\,\frac{\varepsilon}{\Delta x^2} \right) \right] + \right.
$$
$$
\left. + N_{2,j}^n v\,\Delta t \left[\frac{\mu_j}{2\,\Delta x} + \frac{\varepsilon\mu_j^2}{\Delta x^2} \right] - \frac{c}{2} v\,\Delta t \sum_{l=1}^{n_\mu} (Q_l\,N_{1,l}^n) \right]. \tag{6.23}
$$

At $x = a$, $i = I$, $N_{I-1,j}^n = N_{I+1,j}^n$ $\forall n$ from (6.19):

$$
N_{I,j}^{n+1} = \frac{1}{1 - v\sigma\theta\,\Delta t}\left[N_{I,j}^n \left[1 + v\,\Delta t \left(\sigma(1 - \theta) - 2\mu_j^2\,\frac{\varepsilon}{\Delta x^2} \right) \right] + \right.
$$
$$
\left. + 2N_{I-1,j}^n \frac{\varepsilon\mu_j^2}{\Delta x^2} v\,\Delta t - \frac{c}{2} v\,\Delta t \sum_{l=1}^{n_\mu} (Q_l\,N_{I,l}^n) \right]. \tag{6.24}
$$

6.3 Resolution of the System (I). Verification of the Solution (II)

Once ξ has been found from the solution of (II), it remains to verify condition (6.4) by integrating (I).

For this we discretize the system (I) according to the following scheme.

We take

$$\frac{\partial N}{\partial t} \simeq \frac{N^{n+1} - N^n}{\Delta t}.$$

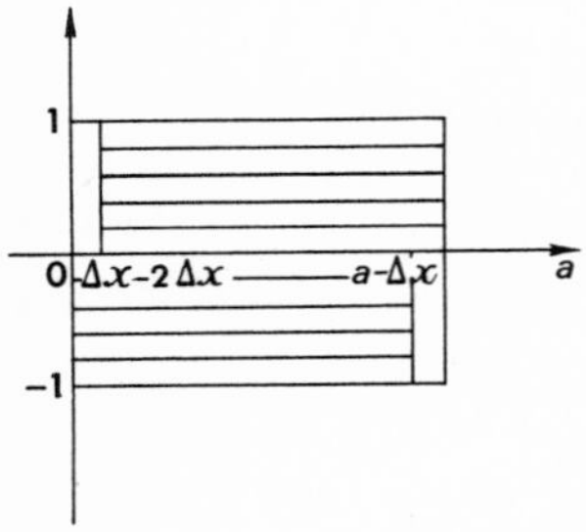

Taking account of (6.2), we write:

$$\frac{\partial N}{\partial x} \simeq \begin{cases} \dfrac{N_i - N_{i-1}}{\Delta x} & \text{for } \mu > 0, \\[2mm] & \quad 1 \le i \le I; \\[2mm] \dfrac{N_{i+1} - N_i}{\Delta x} & \text{for } \mu < 0, \\[2mm] & \quad 0 \le i \le I - 1. \end{cases} \tag{6.25}$$

In this way, no point exterior to the net enters.

Figure 4

Equation (6.1) may then be written:

$$\frac{1}{v} \frac{N_{i,j}^{n+1} - N_{i,j}^n}{\Delta t} + \mu_j \begin{cases} \dfrac{N_{i+1}^n - N_i^n}{\Delta x} \\[2mm] \text{ou} \\[2mm] \dfrac{N_i^n - N_{i-1}^n}{\Delta x} \end{cases} + \sigma[\theta N_{i,j}^{n+1} + (1 - \theta) N_{i,j}^{n+1}]$$

$$- \frac{c}{2} \sum_{l=1}^{n_\mu} Q_l N_{i,l}^n = 0. \tag{6.26}$$

We must therefore solve the following system:

$$N_{i,j}^{n+1} = \left\{ N_{i,j}^n \left[1 - v\,\Delta t\, \frac{\mu_j}{\Delta x} - \sigma(1 - \theta)\, v\,\Delta t \right] + \right.$$

$$\left. + v\,\Delta t \left[-\mu_j \frac{N_{i+1}^n}{\Delta x} + \frac{c}{2} \sum_{l+1}^{n_\mu} Q_l N_{i,l} \right] \right\} \frac{1}{1 + \sigma\theta v\,\Delta t} \quad \text{for } \mu < 0, \tag{III}$$

$$N_{i,j}^n = \left\{ N_{i,j}^n \left[1 - v\,\Delta t\,\frac{\mu_j}{\Delta x} - \sigma(1-\theta)\,v\,\Delta t \right] + \right.$$

$$\left. + v\,\Delta t \left[\mu_j \frac{N_{i+1}^n}{\Delta x} + \frac{c}{2} \sum_{l+1}^{n_\mu} Q_l\,N_{i,l}^n \right] \right\} \frac{1}{1 + \sigma\theta v\,\Delta t} \qquad \text{for} \qquad \mu > 0,$$

(III)
(Cont'd)

with the conditions:

$$N_{i,j}^n = 0 \qquad \forall n \quad \begin{cases} i = 0 \quad \mu > 0 \quad (j \geqslant L+1), \\ i = I \quad \mu < 0 \quad (j \leqslant L), \end{cases}$$

$$N_{i,j}^0 = \xi_{i,j} \qquad \text{previously found (from II)}$$

The solution of this system (III) denoted by $\chi^* = \chi^*(x, \mu) = N(x, \mu, T)$ must satisfy

$$\int_0^a \int_{-1}^1 [\chi^*(x, \mu) - \chi(x, \mu)]^2 \, dx \, d\mu \leqslant \eta^2, \tag{6.27}$$

or in discretized form:

$$\sum_{i=i_1}^{i_2} \sum_{l=1}^{n_\mu} Q_l [\chi_i^*(\mu_1) - \chi_i(\mu_l)]^2 \leqslant \eta^2. \tag{6.28}$$

where:

$$\text{for} \quad \mu_l < 0 \quad (l \leqslant L) \quad i_1 = 0, \quad i_2 = I - 1;$$
$$\text{for} \quad \mu_l > 0 \quad (l > L) \quad i_1 = 1, \quad i_2 = I.$$

6.4 Numerical Results

We carried out two series of numerical applications.

6.4.1 $\chi(x, \mu)$ is given by:

$$\text{for} \quad \mu < 0 \quad \begin{cases} \chi(x, \mu) = 1 & \text{for} \quad x \in [0\,;0.8], \\ \chi(x, \mu) = \cos \pi \left(\dfrac{x - 0.8}{0.4} \right) & \text{for} \quad x \in [0.8\,;1]; \end{cases}$$

$$\text{for} \quad \mu > 0 \quad \begin{cases} \chi(x, \mu) = 1 & \text{for} \quad x \in [0,2 \ ; 1] \ , \\[2mm] \chi(x, \mu) = \cos \pi \left(\dfrac{0.2 - x}{0.4} \right) & \text{for} \quad x \in [0 \ ; 0.2] \ . \end{cases}$$

The results are better[1] (cf. Remark 6.1) for $\theta = 0.5$ (Graphs 23 for χ^* and 23 *bis* for ξ) than for $\theta = 0$ (Graphs 22 and 22 *bis*) or $\theta = 1$ (Graphs 24 and 24 *bis*).

Remark 6.2. Let us suppose that χ is given as equal to 1 everywhere and *nonsmoothed* as above, then the values obtained are also good on $(0; 0.8)$ (for $\mu < 0$), but on $(0.8; 1)$ the functions ξ obtained are very oscillatory. Graphs 25 and 25 *bis* show the results obtained with $\theta = 0.5$ (this value is still better here than $\theta = 0$ or $\theta = 1$).

6.4.2

$$\chi(x, \mu) = \begin{cases} -\mu \cos \dfrac{\pi}{2} x & \mu < 0 \ , \\[4mm] \mu \sin \dfrac{\pi}{2} x & \mu > 0 \ , \end{cases}$$

We recorded these results on Graphs 26 and 26 *bis*.

[1] In the sense (6.4), that is to say, (6.27).

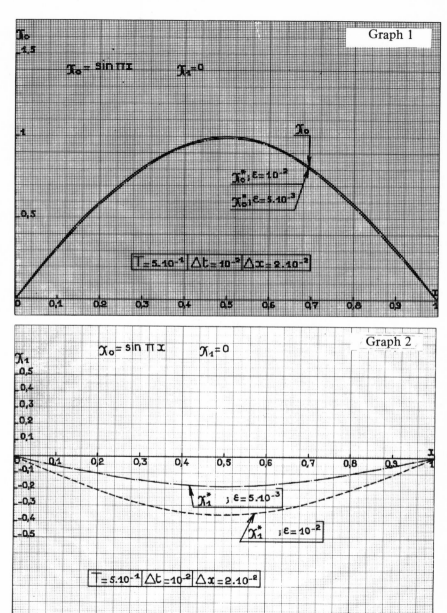

Note: Commas in all graphs represent decimal points.

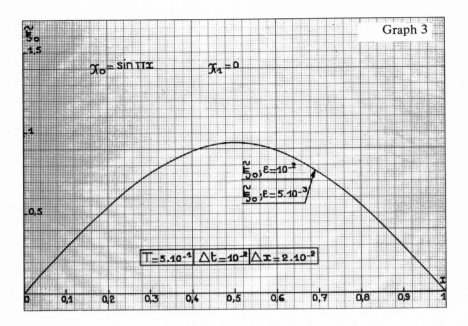

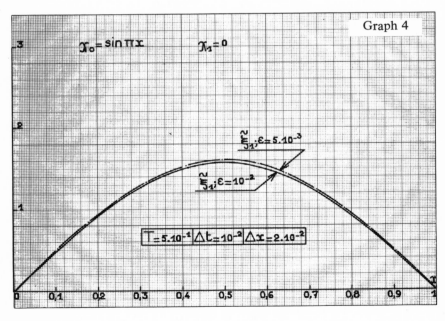

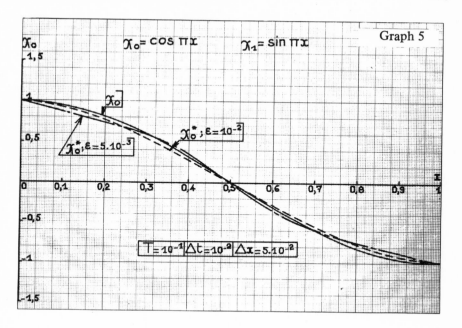

Graph 5

$\mathcal{X}_o = \cos \pi x$ $\mathcal{X}_1 = \sin \pi x$

$\mathcal{X}_o$

$\mathcal{X}_o^*; \varepsilon = 5.10^{-3}$

$\mathcal{X}_o^*; \varepsilon = 10^{-2}$

$T = 10^{-1}$ $\Delta t = 10^{-2}$ $\Delta x = 5.10^{-2}$

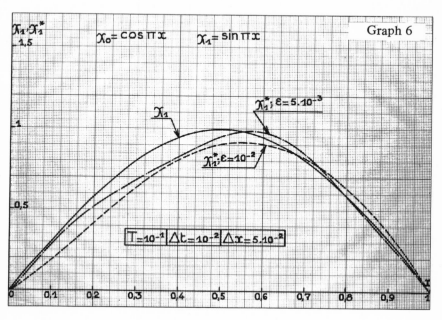

Graph 6

$\mathcal{X}_o = \cos \pi x$ $\mathcal{X}_1 = \sin \pi x$

$\mathcal{X}_1^*; \varepsilon = 5.10^{-3}$

$\mathcal{X}_1$

$\mathcal{X}_1^*; \varepsilon = 10^{-2}$

$T = 10^{-1}$ $\Delta t = 10^{-2}$ $\Delta x = 5.10^{-2}$

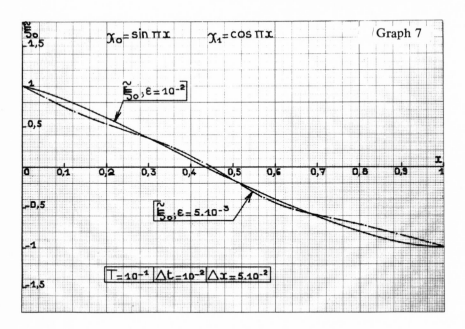

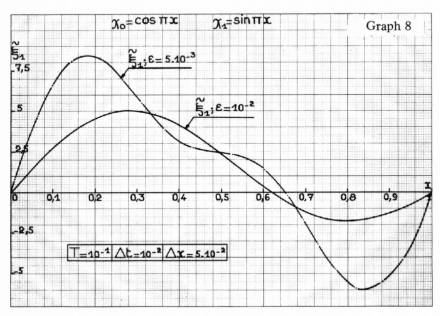

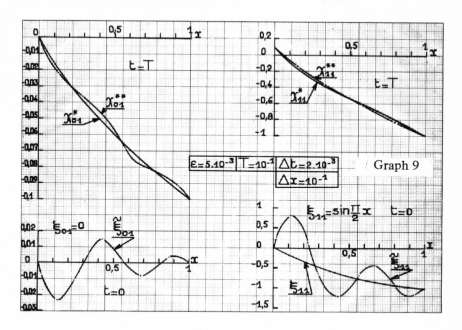

Graph 9

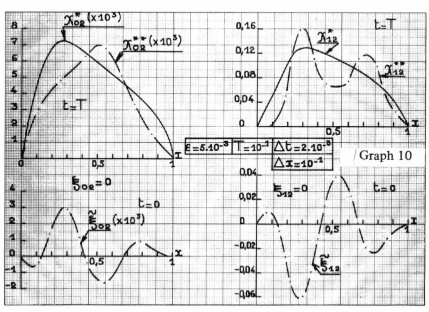

Graph 10

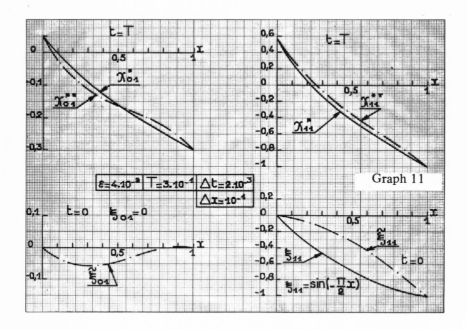

Graph 11

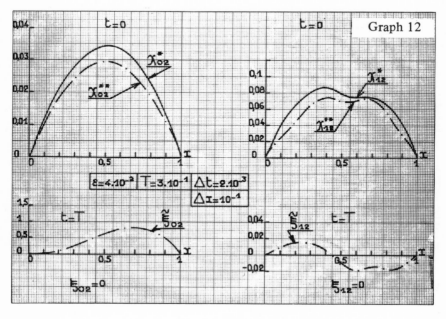

Graph 12

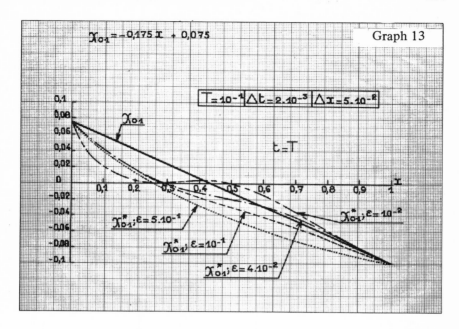

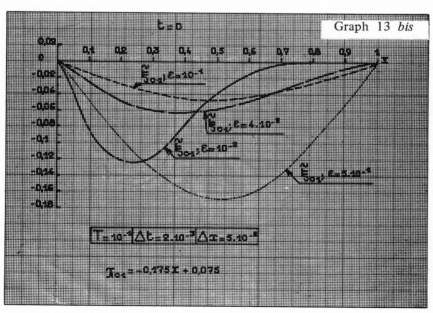

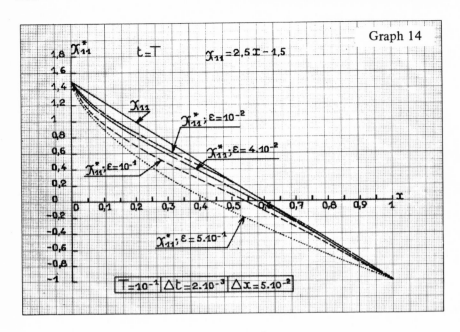

Graph 14

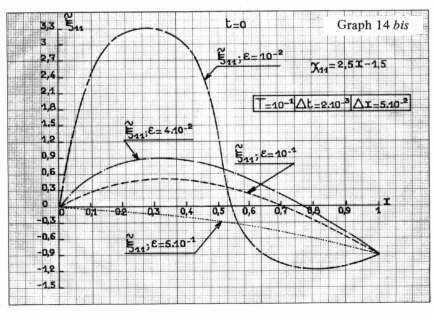

Graph 14 *bis*

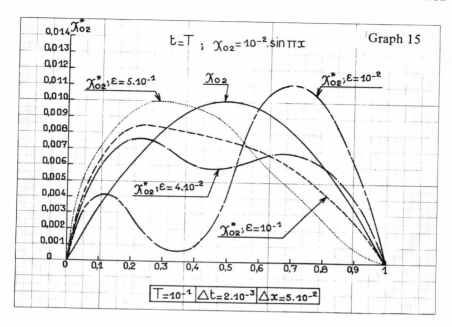

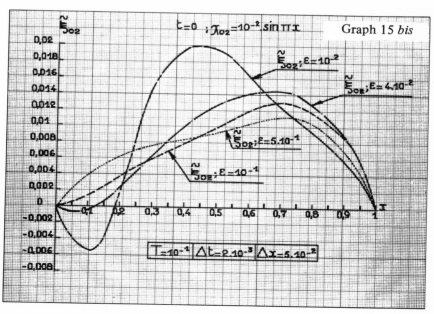

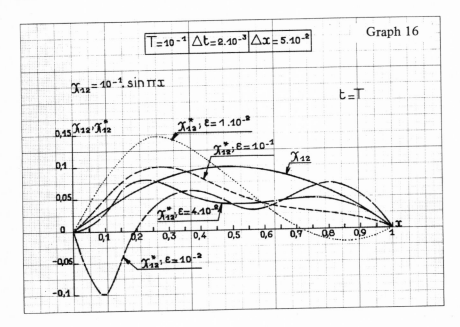

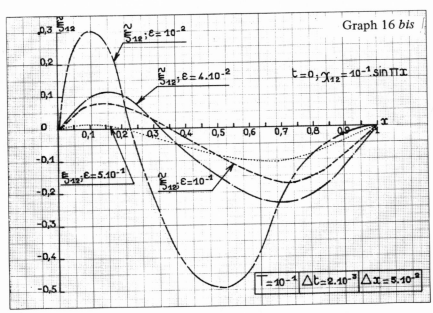

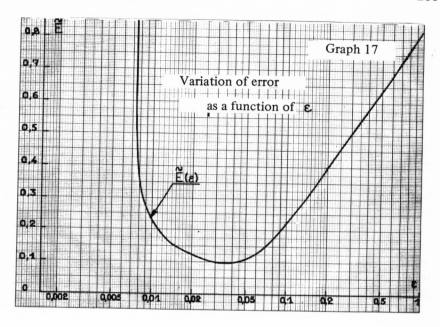

Graph 17

Variation of error

as a function of ε

$\overset{2}{E}(\varepsilon)$

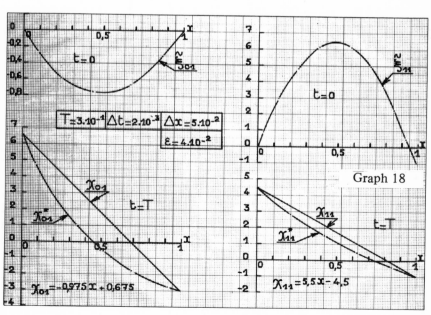

Graph 18

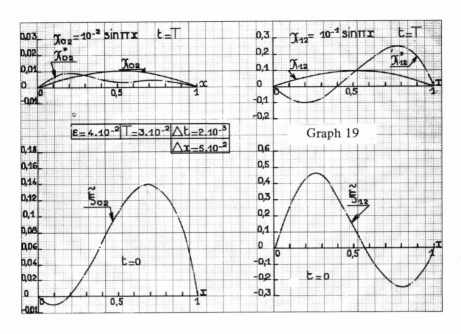

Graph 19

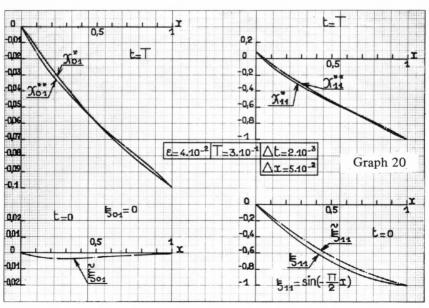

Graph 20

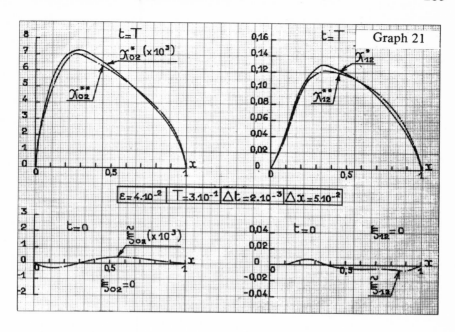

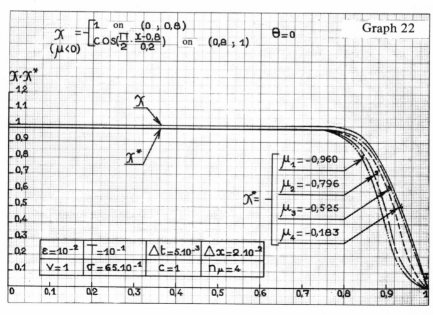

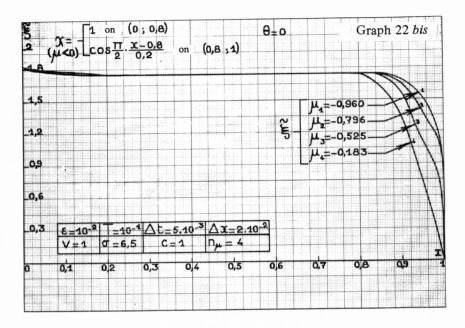

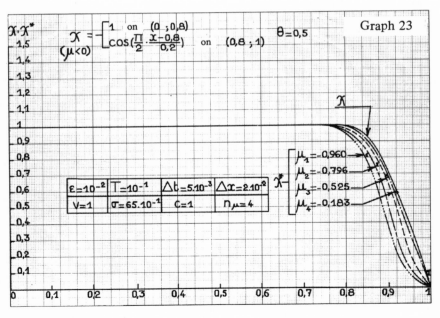

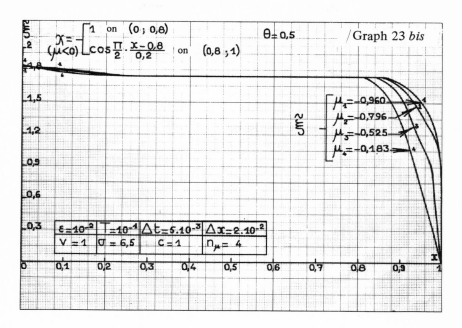

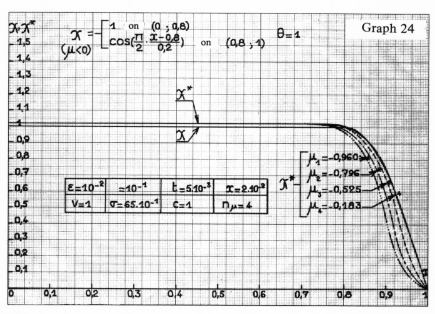

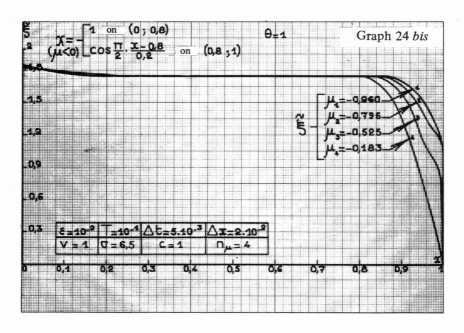

Graph 24 *bis*

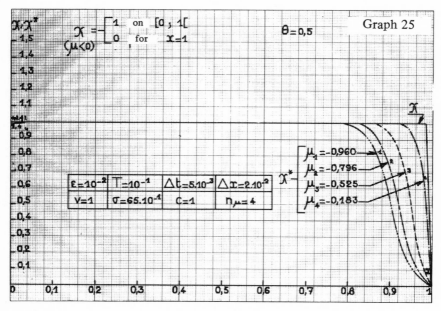

Graph 25

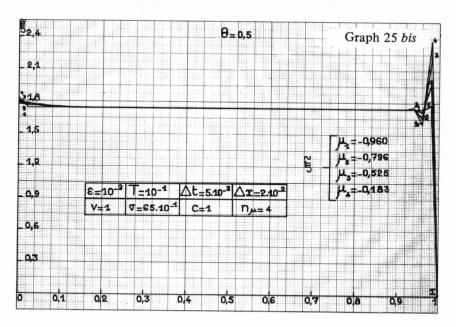

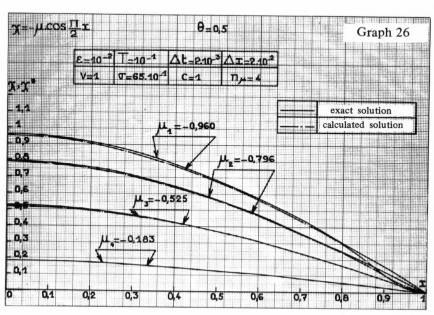

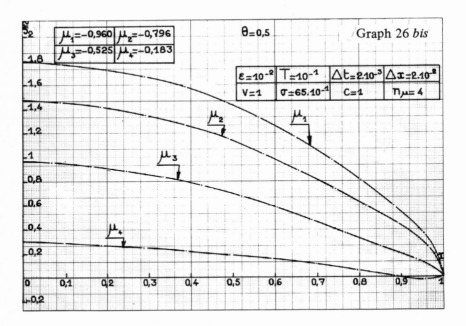

Graph 26 *bis*

Chapter 3

CONTROL IN THE BOUNDARY CONDITIONS

DETAILED OUTLINE

Part One

EQUATIONS OF PARABOLIC TYPE

Part Two

HYPERBOLIC EQUATIONS OF EVOLUTION

Part One

EQUATIONS OF PARABOLIC TYPE

1. FORMULATION OF PROBLEM

1.1 Orientation

Up until now, the "control"—the function at our disposition[1]—was a *given initial quantity*.

We are now going to consider the case (more important for applications, but technically more complicated) where the "control" at our disposition (always without constraints) is *a boundary condition.* ■

It is worth noting that when ξ is a given initial condition, this corresponds physically not to a control, in the usual sense, but rather to a *"sensitivity analysis"*; which is to say, how the solution, in particular at a time T, depends on fluctuations in the initial conditions. On the contrary, when the functions at our disposition are in the boundary conditions, there is really control of the physical process in the sense that we can affect the process at each instant of time in order to achieve a given objective "as well as possible."

We can also exert influence:

—first on ξ

—subsequently on the boundary conditions to attain a given objective "optimally."

1.2 Example

Let Ω be an open region in $\mathbf{R}^n$, with boundary Γ (a variety of dimension $n - 1$, continuously differentiable once, with Ω on one side of Γ).

---

[1] And up to now, *without constraints,* apart from belonging to an appropriate space, for the case of constraints, see Chapter 6.

Let

$$\Sigma = \Gamma \times \,]0, T[$$

be the *lateral boundary* of the cylinder

$$Q = \Omega \times \,]0, T[\subset \mathbf{R}^{n+1} \ ;$$

the notation is as in the preceding chapters: $x \in \Omega$, $t \in (0, T)$. For a function g given on Σ (we will make precise further on the hypotheses on g), we consider *the* function $u = u(x, t)$, solution of:

$$\frac{\partial u}{\partial t} - \Delta u = 0 \quad \text{in} \quad Q \quad (^1) \tag{1.1}$$

$$u(x, 0) = 0 \tag{1.2}$$

$$u(x, t) = g(x, t), \qquad (x, t) \in \Sigma; \quad \text{in short}: \ u|_{\Sigma} = g \ . \tag{1.3}$$

The first form of the problem (approximate) is the following: "Let χ be a given function (in $L_2(\Omega)$): find g so that:

$$J(g) = \int_{\Omega} |\, u(x, T) - \chi(x)\,|^2 \, dx \tag{1.4}$$

is as small as possible—where u denotes in (1.4) the solution of (1.1), (1.2), (1.3) corresponding to g, thus $u = u(x, t) = u(x, t \,; g)$." In fact, we will verify in the following section (under suitable hypotheses concerning g):

Lemma 1.1:

$$\operatorname*{Inf}_{g} J(g) = 0 \ . \tag{1.5}$$

We can therefore pose:

Problem 1.1: Let $\eta > 0$ be given. Find g_η such that

$$J(g_\eta) \leqslant \eta \ . \tag{1.6}$$

Remark 1.1. The problem is of the *same type* as those considered previously, but the "control" g is now [cf. (1.3)] a *boundary function*. ∎

$(^1)$ As already noted, $\Delta u = \sum\limits_{i=1}^{n} \dfrac{\partial^2 u}{\partial x_i^2}$.

Remark 1.2. Naturally, there will exist *an infinity* of g_n answering (1.6); as in previous chapters, our aim is the systematic construction of *certain* g_n satisfying (1.6). ∎

Remark 1.3. Numerous variants are possible. We will briefly indicate some in Section 4. ∎

Remark 1.4. We are going to verify Lemma 1.1 in the following section and make it more precise. Then in Section 3 we present the QR method, with the variants indicated in Section 4. ∎

2. DEMONSTRATION OF LEMMA 1

2.1 The Space Spanned by g

At first, let $H^{1/2}(\mathbf{R}^m)$ be the space of functions $v \in L_2(\mathbf{R}^m)$ whose Fourier transform $\hat{v}(\xi) = \int_{\mathbf{R}^m} e^{-i\xi x} v(x)\, dx$ (a convergent integral in the sense of L_2) satisfies:

$$(1 + |\xi|)^{1/2}\, \hat{v}(\xi) \in L_2(\mathbf{R}^m), \qquad |\xi|^2 = \xi_1^2 + \cdots + \xi_m^2 . \tag{2.1}$$

With the norm:

$$\left(\int_{\mathbf{R}^m} (1 + |\xi|)\, |\hat{v}(\xi)|^2\, d\xi \right)^{1/2} = \| v \|_{H^{1/2}(\mathbf{R}^m)},$$

this is a Hilbert space. *Formally,* $H^{1/2}(\mathbf{R}^m)$ is the space of functions $v \in L_2(\mathbf{R}^m)$ whose *"derivatives of order 1/2"* are in $L_2(\mathbf{R}^m)$.

Let us now define $H^{1/2}(\Gamma)$.

We use local maps, taking Γ (locally) to $\mathbf{R}^{n-1}$. Then $H^{1/2}(\Gamma)$ is the space of functions whose local images are in $H^{1/2}(\mathbf{R}^{n-1})$. More precisely, let $\{\theta_i\}_{i=1}^N$ be a partition of unity continuously differentiable once on Γ, with each θ_i having its support in a local map. Every function v on Γ can then be written:

$$v = \sum_{i=1}^N (\theta_i v) .$$

Let $(\theta_i v)^*$ be the image on $\mathbf{R}^{n-1}$ of $\theta_i v$ by the corresponding local map of Γ. Then:

$$v \in H^{1/2}(\Gamma) \leftrightarrow (\theta_i v)^* \in H^{1/2}(\mathbf{R}^{n-1}) \qquad \forall i . \tag{2.2}$$

This is a Hilbert space with the norm

$$\left(\sum_{i=1}^{N} \| (\theta_i v)^* \|^2_{H^{1/2}(\mathbf{R}^{n-1})} \right)^{1/2} ,$$

independent (up to an equivalence of norms) of the choice of the partition of unity and the system of local maps.

With this defined, *we will suppose that g spans the space*:

$$L_2(0, T ; H^{1/2}(\Gamma)) .$$

Remark 2.1. The interest in $H^{1/2}(\Gamma)$ is provided by the following theorem (cf., for example, Lions—Magenes [1], Chap. 1):

For every function v of $H^1(\Omega)$ we can define in a unique fashion *its trace* on $\Gamma : v_\Gamma$; when v *spans* $H^1(\Omega)$, v_Γ *spans exactly* the space $H^{1/2}(\Gamma)$. (Whence the quite common terminology: $H^{1/2}(\Gamma)$ is a space of traces.) ∎

Remark 2.2. The space $L_2(0, T ; H^{1/2}(\Gamma))$ is *the space of traces on Σ* of function G traversing $L_2(0, T ; H^1(\Omega))$. ∎

Remark 2.3. If Ω is the unit disc in $\mathbf{R}^2$, $H^{1/2}(\Gamma)$ may be identified with the space of functions:

$$v = \sum_{-\infty}^{\infty} \alpha_n e^{in\theta}$$

such that:

$$\left(\sum_{n=-\infty}^{\infty} (1 + |n|) |\alpha_n|^2 \right)^{1/2} < \infty . ∎$$

2.2 Lemma 1.1

First a precise statement:

When g spans $L_2(0, T ; H^{1/2}(\Gamma))$, the function $x \to u(x, T)$ spans a space dense in $L_2(\Omega)$. $\left.\begin{array}{c}\\\\\end{array}\right\}$ (2.3)

Here is the essential part of the proof. The technical difficulty which appears is resolved in 2.3.

Let ψ be a function of $L_2(\Omega)$ such that

$$(u(T), \psi) = 0 \qquad \forall g. \qquad (^1) \tag{2.4}$$

Let w be *the* solution of:

$$\left. \begin{array}{l} -\dfrac{\partial w}{\partial t} - \Delta w = 0, \\[2mm] w(x, T) = \psi, \\[2mm] w = 0 \text{ on } \Sigma; \end{array} \right\} \tag{2.5}$$

this solution exists and is unique; it satisfies:

$$\left. \begin{array}{l} w \in L_2(0, T; H_0^1(\Omega)), \\[2mm] \dfrac{dw}{dt} \in L_2(0, T; H^{-1}(\Omega)). \end{array} \right\} \tag{2.6}$$

Let us then consider the expression:

$$\int_0^T [(u', w) + (u, w')] \, dt = \int_0^T \frac{d}{dt} (u, w) \, dt = (u(T), w(T)) - (u(0), w(0)).$$

Since $w(T) = \psi$ and $(u(T), \psi) = 0$ by hypothesis, and since $u(0) = 0$, we have:

$$\int_0^T [(u', w) + (u, w')] \, dt = 0. \tag{2.7}$$

But $u' = \Delta u$, $w' = -\Delta w$, and thus (2.7) gives:

$$\int_0^T [(\Delta u, w) - (u, \Delta w)] \, dt = 0$$

and, with the aid of Green's formula, this is equivalent to:

$$\int_\Sigma u \frac{\partial w}{\partial n} \, d\Gamma \, dt = 0,$$

$(^1)$ $u(T)$ denotes the function $x \to u(x, T)$; $(f, g) = \displaystyle\int_\Omega fg \, dx$ (we consider, to simplify a bit, real-valued functions).

or

$$\int_{\Sigma} g \frac{\partial w}{\partial n} \, d\Gamma \, dt = 0 \qquad \forall g \,. \tag{2.8}$$

Therefore

$$\frac{\partial w}{\partial n} = 0 \quad \text{on} \quad \Sigma \,. \tag{2.9}$$

But this, combined with (2.5) implies $w = 0$[1] and therefore $\psi = 0$; hence the property of denseness. ∎

But the preceding demonstration is not complete: the expression:

$$\int_0^T (u, w') \, dt$$

is not definite because $u \in L_2(0, T ; H^1(\Omega))$, $w' \in L_2(0, T ; H^{-1}(\Omega))$ and $H^{-1}(\Omega) = (H_0^1(\Omega))'$ *is not* the dual of $H^1(\Omega)$. We can nevertheless "adapt to the situation": this is the aim of the following section.

2.3 Justification of 2.2

The solution w of problem (2.5) satisfies:

$$w \in L_2(0, T - \varepsilon ; H^2(\Omega)), \quad w' \in L_2(0, T - \varepsilon ; H^2(\Omega)), \quad \forall \varepsilon > 0 \,. \;[2] \tag{2.10}$$

Then, with $\varepsilon > 0$ and any fixed quantity:

$$\int_0^{T-\varepsilon} [(u', w) + (u, w')] \, dt$$

has a sense and is equal on one hand to $(u(T - \varepsilon), w(T - \varepsilon))$, an expression which tends as $\varepsilon \to 0$ towards $(u(T), w(T)) = (u(T), \psi) = 0$, and on the other hand to:

$$\int_{\Sigma_\varepsilon} u \frac{\partial w}{\partial n} \, d\Gamma \, dt = \int_{\Sigma_\varepsilon} g \frac{\partial w}{\partial n} \, d\Gamma \, dt \,, \;[3]$$

---

[1] Uniqueness of the Cauchy problem. Cf. also Chapter 5.
[2] Recall that $H^2(\Omega) = \{ v \mid v, D_j v, D_j D_k v \in L_2(\Omega) , \quad \forall j, k \}$.
[3] The integrals exist.

whence

$$\Sigma_\varepsilon = \Gamma \times \,]0, \, T - \varepsilon[\, .$$

Therefore

$$\lim_{\varepsilon \to 0} \int_{\Sigma_\varepsilon} g \, \frac{\partial w}{\partial n} \, d\Gamma \, dt = 0 \, . \tag{2.11}$$

Let us take g zero for $t \geqslant T - \varepsilon_0$, $\varepsilon_0 > 0$ fixed.

Then, taking $\varepsilon < \varepsilon_0$, we deduce from (2.11) that:

$$\int_{\Sigma_{\varepsilon_0}} g \, \frac{\partial w}{\partial n} \, d\Gamma \, dt = 0 \qquad \forall g \in L_2(0, \, T - \varepsilon_0 \, ; \, H^{1/2}(\Gamma)) \, .$$

Conclusion:

$$\frac{\partial w}{\partial n} = 0 \quad \text{on} \quad \Sigma_{\varepsilon_0} \, , \qquad \forall \varepsilon_0 > 0 \, .$$

We deduce that $w = 0$, whence the result.

3. THE QUASI-REVERSIBILITY METHOD

3.1 Heuristic Idea of the Method

The basic idea is to modify the operator $\frac{\partial}{\partial t} - \Delta$ in such a way that we can then define an "optimal approximate solution," by modification of Remark 1.3.

3.2 First Method

We first replace $\frac{\partial}{\partial t} - \Delta$ by the "neighboring" operator:

$$\frac{\partial}{\partial t} - \varepsilon \frac{\partial^2}{\partial t^2} - \Delta \, , \qquad \varepsilon > 0 \text{ "small,"} \tag{3.1}$$

an operator of elliptic type.

We can then, by a modification of Remark 1.3, find U_ε, the solution of:

$$\frac{\partial U_\varepsilon}{\partial t} - \varepsilon \frac{\partial^2 U_\varepsilon}{\partial t^2} - \Delta U_\varepsilon = 0 \,, \quad \text{in} \quad \Omega \times \left]0, T\right[\,, \tag{3.2}$$

with

$$U_\varepsilon(x, 0) = 0 \,, \tag{3.3}$$

$$U_\varepsilon(x, T) = \chi(x) \,.$$

But *this problem admits an infinite number of solutions*, because, for example, the value of U_ε on Σ has not been specified.

Let θ be an open region of $\mathbf{R}^n$, with $\Omega \subset \theta$.

Let U_ε^θ be *the* solution of:

$$\left.\begin{aligned}
&\frac{\partial U_\varepsilon^\theta}{\partial t} - \varepsilon \frac{\partial^2 U_\varepsilon^\theta}{\partial t^2} - \Delta U_\varepsilon^\theta = 0 \quad \text{in} \quad \theta \times \left]0, T\right[\,, \\[2mm]
&U_\varepsilon^\theta(x, 0) = 0 \,, \\[2mm]
&U_\varepsilon^\theta(x, T) = \begin{cases} \chi(x) & \text{in} \quad \Omega \\ 0 & \text{in} \quad \theta - \overline{\Omega} \quad (^1) \end{cases} \\[2mm]
&U_\varepsilon^\theta(x, t) = 0 \quad \text{if} \quad (x, t) \in \partial\theta \times \left]0, T\right[\quad (^2)
\end{aligned}\right\} \tag{3.5}$$

(the existence and uniqueness of a solution—in a suitable sense—is a consequence of 3.4 below).

Then we *can* take

$$U_\varepsilon = \text{restriction of } U_\varepsilon^\theta \text{ to } \Omega \,,$$

and this restriction *depends effectively on* θ (without taking account of the other degrees of freedom in the choice of $U_\varepsilon^\theta(x, T)$ and of $U_\varepsilon^\theta(x, t)$, $x \in \partial\theta$).

Therefore, the choice of g given by

$$g(x, t) = U_\varepsilon(x, t) \,, \qquad \chi \in \Gamma \,,$$

has no sense.

(^{1}) We could take any value in $\theta - \overline{\Omega}$.
(^{2}) Where $U_\varepsilon^\theta(x, t)$ ≖ any given function.

We are thus led to the method presented in 3.3 below.

3.3 The QR Method

The idea of replacing Δ by a "neighboring" Δ_ε *does not require any condition on the boundary values on Γ.* ∎

The function ρ_ε.

We chose $\varphi(\varepsilon)$, $\varphi(\varepsilon) > 0$, $\varphi(\varepsilon) \to 0$ if $\varepsilon \to 0$. We set:

$$d(x, \Gamma) = \text{distance from } x \text{ to } \Gamma.$$

We then take

$$\rho_\varepsilon(x) = \begin{cases} 1 & \text{if} \quad d(x, \Gamma) \geqslant \varphi(\varepsilon), \\ \dfrac{d(x, \Gamma)}{\varphi(\varepsilon)} & \text{if} \quad d(x, \Gamma) < \varphi(\varepsilon). \end{cases} \tag{3.6}$$

The operator Δ_ε.

We set

$$\Delta_\varepsilon v = \sum_{i=1}^{n} \frac{\partial}{\partial x_i} \left(\rho_\varepsilon^2 \frac{\partial v}{\partial x_i} \right). \tag{3.7}$$

We will verify further on that the problem (in a suitable sense);

$$\Delta_\varepsilon v = f, \quad f \text{ given in } \Omega,$$

possesses a unique solution, *without a condition on the boundary values on Γ.* ∎

We will also verify further on *that there exists a unique u_ε,* the solution of

$$\frac{\partial u_\varepsilon}{\partial t} - \varepsilon \frac{\partial^2 u_\varepsilon}{\partial t^2} - \Delta_\varepsilon u_\varepsilon = 0, \tag{3.8}$$

$$u_\varepsilon(x, 0) = 0, \tag{3.9}$$

$$u_\varepsilon(x, T) = \chi(x), \tag{3.10}$$

without a condition on the boundary values on Σ. ∎

The QR method therefore consists of:

1. *Solving* (3.8), (3.9), (3.10);[1]
2. *Taking* (*extrapolating* since u_ε is not defined on $\Gamma \times (0,\ T)$)
   $g(x,\ t) = \dot{u}_\varepsilon(x,\ t)$, $x \in \Gamma$, $t \in (0,\ T)$. ∎

The situation is now as follows:

i) it is necessary to show that the problem (3.8), (3.9), (3.10) possesses a unique solution; this is the purpose of 3.4 below;

ii) can we choose ε (and $\varphi(\varepsilon)$) suitably depending on η such that for the choice in (3.11) we have

$$J(g) \leqslant \eta ?$$

And in addition, can one give this choice *explicitly*? We will only be able to give fragmentary indications as far as this is concerned;[2]

iii) it is evidently necessary to choose a *numerical* method for the solution of (3.8), (3.9) and (3.10), this point is obvious at least as long as we constrain ourselves to applications or to the use of standard methods—as we will do.

3.4   Existence and Uniqueness of the QR Problem

The space V_ε.

We set:

$$V_. = \left\{ v \mid v \in L_2(\Omega),\quad \rho_\varepsilon \frac{\partial v}{\partial x_i} \in L_2(\Omega) \quad \forall i \right\} ; \tag{3.11}$$

with the norm[3]:

$$|v|^2 + \sum_{i=1}^{n} \left| \rho_\varepsilon \frac{\partial v}{\partial x_i} \right|^2 = \| v \|_{V_\varepsilon}^2 , \tag{3.12}$$

---

[1] It remains to choose ε and $\varphi(\varepsilon)$!
[2] Combined with "numerical evidence."
[3] Recall that $|v|^2 = \displaystyle\int_\Omega v^2 \, dx$.

this is a *Hilbert space.* We have

Lemma 3.1: *The space $D(\Omega)$ is dense in V_ε.*

Proof. 1) Let θ_m be a sequence $\in C^1(\bar{\Omega})$; such that:

$$\begin{cases} \theta_m = 0 \quad \text{if} \quad d(x, \Gamma) < \varepsilon_m, \quad \varepsilon_m \to 0 \quad \text{if} \quad m \to \chi \,. \\ \left| \dfrac{\partial \theta_m}{\partial x_i} \right| \leqslant \dfrac{C}{d(x, \Gamma)} \quad \text{for} \quad d(x, \Gamma) \leqslant 2\, \varepsilon_m, \\ \theta_m = 1 \quad \text{if} \quad d(x, \Gamma) \geqslant 2\, \varepsilon_m \end{cases}$$

(such a sequence exists).

Let v be given in V_ε and let:

$$v_m = \theta_m v \,.$$

We have: $v_m \to v$ in V_ε when $m \to \infty$. The only point not completely obvious is to verify that:

$$\rho_\varepsilon \, v \, \frac{\partial \theta_m}{\partial x_i} \to 0 \quad \text{in} \quad L_2(\Omega) \,.$$

Since this function tends towards 0 almost everywhere, and $\left| \rho_\varepsilon \dfrac{\partial \theta_m}{\partial x_i} \right| \leqslant$ *constant* (depending on ε), the result follows by application of the theorem of Lebesgue.

2) It remains to approximate v_m, in the sense of V_ε, by functions in $D(\Omega)$; this is immediate by smoothing. ∎

We denote by V'_ε the dual of V_ε, a distribution space on Ω of the form:

$$f = f_0 + \sum_{i=1}^{n} D_i(\rho_\varepsilon f_i), \qquad f_0, f_i \in L_2(\Omega) \,. \tag{3.13}$$

We have:

Proposition 3.1: *For a given f in V'_ε, there exists a unique u in V_ε which is a solution of:*

$$- \Delta_\varepsilon u + \lambda u = f, \qquad \lambda > 0 \text{ fixed.} \tag{3.14}$$

Proof. 1) For $u, v \in V_\varepsilon$, we set:

$$a_\varepsilon(u, v) = \sum_{i=1}^{n} (\rho_\varepsilon D_i u, \rho_\varepsilon D_i v) . \qquad (3.15)$$

Then, as is easily seen, the statement is equivalent to demonstrating the existence and uniqueness of the solution of:

$$a_\varepsilon(u, v) + \lambda(u, v) = (f, v) \qquad \forall v \in V_\varepsilon , \qquad (3.16)$$

(where (f, v) denotes the duality between V'_ε and V_ε).

2) Now

$$(a_\varepsilon(v, v) + \lambda \mid v \mid^2)^{1/2}$$

is a norm equivalent to $\| v \|_{V_\varepsilon}$, whence the result, by application of the projection lemma. ∎

The space $H^{1/2}(\Omega)$.

We denote by $H^{1/2}(\Omega)$ the space of restrictions to Ω of the functions of $H^{1/2}(\mathbf{R}^n)$ (cf. 2.1 for the definition of this space). ∎

We are now in a position to demonstrate

Theorem 3.1: *Let χ be given in $H^{1/2}(\Omega)$.*[1] *Then, there exists a unique u_ε satisfying:*

$$u_\varepsilon \in L_2(0, T ; V_\varepsilon) , \qquad (3.17)$$

$$u'_\varepsilon \in L_2(0, T ; L_2(\Omega)) ,$$

and conditions (3.8), (3.9), (3.10).[2]

Proof. 1) To begin with, let $\Phi \in L_2(0, T ; H^1(\Omega))$, with $\Phi' \in L_2(0, T ; L_2(\Omega))$, such that:

$$\Phi(x, 0) = 0 , \qquad \Phi(x, T) = \chi(x) .$$

----

[1] We can restrain ourselves, for the sake of simplicity, to the case $\chi \in H^1(\Omega)$. The "best" hypothesis on χ uses the theory of interpolation of Hilbert spaces and is not given here.

[2] Which are meaningful.

We set in what follows:

$$\Phi(t) = \text{function } x \to \Phi(x, t).$$

2) Denote by $\mathcal{V}_\varepsilon$ the space

$$\mathcal{V}_\varepsilon = \{ v \mid v \in L_2(0, T; V_\varepsilon), \quad v' \in L_2(0, T; L_2(\Omega)), \quad v(0) = v(T) = 0 \}$$

and, for $u, v \in \mathcal{V}_\varepsilon$, set:

$$\pi_\varepsilon(u, v) = \int_0^T [\varepsilon(u', v') + (u', v) + a_\varepsilon(u(t), v(t))] \, dt . \tag{3.19}$$

The space $\mathcal{V}_\varepsilon$ is a Hilbert space for

$$\left(\int_0^T (\| v(t) \|_{V_\varepsilon}^2 + | v'(t) |^2) \, dt \right)^{1/2} ;$$

but

$$\int_0^T | v(t) |^2 \, dt \leqslant C \int_0^T | v'(t) |^2 \, dt$$

(since $v(0) = 0$, for example) and therefore on $\mathcal{V}_\varepsilon$ the following norm is equivalent:

$$\| v \|_{\mathcal{V}_\varepsilon} = \left(\int_0^T (a_\varepsilon(v(t), v(t)) + | v'(t) |^2) \, dt \right)^{1/2} . \tag{3.20}$$

Let us note that, since $\displaystyle \int_0^T (v', v) \, dt = 0$, we have:

$$\pi_\varepsilon(v, v) = \int_0^T (a_\varepsilon(v(t), v(t)) + \varepsilon | v'(t) |^2) \, dt . \tag{3.21}$$

Therefore

$$\pi_\varepsilon(v, v) \geqslant \text{Inf}(\varepsilon, 1) \| v \|_{\mathcal{V}_\varepsilon}^2 .$$

Consequently, *for f given in $\mathcal{V}_\varepsilon$, there exists a unique $w_\varepsilon \in \mathcal{V}_\varepsilon$, satisfying*

$$\pi_\varepsilon(w_\varepsilon, v) = (f, v) \qquad \forall v \in \mathcal{V}_\varepsilon , \tag{3.22}$$

(f, v) denotes the scalar product between $\mathcal{V}_\varepsilon'$ and $\mathcal{V}_\varepsilon$.

3) We verify easily that w_ε, the solution of (3.22), satisfies:

$$\frac{\partial w_\varepsilon}{\partial t} - \varepsilon \frac{\partial^2 w_\varepsilon}{\partial t^2} - \Delta_\varepsilon w_\varepsilon = f, \qquad w_\varepsilon(0) = w_\varepsilon(T) = 0 . \tag{3.23}$$

4) Now, $w = u_\varepsilon - \Phi$ satisfies

$$\frac{\partial w}{\partial t} - \varepsilon \frac{\partial^2 w}{\partial t^2} - \Delta_\varepsilon w = - (\Phi - \varepsilon \Phi'' - \Delta_\varepsilon \Phi) , \qquad w(0) = w(T) = 0 . \tag{3.24}$$

But $\Phi' - \varepsilon \Phi'' - \Delta_\varepsilon \Phi$ is in $\mathscr{V}'_\varepsilon$; we can then set

$$f = - (\Phi' - \varepsilon \Phi'' - \Delta_\varepsilon \Phi)$$

and then $w = w_\varepsilon$, which demonstrates the theorem. ∎

4. REMARKS AND VARIANTS

4.1 Initial Data and Second Member Nonzero

We can consider Problem 1.1, but with $u(x, 0)$ a *given nonzero* quantity. More generally, let u be the solution of

$$\frac{\partial u}{\partial t} - \Delta u = f, \qquad f \text{ given and fixed} \tag{4.1}$$

$$u(x, 0) = u_0(x) , \qquad u_0 \text{ given and fixed} \tag{4.2}$$

$$u(x, t) = g(x, t) \qquad \text{on } \Sigma, g \text{ variable} . \tag{4.3}$$

The preceding results may be readily adapted.

4.2 More General Parabolic Operator of the Second Order

Let A be the operator given by:

$$Av = - \sum_{i,j=1}^{n} \frac{\partial}{\partial x_i} \left(a_{ij}(x, t) \frac{\partial v}{\partial x_j} \right) , \tag{4.4}$$

where

$$a_{ij} \in C^3(\overline{\Omega} \times [0, T]) ,$$

$$\sum_{i,j=1}^{n} a_{ij}(x, t)\, \xi_i\, \xi_j \geqslant \alpha(|\,\xi_1\,|^2 + \cdots + |\,\xi_n\,|^2), \quad \forall \xi = \{\,\xi_1, ..., \xi_n\,\} \in \mathbf{R}^n \qquad (4.5)$$

$$\alpha > 0 \quad \text{independent of } x \text{ and } t .$$

Therefore the operator has coefficients depending on x and t; we can write it as

$$A\left(x, t\, \frac{\partial}{\partial x}\right)$$

Let u then be the solution of

$$\frac{\partial u}{\partial t} + Au = 0 \qquad (\text{or } f), \qquad (4.6)$$

$$u(x, 0) = 0 \qquad (\text{or } u_0), \qquad (4.7)$$

$$u(x, t) = g(x, t) \qquad \text{on } \Sigma, g \text{ variable}. \qquad (4.8)$$

All the results of the preceding sections may be adapted to this case.

The operator A_ε (which is to A as $-\Delta_\varepsilon$ is to $-\Delta$) is defined by:

$$A_\varepsilon v = \sum_{i,j=1}^{n} \frac{\partial}{\partial x_i}\left(\rho_\varepsilon^2\, a_{ij}\, \frac{\partial v}{\partial x_j}\right), \qquad (4.9)$$

where ρ_ε is given by (3.6).

4.3 Control of Boundary Conditions, With Boundary Conditions of the Neumann Type

Consider now the solution u of:

$$\frac{\partial u}{\partial t} - \Delta u = 0 , \qquad (4.10)$$

$$u(0) = 0 , \qquad (4.11)$$

$$\left.\frac{\partial u}{\partial v}\right|_{\Sigma} = g, \qquad (^1) \tag{4.12}$$

where g is variable (in a suitable space).

We have again the analogue of Lemma 1.1:

Lemma 4.2: *When g spans the space $L_2(0, T; H^{-1/2}(\Gamma))$ $(^2)$, $u(T)$ spans a space dense in $L_2(\Omega)$.*

Proof. It is analogous to Section 2. Let $\psi \in L_2(\Omega)$, with $(u(T), \psi) = 0$, $\forall g$. We introduce v, the solution of

$$-\frac{\partial v}{\partial t} - \Delta v = 0, \qquad v(T) = \psi, \qquad \left.\frac{\partial v}{\partial v}\right|_{\Sigma} = 0.$$

We consider now the expression (with justification as in Section 2):

$$\int_0^T [(u', v) + (u, v')]\, dt,$$

which is on one hand equal to zero and on the other hand equal to $\int_{\Sigma} \frac{\partial u}{\partial v} v\, d\Sigma$, therefore:

$$\int_{\Sigma} gv\, d\Sigma = 0 \qquad \forall g$$

therefore $v|_{\Sigma} = 0$ and thus $v = 0$, and finally $\psi = 0$. ∎

We consider next the problem analogous to Problem 1.1: *for given $\eta > 0$, find g_η such that:*

$$\int_{\Omega} |u(x, T) - \chi(x)|^2\, dx \leqslant \eta. ∎$$

The QR method is as follows:
We solve (3.8), (3.9), (3.10), then we take

$$g = \left.\frac{\partial u_\varepsilon}{\partial v}\right|_{\Sigma}.$$

$(^1)$ $\frac{\partial}{\partial v}$ = normal derivative to Γ. We can also consider the Neumann problem for (4.6) or other boundary conditions, etc.

$(^2)$ $H^{-1/2}(\Gamma)$ ≈ dual of $H^{1/2}(\Gamma)$.

4.4 Other Geometries

Let us suppose that Ω is given as in Figure 1, with

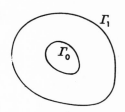

Figure 1

$$\Gamma = \Gamma_0 \cup \Gamma_1,$$

each Γ_j is a variety of type C^1, Ω is on only one side of Γ_j.

Consider next the solution u of:

$$\frac{\partial u}{\partial t} - \Delta u = 0, \tag{4.13}$$

$$u(0) = 0, \tag{4.14}$$

$$\left. \begin{array}{l} u = 0 \text{ on } \Sigma_0 = \Gamma_0 \times]0, T[, \\ u = g \text{ on } \Sigma_1 = \Gamma_1 \times]0, T[, \end{array} \right\} \tag{4.15}$$

where g is variable.

We show again (same type of proof!) that, *when g spans $L_2(0, T; H^{1/2}(\Gamma_1))$, then u(T) spans a space dense in* $L_2(\Omega)$. We can therefore pose a problem analogous to Problem 1.1. ■

Here is the QR solution. The principle is the same; it suffices now to introduce an operator $\tilde{\Delta}_\varepsilon$ "near" Δ, and *degenerating only on Σ_1*. More precisely, we pose:

$$\tilde{\rho}_\varepsilon(x) = \begin{cases} 1 & \text{if } d(x, \Gamma_1) \geqslant \varphi(\varepsilon), \\ \dfrac{d(x, \Gamma_1)}{\varphi(\varepsilon)} & \text{if } d(x, \Gamma_1) < \varphi(\varepsilon), \end{cases} \tag{4.16}$$

then

$$\tilde{\Delta}_\varepsilon v = \sum_{i=1}^n \frac{\partial}{\partial x_i} \left((\tilde{\rho}_\varepsilon)^2 \frac{\partial v}{\partial x_i} \right). \tag{4.17}$$

We then consider u_ε, the solution of:

$$\left. \begin{array}{l} \dfrac{\partial u_\varepsilon}{\partial t} - \tilde{\Delta}_\varepsilon u_\varepsilon - \varepsilon \dfrac{\partial^2 u_\varepsilon}{\partial t^2} = 0, \\ u_\varepsilon(0) = 0, \qquad u_\varepsilon(T) = \chi, \\ u_\varepsilon |_{\Sigma_0} = 0, \end{array} \right\} \tag{4.18}$$

without boundary conditions on Σ_1. ■

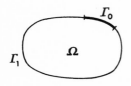

Figure 2

We then take

$$g = u_\varepsilon|_{\Sigma_1}.$$

Remark 4.1. The same type of problem may be posed and the same type of conclusions hold as in the case of the geometry of Figure 2.

4.5 Higher Order

We have up to now, in this chapter, considered operators of the form $\frac{\partial}{\partial t} + A$, where A *is an operator of the second order*. We can consider problems of the same type for a positive elliptic operator of order $2m, m > 1$.[1] We will make this precise by means of an example.

Let u be the solution of:

$$\frac{\partial u}{\partial t} + \Delta^2 u = 0, \qquad (\Delta^2 = \Delta.\Delta), \qquad\qquad (4.19)$$

$$u(0) = 0, \qquad\qquad (4.20)$$

$$\left.\begin{array}{l} u|_\Sigma = g_0, \\[2mm] \dfrac{\partial u}{\partial v}\bigg|_\Sigma = g_1, \end{array}\right\} \qquad\qquad (4.21)$$

where g_0 and g_1 are variable.

We have[1]:

Lemma 4.3: *When g_0 spans $L_2(0, T; H^{3/2}(\Gamma))$ and g_1 spans $L_2(0, T; H^{1/2}(\Gamma))$, then $u(T)$ spans a space dense in $L_2(\Omega)$.*

Proof. Let $\psi \in L_2(\Omega)$, with:

$$(u(T), \psi) = 0 \qquad \forall \{g_0, g_1\}.$$

---

[1] Provided that the uniqueness property of the Cauchy problem holds.

[2] $H^{3/2}(\Gamma)$ = space of functions on Γ whose "derivatives of the first order on Γ" are in $H^{1/2}$; we can provide a Hilbertain structure; cf. the details in Lions–Magenes [1].

We introduce v, the solution of:

$$\left.\begin{array}{l} -\dfrac{\partial v}{\partial t} + \Delta^2 v = 0 , \\[2mm] v(T) = \psi , \\[2mm] v|_\Sigma = \dfrac{\partial v}{\partial \nu}\Big|_\Sigma = 0 . \end{array}\right\} \tag{4.22}$$

The expression (with justification as in Section 2)

$$\int_0^T \big((u', v) + (u, v')\big)\, dt$$

is zero and also is equal to:

$$-\int_\Sigma \frac{\partial u}{\partial \nu}\, \Delta v \, d\Gamma \, dt + \int_\Sigma u \frac{\partial \Delta v}{\partial \nu}\, d\Gamma \, dt .$$

Therefore

$$-\int_\Sigma g_1\, \Delta v \, d\Gamma \, dt + \int_\Sigma g_0 \frac{\partial \Delta v}{\partial \nu}\, d\Gamma \, dt = 0 \qquad \forall g_0, g_1 ,$$

and therefore

$$\Delta v|_\Sigma = 0 , \qquad \frac{\partial \Delta v}{\partial \nu}\Big|_\Sigma = 0 . \tag{4.23}$$

This, combined with (4.22), shows that $v = 0$, therefore $\psi = 0$. ∎

The problem:

To find $g_{0,\eta}, g_{1,\eta}$ such that:

$$\int_\Omega \Big| u(x, T) - \chi(x) \Big|^2 \, dx \leqslant \eta . \ \blacksquare$$

QR Solution

We consider u_ε the solution of:

$$\frac{\partial u_\varepsilon}{\partial t} - \varepsilon \frac{\partial^2 u_\varepsilon}{\partial t^2} + \Delta_\varepsilon^2 u_\varepsilon = 0 \qquad (^1) , \tag{4.24}$$

(¹) Δ_ε defined as in (3.7).

with

$$u_\varepsilon(0) = 0, \qquad u_\varepsilon(T) = \chi, \tag{4.25}$$

without any boundary condition on Σ. This problem admits a unique solution. We then take:

$$g_0 = u_\varepsilon|_\Sigma, \qquad g_1 = \left.\frac{\partial u_\varepsilon}{\partial v}\right|_\Sigma . \blacksquare$$

Remark 4.2. It is naturally interesting to pose the following problem: what happens if, for example, we control only g_1?

$$\left.\begin{aligned}
&\frac{\partial u}{\partial t} + \Delta^2 u = 0, \\
&u(0) = 0, \\
&u|_\Sigma = 0, \qquad \left.\frac{\partial u}{\partial v}\right|_\Sigma = g_1 . \blacksquare
\end{aligned}\right\}$$

4.6 Systems

All that has been said holds for elliptic systems, for example, for the operator:

$$\frac{\partial}{\partial t} + A, \qquad A \quad \textit{the system of elasticity.}$$

4.7 Problems with Time-Lag

We can consider in the same way:

$$\left.\begin{aligned}
&\frac{\partial u}{\partial t} - \Delta u + cu(x, t - \tau_0) = 0, \quad x \in \Omega, \quad t > 0, \tau_0 > 0 \text{ given}, \\
&u(x, t) = 0 \quad \text{for} \quad x \in \Omega, \quad t \in (0, \tau_0), \\
&u(x, t) = g \quad \text{on} \quad \Sigma_{\tau_0},
\end{aligned}\right\} \tag{4.26}$$

where

$$\Sigma_{\tau_0} = \Gamma \times {]}\tau_0, T{[}, \qquad g \text{ variable.} \blacksquare$$

Remark 4.3. We can similarly consider for (1.1), (1.2), (1.3) other functionals than $J(g)$ (as in Chapter 1). ■

5. NUMERICAL APPLICATIONS

5.1 Example

We take the case of Problem 1.1 in one space dimension with $\Omega =]0, 1[$. We apply the general method of Section 3. We take $\rho_\varepsilon(x)$ as in Figure 3.

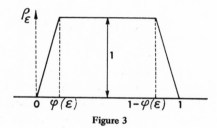

Figure 3

The problem to resolve is then:

$$\left.\begin{array}{l}
\dfrac{\partial u_\varepsilon}{\partial t} - \varepsilon \dfrac{\partial^2 u_\varepsilon}{\partial t^2} - \dfrac{\partial}{\partial x}\left[\rho_\varepsilon^2(x)\dfrac{\partial u_\varepsilon}{\partial x}\right] = 0, \\[2mm]
u_\varepsilon(x, 0) = \xi(x) \quad (^1), \\[2mm]
u_\varepsilon(x, T) = \chi(x),
\end{array}\right\} \qquad (5.1)$$

and we have done what was necessary so that there are no boundary conditions at $x = 0$ and 1.

Remark 5.1. There exists at the moment only heuristic methods for the choice of the numbers ε and $\varphi(\varepsilon)$ (where it is understood that the choice of $\varphi(\varepsilon)$ has only a limited importance). ■

5.2 Discretization of (5.1)

To simplify, we write u in place of u_ε and ρ in place of ρ_ε; u_i^n is the approximation of u at the point $i\,\Delta x$, $n\,\Delta t$;

(^{1}) The general method of Section 3 remains valid when the initial given function is not zero; let us add that the $\xi(x)$ introduced here has not the same role as in preceding chapters.

$$0 \leqslant i \leqslant J, \qquad J \, \Delta x = 1$$

$$0 \leqslant n \leqslant M, \qquad M \, \Delta t = T.$$

We take for the t-derivatives:

$$\left(\frac{\partial u}{\partial t}\right)_i^n \simeq \frac{u_i^{n+1} - u_i^{n-1}}{2 \, \Delta t}$$

$$\left(\frac{\partial^2 u}{\partial t^2}\right)_i^n \simeq \frac{u_i^{n+1} - 2 \, u_i^n + u_i^{n-1}}{\Delta t^2}.$$

The x-derivative at (i, n) of a function v is evaluated as:

$$\left(\frac{\partial v}{\partial x}\right)_i^n \simeq \frac{v_{i+1/2}^n - v_{i-1/2}^n}{\Delta x}.$$

Of course, it is not necessary to have the boundary conditions enter into the numerical integration; the components u_0^n and u_J^n of the solution can thus enter into the discrete formulas with a zero "weight." Hence there are the two following possibilities:

(i) we evaluate:

$$\left(\rho^2 \frac{\partial u}{\partial x}\right)_{i+1/2}^n \qquad \text{by} \qquad \rho_{i+1}^2 \frac{u_{i+1}^n - u_i^n}{\Delta x}$$

and

$$\left(\rho^2 \frac{\partial u}{\partial x}\right)_{i-1/2}^n \qquad \text{by} \qquad \rho_{i-1}^2 \frac{u_i^n - u_{i-1}^n}{\Delta x},$$

which yields:

$$\left[\frac{\partial}{\partial x}\left[\rho^2 \frac{\partial u}{\partial x}\right]\right]_i^n \simeq \rho_{i+1}^2 \frac{u_{i+1}^n - u_i^n}{\Delta x^2} - \rho_{i-1}^2 \frac{u_i^n - u_{i-1}^n}{\Delta x^2}. \tag{5.2}$$

To eliminate the components u_0^n and u_J^n it suffices then to take (cf. Fig. 4):

$$\rho_0 = \rho_J = 0 \, ;$$

(ii) we evaluate:

$$\left(\rho^2 \frac{\partial u}{\partial x}\right)_{i+1/2}^n \qquad \text{by} \qquad \rho_{i+1/2}^2 \frac{u_{i+1}^n - u_i^n}{\Delta x}$$

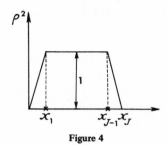

Figure 4

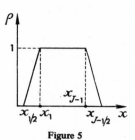

Figure 5

and

$$\left(\rho^2 \frac{\partial u}{\partial x}\right)^n_{i-1/2} \quad \text{by} \quad \rho^2_{i-1/2} \frac{u^n_i - u^n_{i-1}}{\Delta x}$$

which yields:

$$\left[\frac{\partial}{\partial x}\left[\rho^2 \frac{\partial u}{\partial x}\right]\right]^n_i \simeq \rho^2_{i+1/2} \frac{u^n_{i+1} - u^n_i}{\Delta x^2} - \rho^2_{i-1} \frac{u^n_i - u^n_{i-1}}{\Delta x^2} \tag{5.3}$$

and to eliminate the components u^n_0 and u^n_j, it suffices then to take (cf. Fig. 5):

$$\rho_{1/2} = \rho_{J-1/2} = 0.$$

In the discrete form, using the second option above, the equation may be written:

$$\frac{u^{n+1}_i - u^{n-1}_i}{2\,\Delta t} - \varepsilon \frac{u^{n+1}_i - 2\,u^n_i + u^{n-1}_i}{\Delta t^2} - \\ - (\rho^n_{i+1/2})^2 \frac{u^n_{i+1} - u^n_i}{\Delta x^2} - (\rho^n_{i-1/2})^2 \frac{u^n_i - u^n_{i-1}}{\Delta x^2} = 0\,. \tag{5.4}$$

To (5.4) it is necessary naturally to add the corresponding discrete conditions at $t = 0$ and $t = T$, namely:

$$\left.\begin{array}{ll} u^0_i = \xi_i & 0 \leqslant i \leqslant J\,, \\ u^M_i = \chi_i & 0 \leqslant i \leqslant J\,. \end{array}\right\} \tag{5.5}$$

We thus must solve a linear system where the matrix is 5-diagonal and nonsymmetric.

5.3 Solution of the Problem: Obtaining the Boundary Conditions

The preceding method permits us to calculate the solution $u_\varepsilon(x, t)$ in a domain $Q' = \Omega' \times (0, T)$, where $\overline{\Omega}' \subset \Omega$.

To get $u_\varepsilon|_\Gamma$, *we must extend* $u_\varepsilon(x, t)$ *by extrapolation* to $Q = \Omega \times (0, T)$: in the present case, one-dimensional, we denote by g_0^* and g_1^* the functions obtained at $x = 0$ and $x = 1$ in this way.

Remark 5.2. Let us recall, of course, for a given η, there is both *nonuniqueness* and *instability* for g satisfying:

$$J(g) \leqslant \eta .$$

In the absence of *constraints on the* g, there is no privileged choice of g in the set of all admissible solutions g, *which can vary greatly among themselves.* ∎

Finally, *at the end of the verification*, we integrate the direct system

$$\left.\begin{aligned}
&\frac{\partial U}{\partial t} - \frac{\partial^2 U}{\partial x^2} = 0 , \\
&U(x, 0) = \xi(x) , \\
&U(0. t) = g_0^*(t) , \qquad U(1, t) = g_1^*(t) ,
\end{aligned}\right\} \tag{5.6}$$

and we set:

$$\gamma^*(x) = U(x, T) . \tag{5.7}$$

5.4 Numerical Results

5.4.1 USE OF AN ANALYTIC SOLUTION

The system:

$$\left.\begin{aligned}
&\frac{\partial u}{\partial t} - \frac{\partial^2 u}{\partial x^2} = 0 , \quad 0 \leqslant x \leqslant 1 , \\
&\hspace{4.5cm} 0 \leqslant t \leqslant T , \\
&u(1, t) = u(0, t) = 2(T - t) , \\
&u(x, 0) = 2\,T + x(1 - x) ,
\end{aligned}\right\} \tag{5.8}$$

admits for solution:

$$u(x, t) = 2(T - t) + x(1 - x).$$ (5.9)

We therefore take in (5.1):

$$\xi(x) = 2\,T + x(1 - x)$$

and

$$\chi(x) = x(1 - x).$$

We thus calculate $g_0^*(t)$ and $g_1^*(t)$ (cf. 5.3). We note (cf. Graph 1 *bis*) that the functions $g_1^*(t)$ (cf. Remark 5.2) are not "near" the functions:

$$u(0, t) = u(1, t) = g_i(t) = 2(T - t),$$

which correspond to $\eta = 0$. The function χ^* [cf. (3.7)] is represented on Graph 1. ∎

5.4.2 USE OF A SOLUTION CONSTRUCTED NUMERICALLY

The numerical integration of the system:

$$\left. \begin{array}{l} \dfrac{\partial u}{\partial t} - \dfrac{\partial^2 u}{\partial x^2} = 0\,, \\[2mm] u(1, t) = u(0, t) = \dfrac{t}{T}\,(= g_i(t))\,, \\[2mm] u(x, 0) = 0\,, \end{array} \right\}$$ (5.10)

leads to a function $\chi(x)$. We solve the one-dimensional problem with this function $\chi(x)$ as given. The results are given on Graphs 2 and 2 *bis* and give rise to the same remarks as on 5.3 and as for the preceding example.

Part Two

HYPERBOLIC EQUATIONS OF EVOLUTION

1. STATEMENT OF THE PROBLEM

1.1 The Equations of the Problem

We consider, using notation analogous to that of the first part, the solution:

$$u = u(x, t) = u(x, t \; ; g), \qquad x \in \Omega, \qquad t \in]0, T[, \tag{1.1}$$

of the *wave equation*

$$\frac{\partial^2 u}{\partial t^2} - \Delta u = 0, \qquad x \in \Omega, \qquad t \in]0, T[\tag{1.2}$$

with *initial conditions*

$$u(x, 0) = 0, \qquad \frac{\partial u}{\partial t}(x, 0) = 0 \tag{1.3}$$

and *boundary conditions*

$$u(x, t) = g(x, t), \qquad (x, t) \in \Sigma = \Gamma \times]0, T[, \tag{1.4}$$

where g is given (in a suitable class of functions).

The problem in (1.2), (1.3), (1.4) admits, as is well known, a unique solution, which we denote by (1.1). ∎

Remark 1.1. The considerations which follow may be extended without difficulty to the "inhomogeneous" case:

$$\frac{\partial^2 u}{\partial t^2} - \Delta u = f, \qquad f \text{ given} \neq 0, \tag{1.5}$$

$$u(x, 0) = u_0(x), \qquad \frac{\partial u}{\partial t}(x, 0) = u_1(x), \qquad u_0 \text{ and } u_1 \text{ given in } \Omega. \blacksquare \tag{1.6}$$

Remark 1.2. Similarly, all that follows extends readily to the case where instead of $\frac{\partial^2}{\partial t^2} - \Delta$, we consider the hyperbolic operator:

$$\frac{\partial^2}{\partial t} - \sum \frac{\partial}{\partial x_i} \left(a_{ij}(x, t) \frac{\partial}{\partial x_j} \right) + \sum a_i(x, t) \frac{\partial}{\partial x_i} + a_0(x, t) \qquad (1.7)$$

where $a_{ij} = a_{ji}$ and

$$\sum a_{ij}(x, t)\, \xi_i\, \xi_j \geqslant \alpha(\xi_1^2 + \cdots + \xi_n^2), \qquad \alpha > 0 . \blacksquare$$

Remark 1.3. We will note that the method used is general: reducing the whole question to a *problem of uniqueness* (cf. 2 and 3 above); on the other hand it uses in an absolutely fundamental fashion the *linearity* of the problem; for nonlinear cases we can only *adapt heuristically* the QR method proposed below. $\blacksquare$

1.2 The Problem

Let χ_0 and χ_1 be two given functions.
We associate, with each g, the functional:

$$J(g) = \| u(x, T ; g) - \chi_0 \|_0^2 + \| \frac{\partial u}{\partial t} (x, T ; g) - \chi_1 \|_1^2 \qquad (1.8)$$

where

$$\| \quad \|_i \text{ designates a convenient norm.}$$

We take generally:

$$J(g) = \int_O | u(x, T ; g) - \chi_0(x) |^2 \, dx + \int_\Omega \left| \frac{\partial u}{\partial t} (x, T ; g) - \chi_1(x) \right|^2 dx . \qquad (1.9)$$

Remark 1.4. We can also *normalize* (1.9). $\blacksquare$
We are interested in the following questions (evidently of the same *type* as those studied in the first part of the chapter):

QUESTION 1.1

Under what conditions do we have

$$\inf_g J(g) = 0 \ ? \qquad (1.10)$$

QUESTION 1.2

If (1.10) holds, find g_n such that

$$J(g_\eta) \leqslant \eta \cdot \text{κx} \tag{1.11}$$

where η is a given positive quantity. ∎

Remark 1.5. Nature of the control g. In order that there is no difficulty concerning the nature of the spaces in (1.2), (1.3), (1.4) is solved, we suppose that:

$g \in \mathcal{D}(\Sigma)$ = space of functions which are infinitely differen- ⎫
tiable with compact support in $\Sigma = \Gamma \times]0, T]$, ⎬ (1.12)
⎭

supposing that Γ is an infinitely differentiable variety.

For $g \in \mathcal{D}(\Sigma)$, the solution $u(x, t; g)$ is infinitely differentiable in $\bar{\Omega} \times [0, T]$ and therefore

$$u(x, T; g), \ \frac{\partial u}{\partial t}(x, T; g) \in \mathcal{D}(\bar{\Omega}) \qquad \text{(functions } C^\infty \text{ in } \bar{\Omega}). \tag{1.13}$$

It is clear that, in general, even if (1.10) holds, the infimum *is not attained* when $g \in \mathcal{D}(\Sigma)$, since $u(x, T; g)$, $\frac{\partial u}{\partial t}(x, T; g)$ satisfy (1.13) — and therefore if χ_0 and χ_1 are not differentiable, we cannot have $J(g) = 0$.

We can still say that *the image of $\mathcal{D}(\Sigma)$ under the mapping*

$$g \rightarrow \{ u(x, T; g), \frac{\partial u}{\partial t}(x, T; g) \} \tag{1.14}$$

is not closed in the product space $L^2(\Omega) \times L^2(\Omega)$.

(The image of the space spanned by g can be actually closed only when g describes a much more complicated space.) ∎

Remark 1.6. There are no constraints on g. In order to guarantee the existence of a function g attaining the lower ground, it is necessary to *modify* the functional (1.9), replacing it, for example, by

$$J_\lambda(g) = J(g) + \lambda \| g \|^2, \qquad \lambda > 0 \tag{1.15}$$

where $\|g\|$ is a "convenient" norm for g (uniqueness will then hold for the "optimal control," but evidently Inf $J_\lambda(g) \neq 0$!). ∎

Remark 1.7. We can consider in (1.9) that $\{u(x, T; \xi), \frac{\partial u}{\partial t}(x, T; g)\}$ is "the observation." We could also consider as *observation*:

$$\frac{\partial u}{\partial n}(x, t; g) \qquad \text{(normal derivative of } u \text{ on } \Sigma),$$

which would lead to the functional:

$$\tilde{J}(g) = \int_\Sigma \left| \frac{\partial u}{\partial n}(x, t; g) - \chi_2(x, t) \right|^2 d\Sigma \qquad (1.16)$$

(or some combinations of J and $\tilde{J}$). But it is then trivial that

$$\text{Inf } \tilde{J}(g) = 0 .$$

If we consider in effect u as the solution of the well posed problem:

$$\frac{\partial u^2}{\partial t^2} - \Delta u = 0 ,$$

$$u(x, 0) = 0 , \qquad \frac{\partial u}{\partial t}(x, 0) = 0 .$$

$$\left. \frac{\partial u}{\partial n} \right|_\Sigma = \chi_2 ,$$

we will take $g = u(x, t)|_\Sigma$ and evidently $\tilde{J}(g) = 0$. ∎

2. RESULTS OF DENSITY AND NONDENSITY

2.1 Necessary and Sufficient Condition for Density

The question in (1.10) is *equivalent* to seeing if the image of $\mathscr{D}(\Sigma)$ under (1.14) is dense (or not) in $L^2(\Omega) \times L^2(\Omega)$. Let us use for this purpose the Hahn–Banach theorem. Let ψ_0, ψ_1 be given in $L^2(\Omega)$, such that $\{\psi_0, \psi_1\}$ is *orthogonal* to the image of $\mathscr{D}(\Sigma)$ under (1.14):

$$\int_\Omega u(x, T; g) \psi_0(x) \, dx + \int_\Omega \frac{\partial u}{\partial t}(x, T; g) \psi_1(x) \, dx = 0 \qquad \forall g . \qquad (2.1)$$

We wish to see whether or not this implies $\psi_0 = \psi_1 = 0$.

We introduce (compare 2.2, first part of the chapter) w, the solution of:

$$\frac{\partial^2 w}{\partial t^2} - \Delta w = 0 \quad \text{in} \quad \Omega \times {]0, T[}, \tag{2.2}$$

$$w(x, T) = \psi_1(x), \qquad x \in \Omega, \tag{2.3}$$

$$\frac{\partial w}{\partial t}(x, T) = -\psi_0(x), \quad x \in \Omega, \tag{2.4}$$

$$w(x, T) = 0 \quad \text{on} \quad \Sigma \tag{2.5}$$

(this problem certainly admits a unique solution).

Then, using Green's formula:

$$\int_Q \left(\frac{\partial^2 u}{\partial t^2} - \Delta u\right) w \, dx \, dt - \int_Q u \left(\frac{\partial^2 w}{\partial t^2} - \Delta w\right) dx \, dt = 0 =$$

$$= \int_\Omega \frac{\partial u}{\partial t}(x, T; g) \, w(x, T) \, dx - \int_\Omega u(x, T; g) \frac{\partial w}{\partial t}(x, T) \, dx + \int_\Sigma u \frac{\partial w}{\partial n} \, d\Sigma. \tag{2.6}$$

Using (2.3), (2.4) and (2.1), we see that (2.6) reduces to:

$$\int_\Sigma u \frac{\partial w}{\partial n} \, d\Sigma = 0 \tag{2.7}$$

or

$$\int_\Sigma g \frac{\partial w}{\partial n} \, d\Sigma = 0 \qquad \forall g \tag{2.7 bis}$$

therefore

$$\frac{\partial w}{\partial n} = 0 \quad \text{on} \quad \Sigma. \tag{2.8}$$

Thus:

Lemma 2.1: *The necessary and sufficient condition that*

$$\underset{g \in \mathcal{D}(\Sigma)}{\text{Inf}} J(g) = 0$$

is that (2.2), (2.5), (2.8) *imply* $w = 0$. ∎

We deduce from this:

Theorem 2.1: *For T sufficiently large* (depending on the geometry of ii), *we have*

$$\text{Inf}_{g \in \mathscr{D}(\Sigma)} \; J(g) = 0 \, . \blacksquare \tag{2.9}$$

On the other hand, if $T < T_0$, T_0 suitably chosen, we have in general[1]

$$\text{Inf}_{g \in \mathscr{D}(\Sigma)} \; \dot{J}(g) > 0 \, . \blacksquare \tag{2.10}$$

We see here the essential difference between this and the parabolic case studied in the first part of the chapter.

Remark 2.1. We can say that the problem is *controllable if* $\text{Inf}_{g \in \mathscr{D}(\Sigma)} \; J(g) = 0$ and *not controllable otherwise*.

According to this definition, parabolic problems are controllable, while hyperbolic problems are controllable only for sufficiently large T. $\blacksquare$

Remark 2.2. It is necessary to examine the notion of controllability in the preceding remark: the image of g under $u \to u(x, T; g)$, in the *parabolic case, is dense in $L^2(\Omega)$, but it is not identical with $L^2(\Omega)$* (since $u(x, T; g)$ is necessarily, in particular, C^∞ in x). $\blacksquare$

Let us make Theorem 2.1 precise in a particular case.

2.2 The One-Dimensional Case

Let us suppose that

$$\Omega = \,]0, x_0[\, . \tag{2.11}$$

Conditions (2.2), (2.5), (2.8) may be written:

$$\frac{\partial^2 w}{\partial t^2} - \frac{\partial^2 w}{\partial x^2} = 0 \,, \quad 0 < x < x_0 \,, \quad 0 < t < T, \tag{2.12}$$

---

[1] This is to say apart from particular choice of χ_0, χ_1.

$$w(0, t) = w(x_0, t) = 0, \qquad 0 < t < T, \tag{2.13}$$

$$\frac{\partial w}{\partial x}(0, t) = \frac{\partial w}{\partial x}(x_0, t) = 0, \quad 0 < t < T. \tag{2.14}$$

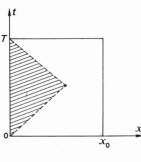

Figure 6

But **(2.12)**, combined with $w(0, t) = 0$, $\frac{\partial w}{\partial x}(0, t) = 0$ implies that $w \equiv 0$ in the shaded region in Figure 6 (bounded by characteristics).

Naturally there is a symmetric conclusion for $x = x_0$, therefore $w \equiv 0$ in the shaded regions in Figures 7, 8, 9 (which represent the three cases of possible figures, without counting the limit cases).

In the case of Figure 7 (and a fortiori in Figure 9) we also deduce that $w \equiv 0$ everywhere in $Q = \Omega \times]0, T[$.

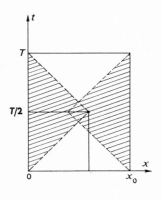

Figure 7

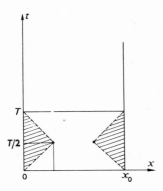

Figure 8

On the other hand, in the case of Figure 8, *we can find* $w \neq 0$, satisfying (2.12), (2.13), (2.14). Consequently

Theorem 2.2: *When* $\Omega =]0, x_0[$, *one-dimensional case, we have*:

$$\underset{g \in \mathscr{D}(\Sigma)}{\text{Inf}} J(g) = 0 \qquad \text{if} \qquad T > x_0 \cdot \tag{2.15}$$

and (in general)

$$\underset{g \in \mathscr{D}(\Sigma)}{\text{Inf }} J(g) > 0 \qquad \text{if} \qquad T < x_0 \, . \blacksquare \qquad \text{(2.16)}$$

Remark 2.3. In the case of Figure 9 ($T > 2\,x_0$) we conclude that $w \equiv 0$ using only one of the shaded triangles. This corresponds to a particular case of 3 below. $\blacksquare$

Remark 2.4. For the case of Theorem 2.2, the problem has been solved independently (together with additional precision) by a direct method (not using Hahn–Banach) by D. L. Russell [1]. $\blacksquare$

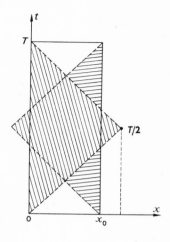

Figure 9

3. SETS OF UNIQUENESS

3.1 One Generalization of the Problem

The generalization which we will give will allow us in addition to make precise Remark 2.3.

Let

$$E = \text{a set contained in } \Sigma, \text{ closed and of positive measure.} \qquad \text{(3.1)}$$

We *define*:

$$\mathscr{D}_E(\Sigma) = \text{the space of functions on } \mathscr{D}(\Sigma) \text{ with support in } E \, . \blacksquare \qquad \text{(3.2)}$$

Example 3.1. Let us reconsider the one-dimensional case (2.11). Then

$$\Sigma = \{\, 0 \,\} \times [0, T] \cup \{\, x_0 \,\} \times [0, T] \, .$$

We can take

$$E = \{\, 0 \,\} \times [0, T] \, . \blacksquare$$

We now consider the case

$$g \in \mathcal{D}_E(\Sigma) . \qquad (3.3)$$

Remark 3.1. The condition (3.3) is a *constraint on the supports.* ∎
We consider $J(g)$ again and we ask if

$$\underset{g\in\mathcal{D}_E(\Sigma)}{\text{Inf}} \ J(g) = 0 . \blacksquare \qquad (3.4)$$

Remark 3.2. We can pose an analogous problem for $\tilde{J}(g)$ defined in (1.16). We see, without difficulty, by the method used in the proof of Theorem 2.1 that if $E \subset \Sigma$ strictly,

$$\underset{g\in\mathcal{D}_E(\Sigma)}{\text{Inf}} \ \tilde{J}(g) > 0 . \blacksquare$$

3.2 Necessary and Sufficient Condition for Density

We make

Definition 3.1: The set $E \subset \Sigma$ is called a *set of uniqueness* if the conditions:

$$\frac{\partial^2 w}{\partial t^2} - \Delta w = 0 \qquad \text{in} \qquad Q = \Omega \times \,]0, T[\, , \qquad (3.5)$$

$$w = 0 \qquad \text{sur} \qquad \Sigma \, , \qquad (3.6)$$

$$\frac{\partial w}{\partial n} = 0 \qquad \text{on} \qquad E \, , \qquad (3.7)$$

imply that

$$w \equiv 0 . \blacksquare \qquad (3.8)$$

Example 3.2. We saw in Section 2 that:

$E = \Sigma$ is a uniqueness set if T is sufficiently large. ∎

Example 3.3. We take $E = \{0\} \times [0, T]$ (cf. Example 3.1). Then (as in Figure 9) the set E *is a uniqueness set if $T > 2\, x_0$.* ∎

We have

Theorem 3.1: *A necessary and sufficient condition that* (3.4) *holds is that E is a uniqueness set.*

Proof. We employ the same principle as in the proof of Lemma 2.1. We introduce w by means of (2.2), (2.3), (2.4), (2.5), and thus to (2.7 bis) with $g \in \mathscr{D}_E(\Sigma)$, namely:

$$\int_\Sigma g \frac{\partial w}{\partial n} \, d\Sigma = 0 \qquad \forall g \text{ with support in } E. \tag{3.9}$$

Therefore $\frac{\partial w}{\partial n} = 0$ on E, whence $w \equiv 0$ if (and only if) E is a set of uniqueness. ∎

3.3 Orientation. QR Method

We will now consider the practical solution of Question 1.2 by the QR method.

As might be imagined, the "approximation of χ_1" is much more difficult to carry out numerically than the approximation of χ_0.

This is why we are going to present three possibilities of application of the QR method, showing how to improve (partially) the approximation of χ_1.

4. THE QR METHOD—FIRST POSSIBILITY

4.1 Formal Statement

We introduce the operator Δ_ε, as in (3.7), first part. We take u_ε the solution of the equation (*of "elliptic" type*):

$$\varepsilon u_\varepsilon^{(4)} + \varepsilon \Delta_\varepsilon u_\varepsilon'' - u_\varepsilon''' + \Delta_\varepsilon u_\varepsilon' = 0 \quad (^1),$$

where the prime denotes $\frac{\partial}{\partial t}$, with boundary conditions

---

(^{1}) This is an "elliptic regularization": cf. W. Strauss [1].

$$u_\varepsilon(0) \ = u'_\varepsilon(0) = 0 \tag{4.2}$$

$$u_\varepsilon(T) = \chi_0$$

$$u'_\varepsilon(T) = \chi_1 . \tag{4.3}$$

We will show further on that u_ε exists and is unique (in a suitable space). We then take

$$g = u_r |_\Sigma \tag{4.4}$$

(after extrapolation, cf. the first part). ∎

Remark 4.1. If we "neglect" in (4.1) terms having ε as a factor, there remains

$$-\frac{\partial}{\partial t}\left(\frac{\partial^2 u_\varepsilon}{\partial t^2} - \Delta_\varepsilon u_\varepsilon\right) = 0 .$$

It therefore comes down to an approximation of

$$\frac{\partial}{\partial t}\left(\frac{\partial^2}{\partial t^2} - \Delta\right) .$$

When $\varepsilon \to 0$, u_ε *cannot* converge towards a u satisfying

$$u'' - \Delta u = 0 , \qquad u(0) = u'(0) = 0 , \qquad u(T) = \chi_0 , \qquad u'(T) = \chi_1 ,$$

since such a function does not exist in general. ∎

4.2 Solution of the QR Problem

We introduce the space V_ε as in the first part (Section 3., 3.4). Then

Theorem 4.1: *We suppose χ_0 and χ_1 given in $H^2(\Omega)$.*[1] *There exists one and only one function u_ε satisfying*

---

[1] These are not optimal hypotheses! But optimal hypotheses require use of the theory of interpolation of Hilbert spaces.

$$u_\varepsilon \in L_2(0, T; V_\varepsilon),$$
$$u'_\varepsilon \in L_2(0, T; V_\varepsilon),$$
$$u''_\varepsilon \in L_2(0, T; L_2(\Omega))$$

$\left.\begin{array}{}\\\\\\\end{array}\right\}$ (4.5)

and conditions (4.1), (4.2), (4.3).

Remark 4.2. (Compare with Remark 3.1, first part.)
We choose for practical reasons

$$g = u_\varepsilon|_{\Sigma'}, \qquad \Sigma' \text{ a surface "near" } \Sigma$$

(since $u_\varepsilon|_{\Sigma'}$ is not defined a priori). ∎

Proof. 1) Let to begin with $\Phi \in L_2(0, T; H^2(\Omega))$, Φ', $\Phi'' \in L_2(0, T; H^2(\Omega))$, with $\Phi(0) = \Phi'(0) = 0$, $\Phi(T) = \chi_0$, $\Phi'(T) = \chi_1$. Then the function

$$w = u_\varepsilon - \Phi$$

satisfies

$$\varepsilon w^{(4)} + \varepsilon \Delta_\varepsilon w'' - w''' + \Delta_\varepsilon w' = f, \qquad (4.6)$$

where

$$f = -\varepsilon \Phi^{(4)} - \varepsilon \Delta_\varepsilon \Phi'' + \Phi''' - \Delta_\varepsilon \Phi'$$

and therefore f can be written

$$f = f'_0 + f'_1 + f_2, \qquad f_i \in L_2(0, T; L_2(\Omega)). \qquad (4.7)$$

In addition

$$w(0) = w'(0) = w(T) = w'(T) = 0. \qquad (4.8)$$

We will have the theorem if we show the existence and uniqueness of w satisfying (3.6), (3.8) and

$$w \in \mathscr{V} \qquad (4.9)$$

where

$$\mathscr{V} = \{ v \mid v \in L_2(0, T; V_\varepsilon), \quad v' \in L_2(0, T; V_\varepsilon), \quad v'' \in L_2(0, T; L_2(\Omega)),$$
$$v(0) = v'(0) = v(T) = v'(T) = 0 \}. \qquad (4.10)$$

2) We introduce $a_\varepsilon(u, v)$ as in the first part (3.15) and write

$$a(u, v) = a_\varepsilon(u, v).$$

We also write, to simplify,

$$a(v, v) = a(v).$$

For $u, v \in \mathscr{V}$, we put:

$$\pi(u, v) = \int_0^T \left[\varepsilon(u'', v'') + \varepsilon a(u', v') + (u'', v') - a(u', v) \right] dt. \qquad (4.11)$$

We verify easily that

$$f \text{ given by (4.7) is in } \mathscr{V}', \text{ dual space of } \mathscr{V}. \qquad (4.12)$$

In addition:

Lemma 4.1: *The conditions* (4.6), (4.8), (4.9) *are equivalent to*

$$\begin{cases} w \in \mathscr{V}, \\ \pi(w, v) = (f, v) \qquad \forall v \in \mathscr{V} \end{cases} \qquad (4.13)$$

(where (f, v) denotes the scalar product of $f \in \mathscr{V}'$ and $v \in \mathscr{V}$).

Proof. 1) Since the space $\mathscr{D}(]0, T[; V_\varepsilon)$ of functions C^∞ with values in V_ε, with compact support in $]0, T[$, is *dense* in $\mathscr{V}$, (4.13) is equivalent to the same condition $\forall v \in \mathscr{D}(]0, T[; V_\varepsilon)$ and even, since $\mathscr{D}(\Omega)$ is dense in V_ε, $\forall v \in \mathscr{D}(Q)$.

2) Now, for $v \in \mathscr{D}(Q)$,

$$\pi(w, v) = \langle \varepsilon w^{(4)} - \varepsilon A w'' - w''' - A w', \ v \rangle$$

(where $A = -\Delta_\varepsilon$), the bracket denoting the duality between $\mathscr{D}'(Q)$ and $\mathscr{D}(Q)$. Hence (4.6). ∎

The theorem will then be a consequence of

$$\text{there exists a unique } w \text{ satisfying (4.13).} \qquad (4.14)$$

The last point is deduced from:

  i) the fact that $\pi(u, v)$ is a continuous bilinear form on $\mathscr{V}$;
  ii) the fundamental inequality

$$\pi(v, v) \geqslant \alpha \| v \|_{\mathscr{V}}^2, \qquad \forall v \in \mathscr{V}, \qquad \alpha > 0. \tag{4.15}$$

  3) *Proof of* (4.15). We have:

$$\int_0^T (v'', v')\, dt = \frac{1}{2} \int_0^T \frac{d}{dt} |v'(t)|^2\, dt = 0,$$

$$\int_0^T a(v', v)\, dt = \frac{1}{2} \int_0^T \frac{d}{dt} a(v(t), v(t))\, dt = 0.$$

Therefore

$$\pi(v, v) = \int_0^T \left[\varepsilon \, | v''(t) |^2 + \varepsilon a(v'(t), v'(t)) \right] dt. \tag{4.16}$$

But, in addition:

$$\int_0^T (| v(t) |^2 + | v'(t) |^2)\, dt \leqslant C_1 \int_0^T | v''(t) |^2\, dt,$$

$$\int_0^T a(v(t), v(t))\, dt \leqslant C_2 \int_0^T a(v'(t), v'(t))\, dt,$$

and thus (3.16) implies (4.15).

This finishes the proof of the theorem. ∎

5. THE QR METHOD–SECOND POSSIBILITY: METHOD OF PENALIZATION

5.1 Formal Statement

We now take u_ε as a solution of the equation (still of "elliptic type") (compare with 4.1):

$$\frac{1}{\varepsilon}(D^2 - \Delta_\varepsilon)^2 u - D(D^2 - \Delta_\varepsilon) u_\varepsilon = 0 \qquad (D = \partial/\partial t) \tag{5.1}$$

the boundary conditions are unchanged (4.2), (4.3).

We will briefly indicate further on that the problem is well-posed. We then take $g = u_\varepsilon|_\Sigma$ as in (4.4). ∎

Remark 5.1. The first term in (5.1) is a *penalty term*[1] since ε is "small."

(i) If $(D^2 - \Delta_\varepsilon) u_\varepsilon$ is "small," we therefore remain close to the original equation.

(ii) If this is not the case, two possibilities arise:

1. $D(D^2 - \Delta_\varepsilon) u_\varepsilon$ is nonnegligible; then the penalization term in (5.1) implies that $(D^2 - \Delta_\varepsilon) u_\varepsilon$ must be of "the order" of $\sqrt{\varepsilon}$, which leads to (i).

2. There remains the case where $D(D^2 - \Delta_\varepsilon) u_\varepsilon = 0$ without forcing $(D^2 - \Delta_\varepsilon) u_\varepsilon$ to be zero; we would then have $(D^2 - \Delta_\varepsilon) u_\varepsilon = h(x)$ (independent of t) so that (5.1) reduces to

$$\frac{1}{\varepsilon} (D^2 - \Delta_\varepsilon) h = 0, \quad \text{that is,} \quad \Delta_\varepsilon h = 0,$$

therefore $h = 0$, whence a contradiction.

This constitutes a formal justification of the introduction of equations of the type (5.1). ∎

5.2 Solution of the QR Problem

The general principle is the following:

1) We first introduce a function Φ satisfying:

$$\Phi(0) = \Phi'(0) = 0,$$

$$\Phi(T) = \chi_0, \quad \Phi'(T) = \chi_1.$$

Then the function

$$w_\varepsilon = u_\varepsilon - \Phi$$

[1] This method of penalization plays an important role in problems considered in Chapters 4 and 5. See also, in a different context, Lions [5], Chapter 5, and Benscrisson–Kenneth [1], Tartar [1].

must satisfy:

$$\left.\begin{aligned}
\frac{1}{\varepsilon} (D^2 - \Delta_\varepsilon)^2 \, w_\varepsilon - D(D^2 - \Delta_\varepsilon) \, w_\varepsilon &= f \qquad (f \text{ known}) , \\
w_\varepsilon(0) &= w_\varepsilon'(0) = 0 , \\
w_\varepsilon(T) &= w_\varepsilon'(T) = 0 .
\end{aligned}\right\} \tag{5.2}$$

2) We introduce [compare with (4.11)]:

$$\pi(u, v) = \frac{1}{\varepsilon} ((D^2 - \Delta_\varepsilon) \, u, (D^2 - \Delta_\varepsilon) \, v) + ((D^2 - \Delta_\varepsilon) \, u, Dv) , \tag{5.3}$$

where

$$(f, g) = \int_0^T \int_\Omega fg \, dx \, dt .$$

The space $\mathscr{V}$ introduced in (4.10) is now replaced by:

$$\hat{\mathscr{V}} = \{ \, v \mid v \in L_2(0, T ; V_\varepsilon) , \quad v' \in L_2(0, T ; L_2(\Omega)) ,$$

$$(D^2 - \Delta_\varepsilon) \, v \in L_2(0, T ; L_2(\Omega)) ,$$

$$v(0) = v'(0) = 0 , \quad v(T) = v'(T) = 0 \, \} .$$

We verify that for all $v \in \mathscr{V}$:

$$\pi(v, v) = \frac{1}{\varepsilon} \| (D^2 - \Delta_\varepsilon) \, v \|_{L_2(Q)}^2 \qquad (\text{where } Q = \Omega \times (0, T)) ;$$

then, for functions satisfying the conditions $v(0) = 0, v'(0) = 0$, we have:

$$\| (D^2 - \Delta_\varepsilon) \, v \|_{L_2(Q)} \geqslant C \, [\| \, v \, \|_{L_2(0,T;V)} + \| \, v' \, \|_{L_2(Q)}] \quad (C \text{ constant})$$

Whence we have an inequality analogous to (4.15). This permits us to demonstrate, as in 4, *that the QR problem* (5.1), (5.2), (5.3) *is well posed.* ∎

6. THE QR METHOD–THIRD POSSIBILITY: MATRIX METHOD

6.1 Orientation

It is quite natural to write, as is classic, the hyperbolic equation as a *system of the first order*.

We will make this precise in the one-dimensional so as not to obscure the exposition. ∎

6.2 Matrix QR Method

The equation (2.1) is equivalent (one-dimensional case, $\Omega =]0, x_0[$) to the system:

$$\frac{\partial u}{\partial t} - \frac{\partial v}{\partial x} = 0 ,$$

$$\frac{\partial v}{\partial t} - \frac{\partial u}{\partial x} = 0 ,$$

(6.1)

or, setting

$$u = \{ u, v \} , \qquad A = \begin{pmatrix} 0 & -\partial/\partial x \\ -\partial/\partial x & 0 \end{pmatrix}$$

(6.2)

$$\frac{\partial u}{\partial t} + Au = 0 .$$

(6.3)

We now proceed by analogy with the first part of the chapter.

Let A_ε be a differential system "near" A and "degenerating at the boundary" (we will make this precise later on). Then we consider the following QR problem:

$$\frac{\partial u_\varepsilon}{\partial t} - \varepsilon \frac{\partial^2 u_\varepsilon}{\partial t^2} + A_\varepsilon u_\varepsilon = 0 , \qquad \varepsilon > 0, x \in \Omega, t \in]0, T[,$$

(6.4)

$$u_\varepsilon(x, 0) = 0 ,$$

(6.5)

$$u_\varepsilon(x, T) = \chi(x) ,$$

(6.6)

where we have put

$$\chi(x) = \{ \chi_0(x) , \chi_1(x) \} .$$

We then take

$$g(x, t) = u_\varepsilon(x, t) , \qquad x \in \Gamma ,$$

(6.8)

with

$$u_\varepsilon = \{ u_\varepsilon, v_\varepsilon \} . \blacksquare$$

It remains to make A_ε precise.

We introduce ρ_ε as in Figure 3 (Section 5, first part) and we put:

$$A_\varepsilon = \begin{pmatrix} 0 & -\dfrac{\partial}{\partial x}(\rho_\varepsilon \,\cdot) \\ -\rho_\varepsilon \dfrac{\partial}{\partial x} & 0 \end{pmatrix}. \tag{6.9}$$

The choice of A_ε is justified by the following formula; if $u = \{u, v\}$ we have

$$(A_\varepsilon\, u,\, u) = 0. \tag{6.10}$$

In effect:

$$(A_\varepsilon\, u,\, u) = -\int_0^1 \frac{\partial}{\partial x}(\rho_\varepsilon\, v).u \,\mathrm{d}x - \int_0^1 \rho_\varepsilon \frac{\partial u}{\partial x}.v \,\mathrm{d}x =$$

$$= -\int_0^1 \frac{\partial}{\partial x}(\rho_\varepsilon\, uv)\, \mathrm{d}x = 0,$$

because

$$\rho_\varepsilon(0) = \rho_\varepsilon(1) = 0$$

(and therefore (6.10) holds *without boundary conditions on* u). ∎

Remark 6.1. We can make (6.4) explicit:

$$\left(\frac{\partial}{\partial t} - \varepsilon \frac{\partial^2}{\partial t^2}\right) u_\varepsilon - \frac{\partial}{\partial x}(\rho_\varepsilon\, v_\varepsilon) = 0,$$

$$\left(\frac{\partial}{\partial t} - \varepsilon \frac{\partial^2}{\partial t^2}\right) v_\varepsilon - \rho_\varepsilon \frac{\partial u_\varepsilon}{\partial x} = 0,$$

therefore, eliminating v_ε:

$$\left(\frac{\partial}{\partial t} - \varepsilon \frac{\partial^2}{\partial t^2}\right)^2 u_\varepsilon - \frac{\partial}{\partial x}\left(\rho_\varepsilon^2 \frac{\partial u_\varepsilon}{\partial x}\right) = 0,$$

$$u_\varepsilon(x, 0) = 0, \qquad \frac{\partial u_\varepsilon}{\partial t}(x, 0) = 0,$$

$$u_\varepsilon(x, T) = \chi_0, \qquad \frac{\partial u_\varepsilon}{\partial t}(x, T) = \chi_1. \,\blacksquare$$

Remark 6.2. We could also use a *parabolic regularization* on system (6.3) (cf. Lions–Magenes [1]). ∎

Remark 6.3. We will see by means of numerical examples that this method does not improve "the approximation of χ_1." It is probably possible to improve the approximation of χ_1 introducing a *weighting factor for* χ_1 but we do not know how to do this systematically in the QR method. ∎

7. NUMERICAL RESULTS (I)

7.1 Generalities

We take a one-dimensional case with:

$$\Omega = \,]0, 1[\, . \tag{7.1}$$

Given any solution u whatever of (1.2), we *calculate*:

$$\chi_0(x) = u(x, T) \, , \qquad \chi_1(x) = \frac{\partial u}{\partial t}(x, T)$$

(in the hyperbolic case this does not restrain the generality of the problem, since there is no "regularization," because the hyperbolic operator is not hypoelliptic). Then we apply the QR method with the three possibilities given in 4, 5, 6.

The corresponding QR problem is solved by discretization. We utilized standard methods; hence we do not give the details of the discretization methods. This gives us $g = u_\varepsilon|_\Sigma$ (actually calculated by extrapolation); we then solve the direct problem:

$$\frac{\partial^2 U_\varepsilon}{\partial t^2} - \frac{\partial^2 U_\varepsilon}{\partial x^2} = 0 \, ,$$

$$U_\varepsilon(x, 0) = 0 \, , \quad \frac{\partial U_\varepsilon}{\partial t}(x, 0) = 0 \, , \tag{7.2}$$

$$U_\varepsilon|_\Sigma = u_\varepsilon|_\Sigma \, ;$$

naturally we solve (7.2) approximately by discretization (standard). We are thus led to approximations of $U_\varepsilon(x, T)$ and $\frac{\partial U_\varepsilon}{\partial t}(x, T)$ which we denote respectively by χ_0^* and χ_1^*.

We have drawn graphs of χ_i and χ_i^* at the end for comparison purposes. ■

We took:

$$T = 2 \quad \text{(that is,} \quad T = 2\,x_0) . \tag{7.3}$$

We also carried out calculations, not given here, with $T = 3/2$; the results are basically of the same nature (the *limit* case $T = x_0$ is given in Section 8). ■

7.2 The Case $u(x, t) = e^{-(x+t)}$

The results of three possibilities:

- without penalization (Section 4);
- with penalization (Section 5);
- matrix formulation (Section 6);

are indicated respectively on Graphs 3, 4, 5 and 6. ■

On these graphs ε_1 denotes the parameter ε introduced in the *factors* of the equations and ε_0 denotes the less important parameter appearing in the functions ρ_ε. ■

As already indicated, the results are bad,[1] for the approximation of χ_1. We have also verified that the results are not essentially affected by:

- other choice of parameters ε_i,
- other methods of discretization,
- a change of the discretization step size.

(This remark is valid for all examples presented here.)

The difficulty met in approximating χ_1 is therefore of a fundamental nature; cf. Remark 6.3. ■

7.3 The Case $u(x, t) = \cos \pi x \cos \pi t$

The results of the three possibilities are indicated on Graphs 7, 8, 9, 10 and 11. ■

---

[1] Note that we have not met this difficulty in the approximation of χ_1 for the same functional J when we control the initial conditions rather than the boundary conditions; cf. Chapter 2, Section 2.5 and 4.4.4.

8. NUMERICAL RESULTS (II)

We carried out a series of numerical trials with Ω still given by (7.1) and this time with:

$$T = 1 . \tag{8.1}$$

Then (cf. Theorem 2.2) we are in the limiting case, and it can be predicted that the QR method gives results which are much worse, even when χ_0 and χ_1 are given in the image of the g-space under the mapping:

$$g \rightarrow \{ u(x, T\,;g)\,, \qquad \frac{\partial u}{\partial t}\,(x, T\,;g)\ \} .$$

This is confirmed in Graphs 12 to 15. ∎

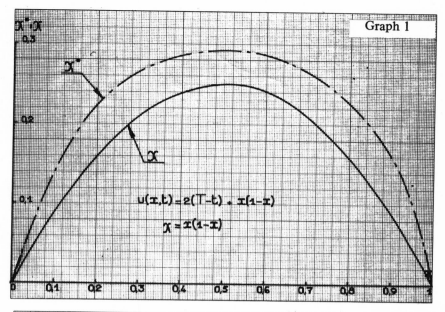

Graph 1

$$u(x,t) = 2(T-t) + x(1-x)$$

$$x = x(1-x)$$

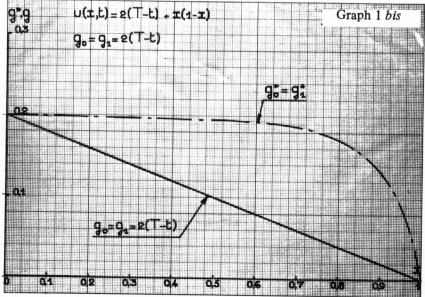

Graph 1 *bis*

$$u(x,t) = 2(T-t) + x(1-x)$$

$$g_0 = g_1 = 2(T-t)$$

$$g_0^* = g_1^*$$

$$g_0 = g_1 = 2(T-t)$$

NOTE: In all graphs commas represent decimal points.

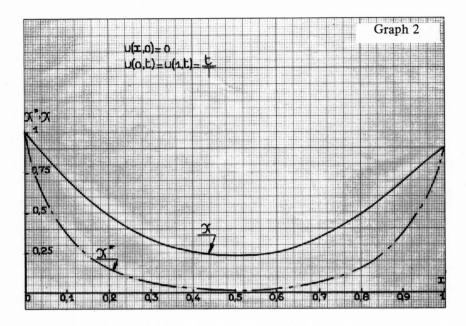

Graph 2

$$u(x,0) = 0$$
$$u(0,t) = u(1,t) = \frac{t}{T}$$

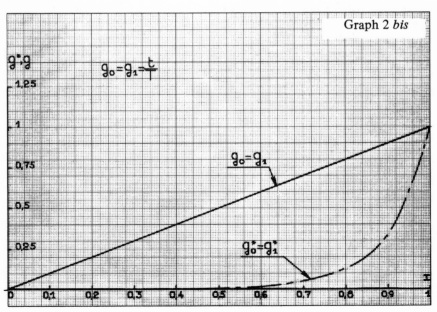

Graph 2 *bis*

$$g_0 = g_1 = \frac{t}{T}$$

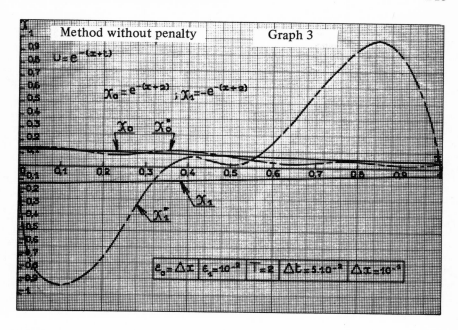

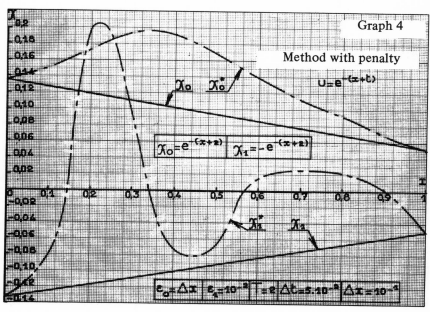

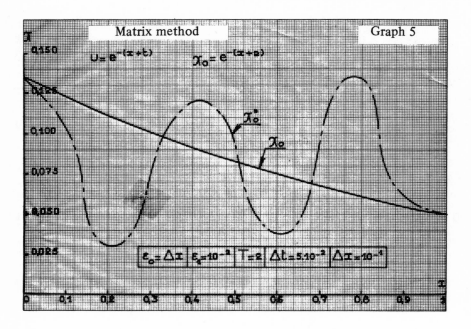

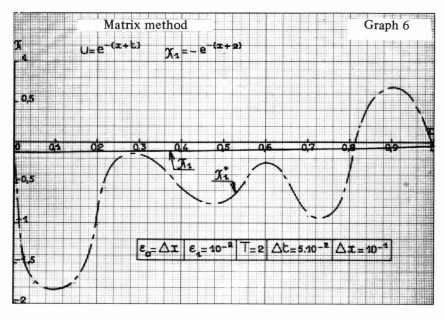

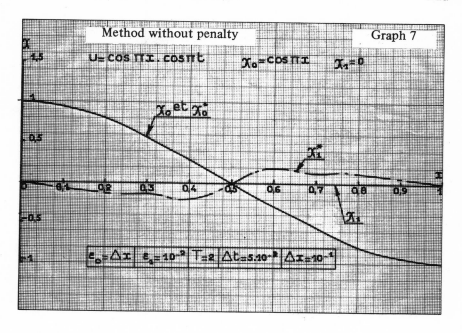

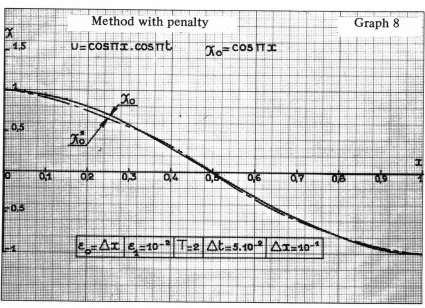

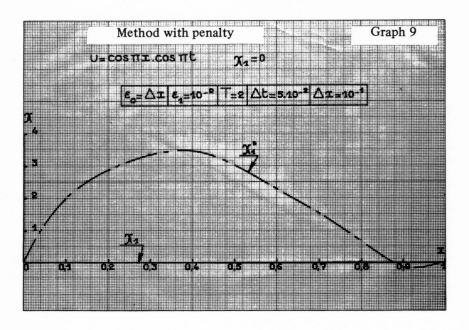

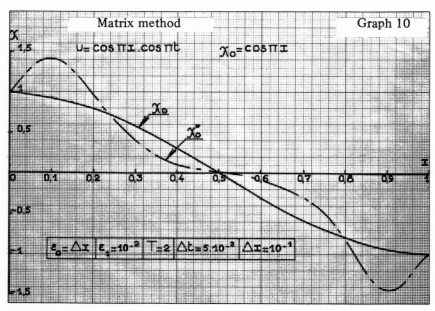

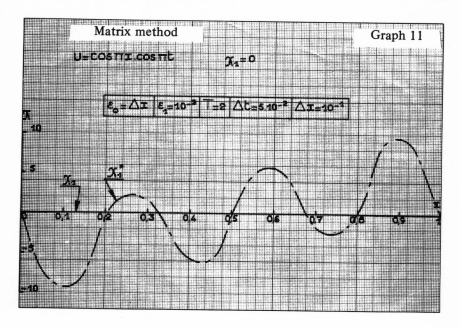

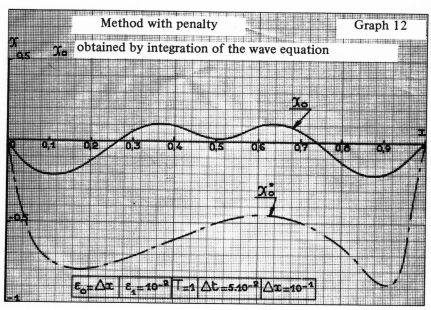

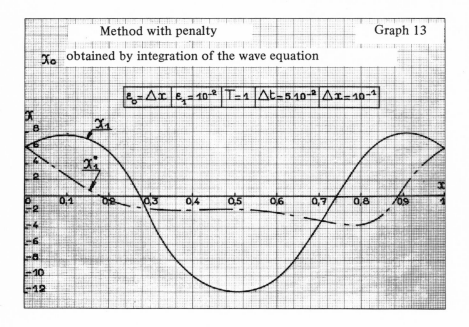

Method with penalty — Graph 13

$\mathcal{X}_0$ obtained by integration of the wave equation

$\varepsilon_0 = \Delta x \mid \varepsilon_1 = 10^{-2} \mid T = 1 \mid \Delta t = 5.10^{-2} \mid \Delta x = 10^{-1}$

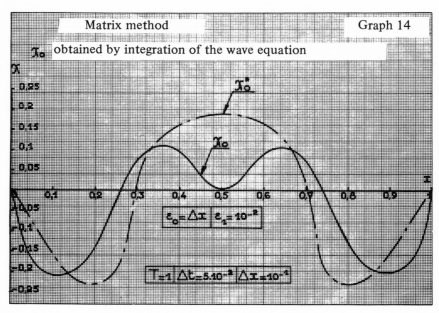

Matrix method — Graph 14

$\mathcal{X}_0$ obtained by integration of the wave equation

$\varepsilon_0 = \Delta x \mid \varepsilon_1 = 10^{-2}$

$T = 1 \mid \Delta t = 5.10^{-2} \mid \Delta x = 10^{-1}$

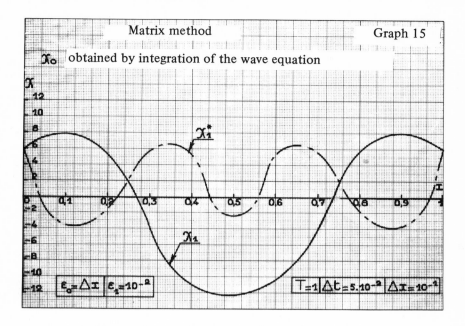

Chapter 4

QUASI-REVERSIBILITY AND ANALYTIC CONTINUATION OF SOLUTIONS OF ELLIPTIC EQUATIONS

DETAILED OUTLINE

1. FORMULATION OF THE PROBLEM

1.1 Orientation

Up until now we have applied the QR method to problems of *evolution.* In this chapter, we are going to show how the same *type* of method may be applied to *stationary* problems.

1.2 The Problem

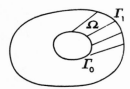

Figure 1

Let Ω be an open domain in $\mathbf{R}^n$, with boundary Γ_0 or Γ_1 (cf. Fig. 1);[1] more precisely: Γ_j is a variety continuously differentiable once of dimension $(n-1)$, and Ω is on one side of Γ_j, $j = 0, 1$.

Let A be an *elliptic* differential operator of the *second order* in Ω [2]:

$$Au = -\sum_{i,j=1}^{n} \frac{\partial}{\partial x_i}\left(a_{ij}(x)\frac{\partial u}{\partial x_j}\right) + a_0 u, \tag{1.1}$$

where

$$a_{ij} \in C^3(\overline{\Omega}), a_0 \in C^0(\overline{\Omega}) \quad \text{(real values for } simplicity) \tag{1.2}$$

$$\sum_{i,j=1}^{n} a_{ij}(x)\,\xi_i\,\xi_j \geqslant \alpha_1(\xi_1^2 + \cdots + \xi_n^2), \quad \alpha_1 > 0, \tag{1.3}$$

$$a_0(x) \geqslant \alpha_0 > 0. \tag{1.4}$$

Let u be a function in Ω which is "A-harmonic," i.e., a solution of:

$$Au = 0 ; \tag{1.5}$$

[1] Later on we will give some variants of this solution.
[2] We will see in Section 4 how this extends to higher orders.

let us set[1]:

$$g_0 = u|_{\Gamma_0} \tag{1.6}$$

$$g_1 = \frac{\partial u}{\partial v_A}\bigg|_{\Gamma_0} \quad [2]. \tag{1.7}$$

The problem is the following: *"calculate" u, given g_0 and g_1.*

Remark 1.1. If g_0 and g_1 are two bounded functions *whatsoever*, then the *Cauchy problem*:

$$Au = 0, \quad u|_{\Gamma_0} = g_0, \quad \frac{\partial u}{\partial v_A}\bigg|_{\Gamma_0} = g_1 \tag{1.8}$$

is *generally badly posed*.

If $\{ g_0, g_1 \}$ is a compatible pair, i.e., for which u *exists*, then the problem no longer in general has a solution for $\{ g_0 + \varepsilon_0, g_1 + \varepsilon_1 \}$, where ε_0 and ε_1 are arbitrarily small functions; consequently, even if $\{ g_0, g_1 \}$ is a compatible pair, the direct treatment of (1.8) will be numerically *unstable* (i.e., impossible). ∎

Remark 1.2. The practical situation is the following: g_0 and g_1 are given "experimentally"; let g_0^*, g_1^* be the values given for g_0 and g_1 [3]: we *know in addition* that the pair $\{ g_0, g_1 \}$ (to which $\{ g_0^*, g_1^* \}$ is an *approximation*) is *compatible* with A; the problem is to calculate u (approximately), for instance on Γ_1. ∎

Remark 1.3. The function u is *defined in a unique fashion* by (1.5), (1.6), (1.7); in effect, if u_1 and u_2 are two actual solutions of the problem, the difference

$$w = u_1 - u_2$$

satisfies

$$\begin{cases} Aw = 0, \\ w|_{\Gamma_0} = 0, \quad \dfrac{\partial w}{\partial v_A}\bigg|_{\Gamma_0} = 0, \end{cases}$$

[1] The hypotheses on u are made precise below.
[2] $\dfrac{\partial}{\partial v_A}$ = conormal derivative, i.e.,

$$\frac{\partial u}{\partial v_A} = \sum_{i,j} a_{ij}(x) \frac{\partial u}{\partial x_j} \cos(v, x_i),$$

v = normal to Γ_0, exterior to Ω.
[3] Naturally in general on a discrete set.

and therefore *(uniqueness* of the Cauchy problem), $w = 0$. (Cf. a discussion of this matter in N. Landis [1].) ∎

Remark 1.4. Another type of problem, important in applications, is the following: let again $\{ g_0^*, g_1^* \}$ be given (experimentally), corresponding to a pair compatible with A; but not only is u not known but not even the domain Ω where u, a solution of $Au = 0$, is defined is known. In other words, in the case of Figure 1 Γ_1 *is not known.* We will give some ideas concerning this problem later on.

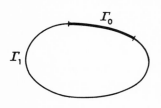

Figure 2

Remark 1.5. The same type of problem presents itself in the case of Figure 2.

We will give subsequently some variants to solve the corresponding problem. ∎

2. ASSOCIATED QR PROBLEMS

2.1 Orientation

The present section fixes the notation and gives the proposed QR methods. We indicate the principal results which are made precise and demonstrated in the following section. A third QR method will be indicated in Section 8. ∎

2.2 Notation

We set:

$$d(x, \Gamma_1) = \text{distance from } x \text{ to } \Gamma_1, \qquad x \in \Omega . \tag{2.1}$$

Let ε_0 be a positive number ("small"). We set

$$M_{\varepsilon_0}(x) = \begin{cases} 1 & \text{if} \quad d(x, \Gamma_1) \geqslant 2\,\varepsilon_0 \\ & \qquad\qquad\qquad\qquad x \in \Omega \\ 0 & \text{if} \quad d(x, \Gamma_1) < \varepsilon_0 \\ \\ \text{continuous in } \Omega, \text{ bounded by } 0 \text{ and } 1. \end{cases} \tag{2.2}$$

and

$$\rho_{\varepsilon_0}(x) = \begin{cases} 1 & \text{if} \quad d(x, \Gamma_1) \geqslant \varepsilon_0 \\ \dfrac{d(x, \Gamma_1)}{\varepsilon_0} & \text{if} \quad d(x, \Gamma_1) \leqslant \varepsilon_0 . \end{cases} \qquad (2.3)$$

2.3 The Associated QR Problem

2.3.1 First Method

Denote by Ω' or Ω_{ε_0} the open domain Ω from which we extract the band $\{ x \mid d(x, \Gamma_1) \leqslant \varepsilon_0 \}$. We will define u_ε as *the* solution of

$$A^*(M_{\varepsilon_0}^2 \, A u_\varepsilon) = 0 \qquad (2.4)$$

(where A^* denotes the adjoint of A), with the boundary conditions

$$u_\varepsilon|_{\Gamma_0} = g_0 , \qquad \frac{\partial u_\varepsilon}{\partial \nu_A}\bigg|_{\Gamma_0} = g_1 . \qquad (2.5)$$

We will show in Section 3 that (2.4), (2.5) defines u_ε in a unique fashion, the problem is *well posed*. ∎

Remark 2.1. We note that: (i) the operator appearing in (2.4) is of the fourth degree "inside" the domain $\Omega' = \Omega_{\varepsilon_0}$, (ii) the operator is degenerate on the boundary part $d(x, \Gamma_1) = \varepsilon_0$, in such a way *that there is no need of any other conditions apart from* (2.5). ∎

Remark 2.2. We could also work here with ρ_{ε_0} *in place of* M_{ε_0}; we use M_{ε_0} so that there is uniformity of notation with 2.3.2. We will show in Section 3 that

$$u_\varepsilon = u \quad \text{in} \quad \Omega' . \qquad (2.6)$$

We have therefore (in agreement with the general idea of Quasi-Reversibility) *replaced an improperly posed problem by a family of properly posed problems.* ∎

Remark 2.3. It is essential, in numerous problems, to introduce *penalty terms* (cf. Chapter 3, second part, Remark 4.1, and Chapter 5). This is why we are going to propose a second QR method, mathematically more complicated, but operationally quite close to the first. *We emphasize, both theoretically and numerically, that penalty terms are not essential in the problems of the present chapter, but they are important in the extension problems of chapter 5.* ■

2.3.2 SECOND METHOD

We now define u_ε as the solution of

$$\frac{1}{\varepsilon_1^2} A^*(M_{\varepsilon_0}^2 A u_\varepsilon) - \sum_{i,j=1}^n \frac{\partial}{\partial x_i} \left(\rho_{\varepsilon_0}^2 a_{ij} \frac{\partial u_\varepsilon}{\partial x_j} \right) + a_0 u_\varepsilon = 0 , \qquad (2.7)$$

where

$$\varepsilon_1 > 0, \text{ ``small'' (designed to tend to zero)}$$

$$u_\varepsilon = u_{\varepsilon_0, \varepsilon_1}, \quad \varepsilon = \{ \varepsilon_0, \varepsilon_1 \} ,$$

with the *boundary conditions of* (2.5). ■

We will show in Section 3 that (2.7), (2.5) define u_ε in a unique fashion. The essence of Remark 2.1 applies here.

We will show in Section 3 that

$$u_\varepsilon \to u \quad \text{when} \quad \varepsilon_0, \varepsilon_1 \to 0 . \ (^1) \qquad (2.8)$$

We have therefore again (following the general idea of quasi-reversibility) *approached an improperly posed problem by a family of well posed problems.* ■

Remark 2.4. Let us set:

$$A_{\varepsilon_0} \varphi = - \sum_{i,j=1}^n \frac{\partial}{\partial x_i} \left(\rho_{\varepsilon_0}^2 a_{ij} \frac{\partial \varphi}{\partial x_j} \right) + a_0 \varphi ;$$

A_{ε_0} is an ``approximation'' to $A$ (``degenerate'' on Γ) and u_ε satisfies

$$\frac{1}{\varepsilon_1^2} A^*(M_{\varepsilon_0} A u_\varepsilon) + A_{\varepsilon_0} u_\varepsilon = 0 .$$

$(^1)$ Making precise what sense this convergence has!

The limit function u should satisfy

$$Au = 0 .$$

The term $\frac{1}{\varepsilon_1^2} A^*(M_{\varepsilon_0} Au_\varepsilon)$ appears therefore as the "penalty term," discussed in Remark 2.3. ∎

Remark 2.5. (Valid for the two methods.) From the practical point of view, once the ε are chosen, we solve the systems (2.4), (2.5) or (2.7), (2.5) directly—by discretization. ∎

3. CONVERGENCE THEOREMS

We consider first the more complicated case (Method 2.3.2); the first method is studied in 3.5.

3.1 Function Spaces

We set:

$$W_{\varepsilon_0} = \left\{ v \mid v \in L_2(\Omega) \; ; \; \rho_{\varepsilon_0} D_j v \in L_2(\Omega) \; ; \; j = 1, ..., n \; ; \; D_j = \frac{\partial}{\partial x_j} \; ; \right.$$
$$\left. M_{\varepsilon_0} Av \in L_2(\Omega) \right\}. \tag{3.1}$$

More precisely, we suppose that Av is *locally* in $L_2(\Omega)$ and that $M_{\varepsilon_0} Av \in L_2(\Omega)$. If we set

$$|f| = \left(\int_\Omega f(x)^2 \, dx \right)^{1/2} \text{(norm in} \quad L_2(\Omega)) ,$$

we define the norm in W_{ε_0} by:

$$\| v \|_{W_{\varepsilon_0}} = \left(| v |^2 + \sum_{j=i}^{n} | \rho_{\varepsilon_0} D_j v |^2 + | M_{\varepsilon_0} Av |^2 \right)^{1/2} . \tag{3.2}$$

Lemma 3.1: *With the norm of (3.2), W_{ε_0} is a Hilbert space.*

Proof. Let v_m be a Cauchy sequence for the norm 3.2. Then $v_m \to v$ in $L_2(\Omega)$, $\rho_{\varepsilon_0} D_j v_m \to \theta_j$ in $L_2(\Omega)$ and

$$M_{\varepsilon_0} A v_m \to \chi \quad \text{in} \quad L_2(\Omega). \qquad (3.3)$$

But then $D_j v_m \to D_j v$ in $H^{-1}(\Omega),$[1] therefore:

$$\rho_{\varepsilon_0} D_j v \to \rho_{\varepsilon_0} D_j v \quad \text{in} \quad H^{-1}(\Omega)$$

and therefore

$$\theta_j = \rho_{\varepsilon_0} D_j v.$$

It remains to show that $\chi = M_{\varepsilon_0} A v$. Now let $\varphi \in D(\Omega)$; then $\langle \; , \; \rangle$ denote the duality between $D'(\Omega)$ (distributions on Ω) and $D(\Omega)$:

$$\langle M_{\varepsilon_0} A v_m, \varphi \rangle \to \langle \chi, \varphi \rangle$$

and

$$\langle M_{\varepsilon_0} A v_m, \varphi \rangle = \langle A v_m, M_{\varepsilon_0} \varphi \rangle = \sum_{i,j=1}^{n} \int_{\Omega} a_{ij} \frac{\partial v_m}{\partial x_j} \frac{\partial}{\partial x_i} (M_{\varepsilon_0} \varphi) \, dx.$$

Since $\rho_{\varepsilon_0} = 1$ on Ω_{ε_0}, we know that $\dfrac{\partial v_m}{\partial x_j} \to \dfrac{\partial v}{\partial x_j}$ in $L_2(\Omega_{\varepsilon_0})$ and since

$$\int_{\Omega} a_{ij} \frac{\partial v_m}{\partial x_j} \frac{\partial}{\partial x_i} (M_{\varepsilon_0} \varphi) \, dx = \int_{\Omega_{\varepsilon_0}} a_{ij} \frac{\partial v_m}{\partial x_j} \frac{\partial}{\partial x_i} (M_{\varepsilon_0} \varphi) \, dx,$$

we see that

$$\langle M_{\varepsilon_0} A v_m, \varphi \rangle \to \sum_{i,j=1}^{n} \int_{\Omega_{\varepsilon_0}} a_{ij} \frac{\partial v}{\partial x_j} \frac{\partial}{\partial x_i} (M_{\varepsilon_0} \varphi) \, dx = \langle M_{\varepsilon_0} A v, \varphi \rangle$$

therefore

$$\langle \chi, \varphi \rangle = \langle M_{\varepsilon_0} A v, \varphi \rangle \qquad \forall \varphi \in D(\Omega)$$

whence the result. ∎

[1] Recall that $H^{-1}(\Omega)$ = the dual of $H_0^1(\Omega)$, with $H_0^1(\Omega)$ the space of functions $u \in L_2(\Omega)$ with $\dfrac{\partial u}{\partial x_i} \in L_2(\Omega)$, $i = 1, \ldots, n$ and $u = 0$ on the boundary of Ω.

We next introduce

$$V_{\varepsilon_0} = \text{adherence of } D(\Omega) \text{ in } W_{\varepsilon_0}. \tag{3.4}$$

Lemma 3.2: *The space V_{ε_0} coincides with the subspace of W_{ε_0} of functions v such that*

$$v|_{\Gamma_0} = 0, \qquad \frac{\partial v}{\partial v_A}\bigg|_{\Gamma_0} = 0. \tag{3.5}$$

Proof. 1) Let $v \in V_{\varepsilon_0}$. Then *in* Ω_{ε_0}, we have $v = \lim \varphi_m$, $\varphi_m \equiv 0$ in a (variable) neighborhood of Γ_0, with the limit in the sense:

$$\varphi_m \to v \quad \text{in} \quad L_2(\Omega_{\varepsilon_0}), \qquad D_j \varphi_m \to D_j v \quad \text{in} \quad L_2(\Omega_{\varepsilon_0})$$

and

$$A\varphi_m \to {}^A v \quad \text{in} \quad L_2(\Omega_{\varepsilon_0}).$$

Therefore we have (3.5) (in effect, from the regularity of elliptic problems—cf., for example, Magenes, Stampacchia [1], or Lions, Magenes [1]:

$$D_i D_j \varphi_m \to D_i D_j v \quad \text{in} \quad L_2(\Omega_{\varepsilon_0}).$$

2) The interesting point is that belonging to V_{ε_0} *does not imply any conditions on* Γ_1. To that end, let us introduce a sequence θ_m of functions having the following properties:

$$\left.\begin{array}{l} \theta_m \in C^2(\overline{\Omega}), \\ \theta_m(x) = 0 \quad \text{if} \quad d(x, \Gamma_1) \leqslant \rho_m, \rho_m \to 0 \quad \text{if} \quad m \to \infty, \\ |D_j \theta_m(x)| \leqslant C/\rho_m, \quad j = 1, ..., n, \\ \theta_m(x) = 1 \quad \text{if} \quad d(x, \Gamma_1) \geqslant 2\rho_m. \end{array}\right\} \tag{3.6}$$

Such a sequence exists. Let then v be in W_{ε_0}; we are going to verify that

$$\theta_m v \to v \quad \text{in} \quad W_{\varepsilon_0}. \tag{3.7}$$

Since $\theta_m v = v$ on Ω_{ε_0} for m sufficiently large and since $\theta_m v \to v$ in $L_2(\Omega)$, the only thing to verify is that

$$\rho_{\varepsilon_0} D_j(\theta_m v) \to \rho_{\varepsilon_0} D_j v \quad \text{in} \quad L_2(\Omega),$$

i.e., that

$$\rho_{\varepsilon_0}(D_j \, \theta_m) \, v \to 0 \quad \text{in} \quad L_2(\Omega) \, .$$

Now, from (3.6), $\rho_{\varepsilon_0}(D_j \, \theta_m)$ is bounded and (3.8) is a result of Lebesque's theorem (since $D_j \, \theta_m = 0$ if $d(x, \Gamma_1) > 2 \rho_m$, therefore $\rho_{\varepsilon_0}(D_i \, \theta_m) \, v \to 0$ a.e.).

3) To finish the proof of the lemma, it suffices to introduce a sequence of functions ψ_m having properties analogous to those of θ_m in the neighborhood of Γ_1 *and* of Γ_0. Then, if v is in W_{ε_0} *and satisfies* (3.5), we have:

$$\psi_m \, v \to v \quad \text{in} \quad W_{\varepsilon_0} \, .$$

Therefore v is the limit in W_{ε_0} of functions of compact support in Ω. We are thus led to the case where v is of compact support in Ω; but then v is the limit (in W_{ε_0}) of elements of $D(\Omega)$ by regularization. ∎

3.2 Bilinear Forms

For u and v, given functions in Ω. we set (when the expressions have meaning):

$$a_{\varepsilon_0}(u, v) = \sum_{i,j=1}^{n} (a_{ij} \, \rho_{\varepsilon_0} \, D_j \, u, \, \rho_{\varepsilon_0} \, D_i \, v) + (a_0 \, u, v) \, , \qquad (3.9)$$

where

$$(f, g) = \int_\Omega f(x) \, g(x) \, dx \, ,$$

and

$$\pi_\varepsilon(u, v) = \frac{1}{\varepsilon_1^2} (M_{\varepsilon_0} \, Au, \, M_{\varepsilon_0} \, Av) + a_{\varepsilon_0}(u, v) \, . \, \blacksquare \qquad (3.10)$$

We now make precise the hypotheses on u, satisfying (1.5), (1.6), (1.7). We suppose that:

$$u \in H^1(\Omega) \, , \qquad Au = 0 \, ; \qquad (3.11)$$

then

$$u \mid_{\Gamma_0} = g_0 \in H^{1/2}(\Gamma_0) \quad (^1). \tag{3.12}$$

$$\left. \frac{\partial u}{\partial v_A} \right|_{\Gamma_0} = g_1 \in H^{-1/2}(\Gamma_0) \quad (^2). \tag{3.13}$$

Lemma 3.3: *For u given satisfying* (3.11), *there exists a unique* $w_\varepsilon \in V_\varepsilon$ *with*

$$\pi_\varepsilon(w_\varepsilon, v) = a_{\varepsilon_0}(u, v), \qquad \forall v \in V_\varepsilon. \tag{3.14}$$

Proof. It is sufficient to verify the *coerciveness* of π_ε:

$$\pi_\varepsilon(v, v) = \frac{1}{\varepsilon_1^2} \mid M_{\varepsilon_0} Av \mid^2 + a_{\varepsilon_0}(v, v) \geqslant \frac{1}{\varepsilon_1^2} \mid M_{\varepsilon_0} Av \mid^2 +$$
$$+ \alpha_0 \mid v \mid^2 + \alpha_1 \sum_{j=1}^{n} \mid \rho_{\varepsilon_0} D_j v \mid^2 . \blacksquare \tag{3.15}$$

A further consequence of (3.15) is that we have:

Lemma 3.4: *There exists a constant C such that*

$$\mid M_{\varepsilon_0} Aw_\varepsilon \mid \leqslant C\varepsilon_1, \tag{3.16}$$

$$\mid w_\varepsilon \mid + \sum_{j=1}^{n} \mid \rho_{\varepsilon_0} D_j w_\varepsilon \mid \leqslant C. \tag{3.17}$$

Proof. In effect, $\mid a_{\varepsilon_0}(u, v) \mid \leqslant C_1 \parallel v \parallel$, whence the result follows upon taking $v = w_\varepsilon$ in (3.14) and using (3.15). $\blacksquare$

3.3 First Consequence of Convergence

Proposition 3.1: *When* $\varepsilon = \{ \varepsilon_0, \varepsilon_1 \} \to 0$ *(in* $\mathbf{R}^2$), $w_\varepsilon \to 0$ *in the sense*

$$w_\varepsilon \to 0 \quad \text{in } L_2(\Omega) \text{ (strongly)}, \tag{3.18}$$

$$\rho_{\varepsilon_0} D_j w_\varepsilon \to 0 \quad \text{in } L_2(\Omega) \text{ (strongly) } \forall j. \tag{3.19}$$

(^{1}) Space of Sobolev of fractional order 1/2; cf. Lions–Magenes [1].
(^{2}) Dual space of $H^{1/2}(\Gamma_0)$.

Proof. In view of (3.16), (3.17), we can extract from w_ε a sequence $W_{\varepsilon'}, \varepsilon' = \{\varepsilon_0', \varepsilon_1'\}$ such that

$$w_{\varepsilon'} \to w \quad \text{in } L_2(\Omega) \text{ weakly,} \tag{3.20}$$

$$\rho_{\varepsilon_0} D_j w_{\varepsilon'} \to \theta_j \quad \text{in } L_2(\Omega) \text{ weakly,} \tag{3.21}$$

$$M_{\varepsilon_0'} A w_{\varepsilon'} \to 0 \quad \text{in } L_2(\Omega) \text{ strongly.} \tag{3.22}$$

We see, as in the proof of Lemma 3.1 that

$$\theta_j = D_j w. \tag{3.23}$$

Therefore

$$w \in H^1(\Omega)$$

and w is the limit in $H^1(\Omega_{\varepsilon_0})$ (ε_0 fixed) of functions which are zero on Γ_0, therefore

$$w \in H^1(\Omega), \qquad w|_{\Gamma_0} = 0. \tag{3.24}$$

Let us note next that (3.22) implies

$$A w_{\varepsilon'} \to A w = 0 \text{ dans } L_2(\Omega_{\varepsilon_0}), \quad (\varepsilon_0' < \varepsilon_0) \tag{3.25}$$

and then, after Lions-Magenes [1],

$$\left.\frac{\partial w_{\varepsilon'}}{\partial v_A}\right|_{\Gamma_0} \to \left.\frac{\partial w}{\partial v_A}\right|_{\Gamma_0} \quad \text{in} \quad H^{-1/2}(\Gamma_0) \, ;$$

since (Lemma 3.2), $\left.\dfrac{\partial w_{\varepsilon'}}{\partial v_A}\right|_{\Gamma_0} = 0$, we have therefore

$$\left.\frac{\partial w}{\partial v_A}\right|_{\Gamma_0} = 0. \tag{3.26}$$

But $Aw = 0$ and that combined with (3.24), (3.26) implies that $w = 0$. But then (3.20), (3.21) imply (3.18), (3.19) with *weak* convergence (without the necessity of extracting a subsequence).

   2) It remains only to show the *strong* convergence. We deduce from (3.14) that

$$\pi_\varepsilon(w_\varepsilon, w_\varepsilon) = a_{\varepsilon_0}(u, w_\varepsilon) \to a(u, w) = 0 \tag{3.27}$$

therefore using (3.15), we have the result. ∎

Remark 3.1. We deduce from (3.27) that:

$$| M_{\varepsilon_0} A w_\varepsilon | = o(\varepsilon_1), \quad \text{where} \quad \frac{o(\lambda)}{\lambda} \to 0 \quad \text{if} \quad \lambda \to 0 . \blacksquare \qquad (3.28)$$

3.4 Convergence of the Second QR Method

We are now in a position to show

Theorem 3.1: *Let A be given with (1.1), . . . , (1.4). Let $\{ g_0, g_1 \}$ be given functions , $g_0 \in H^{1/2}(\Gamma_0), g_1 \in H^-$ (Γ_0), compatible with A (i.e., there exists a unique u, in $H(\Omega)$, satisfying $Au = 0$, $u \vert_{\Gamma_0} = g_0$, $\dfrac{\partial u}{\partial v_A}\Big\vert_{\Gamma_0} = g_1$). Then:*

i) *there exists a unique $u_\varepsilon \in w_\varepsilon$ satisfying*

$$u_\varepsilon \vert_{\Gamma_0} = g_0 , \quad \frac{\partial u_\varepsilon}{\partial v_A}\Big\vert_{\Gamma_0} = g_1 , \qquad (3.29)$$

and

$$\frac{1}{\varepsilon_1^2} A^*(M_{\varepsilon_0}^2 A u_\varepsilon) - \sum \frac{\partial}{\partial x_i}\left(\rho_{\varepsilon_0}^2 a_{ij} \frac{\partial u_\varepsilon}{\partial x_j} \right) + a_0 u_\varepsilon = 0 \qquad (3.30)$$

[*derivatives are in the sense of distributions; M_{ε_0}, ρ_{ε_0} as defined in (2.2), (2.3)*];

ii) *when $\varepsilon = \{ \varepsilon_0, \varepsilon_1 \} \to 0$, we have:*

$$u_\varepsilon \to u \quad \text{in} \quad L_2(\Omega) , \quad \rho_{\varepsilon_0} D_j u_\varepsilon \to D_j u$$

in $L_2(\Omega)$, $\forall j$;

iii) *we have, when $\varepsilon \to 0$:*

$$\| M_{\varepsilon_0} A u_\varepsilon \|_{L_2(\Omega)} = o(\varepsilon_1) . \qquad (3.31)$$

Remark 3.2. This refines, as stated, the result of (2.8) and justifies, in this case, the second QR method. $\blacksquare$

Proof. If w_ε is defined as in Lemma 3.3, we put (for the moment; we will see that $\varphi_\varepsilon = u_\varepsilon$):

$$\varphi_\varepsilon = u - w_\varepsilon . \qquad (3.32)$$

Then $\varphi_\varepsilon \in H^1(\Omega_{\varepsilon_0})$ and $A\varphi_\varepsilon \in L_2(\Omega_{\varepsilon_0})$, therefore $\varphi_\varepsilon|_{\Gamma_0}$ and $\dfrac{\partial \varphi_\varepsilon}{\partial v_A}\bigg|_{\Gamma_0}$ are defined and (3.32) and Lemma 3.2 show that:

$$\varphi_\varepsilon|_{\Gamma_0} = u|_{\Gamma_0} = g_0 , \quad \frac{\partial \varphi_\varepsilon}{\partial v_A}\bigg|_{\Gamma_0} = \frac{\partial u}{\partial v_A}\bigg|_{\Gamma_0} = g_1 . \tag{3.33}$$

In addition, if $v \in D(\Omega)$:

$$\pi_\varepsilon(\varphi_\varepsilon, v) = \pi_\varepsilon(u, v) - \pi_\varepsilon(w_\varepsilon, v) ;$$

but since $Au = 0$, $\pi_\varepsilon(u, v) = a_{\varepsilon_0}(u, v)$; and therefore from (3.14):

$$\pi_\varepsilon(\varphi_\varepsilon, v) = 0 \qquad \forall v \in D(\Omega) ,$$

which is *equivalent* to

$$\frac{1}{\varepsilon_1^2} A^*(M_{\varepsilon_0}^2 A\varphi_\varepsilon) - \sum \frac{\partial}{\partial x_i}\left(\rho_{\varepsilon_0}^2 a_{ij} \frac{\partial \varphi_\varepsilon}{\partial x_j} \right) + a_0 \varphi_\varepsilon = 0 . \tag{3.34 bis}$$

Reciprocally, if $\varphi_\varepsilon \in W_\varepsilon$ and satisfies (3.33), (3.34), then $u - \varphi_\varepsilon = w_\varepsilon$ satisfies (3.14). This shows that $\varphi_\varepsilon = u_\varepsilon$ and is unique. Therefore (i) is proven.

   2) The result (ii) is then an immediate consequence of Proposition 3. ∎

3.5 Study of the First QR Method

   We denote by W the space

$$W = \left\{ v \mid v \in H^2(\Omega_{\varepsilon'}) \;\; \forall \varepsilon' > \varepsilon_0, v|_{\Gamma_0} = \frac{\partial v}{\partial v_A}\bigg|_{\Gamma_0} = 0, \\ M_{\varepsilon_0} Av \in L_2(\Omega_{\varepsilon_0}) \right\} ; \tag{3.35}$$

let

$$\| v \|_W = | M_{\varepsilon_0} Av | \quad (\text{norm in} \quad L_2(\Omega_{\varepsilon_0})) . \tag{3.36}$$

The space W with the norm (3.36) is a Hilbert space; in effect, if ε', ε_0, $M_{\varepsilon_0} \geqslant k > 0$ on $\Omega_{\varepsilon'}$, therefore:

$$| M_{\varepsilon_0} Av | \geqslant k | Av |_{L_2(\Omega_{\varepsilon'})} ;$$

since $v|_{\Gamma_0} = 0$ (and the boundary Γ_0 is supposed regular) we have (cf. L. Nirenberg [1])

$$|Av|_{L_2(\Omega_{\varepsilon'})} \geqslant C \|v\|_{H^2(\Omega_{\varepsilon''})}, c > 0, \varepsilon'' \text{ arbitrarily fixed } > \varepsilon',$$

whence the result follows. ∎

If we introduce

$$u_\varepsilon - u = z_\varepsilon,$$

we see that (2.4), (2.5) is equivalent to

$$A^*(M_{\varepsilon_0}^2 Az_\varepsilon) = 0, \tag{3.38}$$

$$z_\varepsilon|_{\Gamma_0} = 0 \qquad \frac{\partial z_\varepsilon}{\partial \nu_A}\bigg|_{\Gamma_0} = 0. \tag{3.39}$$

But if v_1 and $v_2 \in W$, we have:

$$(A^* M_{\varepsilon_0}^2 Av_1, v_2) = (M_{\varepsilon_0} Av_1, M_{\varepsilon_0} Av_2)$$

and therefore (3.38) and $z_\varepsilon|_{\Gamma_0} = 0$ imply that

$$|M_{\varepsilon_0} Az_\varepsilon|^2 = 0;$$

therefore $Az_\varepsilon = 0$ in Ω_{ε_0} which, combined with (3.39) implies that $z_\varepsilon = 0$, whence

Theorem 3.2: *In the QR method of 2.3.1, we have:*

$$u_\varepsilon = u \quad \text{in} \quad \Omega_{\varepsilon_0}.$$

3.6 Second Term Nonzero

Let us suppose that u satisfies in Ω [compare (1.5), (1.6), (1.7)]:

$$Au = f, \quad f \text{ not now} \equiv 0, \tag{3.40}$$

$$u|_{\Gamma_0} = g_0, \quad \frac{\partial u}{\partial \nu_A}\bigg|_{\Gamma_0} = g_1. \tag{3.41}$$

The problem of 1.2 is now: *"to calculate" u from the given func-tions f, g_0 and g_1.* (Remarks 1.1 to 1.5 remain valid.) The first method of 2 adapts in the following fashion: we define u_ε as the solution of the properly posed problem:

$$A^*(M_{\varepsilon_0}^2 A u_\varepsilon) = A^* M_{\varepsilon_0}^2 f, \tag{3.42}$$

$$u_\varepsilon|_{\Gamma_0} = g_0, \qquad \frac{\partial u_\varepsilon}{\partial v_A}\bigg|_{\Gamma_0} = g_1 \cdot \blacksquare \tag{3.41}$$

Remark 3.3. We do not know if the second method of Section 2 can be adapted to the situation "$f \not\equiv 0$." $\blacksquare$

4. SOME VARIANTS

4.1 Analogous Problem in a Different Geometry

Let us now consider the analogous problem in the case where the geometry is of the type indicated in Figure 2, Remark 1.5.

We again introduce the functions ρ_{ε_0} and M_{ε_0}; ρ_{ε_0} is chosen *for example*, with the same analytic definition as in (2.3); the graph of ρ_{ε_0} evidently changes its aspect but the preceding considerations remain valid. The same applies to M_{ε_0}. $\blacksquare$

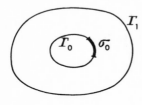

Figure 3

Remark 4.1. When the configuration is that of Figure 1, but g_0 and g_1 become "too large" on a portion σ_0 of Γ_0, *it is of interest not to take account of the data on σ_0* by "suppressing" σ_0 with the aid of weight functions ρ_{ε_0}, M_{ε_0} which are *zero or "small"* on σ_0 *(and in the neighborhood of σ_0)*; cf. the numerical application to Sections 6 and 7. $\blacksquare$

4.2 Problem of Higher Order

The fact that the operator A is of the second order plays no essential role.[1] Let us constrain ourselves to an *example*, in

---

[1] Provided that the property of uniqueness for the Cauchy problem holds (this need not be the case for operators of order superior to 2).

order not to make our exposition too dense! Let u be a *biharmonic* function in Ω:

$$\Delta^2 u = 0 ; \qquad (4.1)$$

set

$$g_0 = u|_{\Gamma_0} ,$$

$$g_1 = \frac{\partial u}{\partial v}\bigg|_{\Gamma_0} ,$$

$$g_2 = \Delta u|_{\Gamma_0} , \qquad (4.2)$$

$$g_3 = \frac{\partial}{\partial v} \Delta u|_{\Gamma_0} .$$

The problem is to calculate (approximately) u starting with the quadruple $\{ g_0, g_1, g_2, g_3 \}$ (compatible in a sense analogous to that of Remark 1.1).

Let us give the principle of the second method (without all the technical details).

We introduce M_{ε_0}, ρ_{ε_0} as in (2.2), (2.3). We set

$$\Delta_{\varepsilon_0} v = \sum_{i=1}^{n} \frac{\partial}{\partial x_i} \left(\rho_{\varepsilon_0}^2 \frac{\partial v}{\partial x_i} \right). \qquad (4.3)$$

We show then that there exists a unique u_ε satisfying[1]

$$\frac{1}{\varepsilon_1^2} \Delta^2 (M_{\varepsilon_0}^2 \Delta^2 u_\varepsilon) + \Delta_{\varepsilon_0}^2 u_\varepsilon = 0 , \qquad (4.4)$$

$$u_\varepsilon|_{\Gamma_0} = g_0 , \quad \frac{\partial u_\varepsilon}{\partial v}\bigg|_{\Gamma_0} = g_1 , \quad \Delta u_\varepsilon|_{\Gamma_0} = g_2 , \quad \frac{\partial \Delta}{\partial v} u_\varepsilon|_{\Gamma_0} = g_3 \quad [2] \qquad (4.5)$$

and

$$u_\varepsilon \to u \quad \text{when} \quad \varepsilon_0, \varepsilon_1 \to 0 \qquad (4.6)$$

(in the topology corresponding to the conditions $\Rightarrow$ of [1]). ∎

---

[1] $u_\varepsilon \in L_2(\Omega)$, $\quad \rho_{\varepsilon_0} D_j u_\varepsilon \in L_2(\Omega) \quad \forall j$, $\quad \Delta_{\varepsilon_0} u_\varepsilon \in L_2(\Omega)$, $\quad M_{\varepsilon_0} \Delta^2 u_\varepsilon \in L_2(\Omega)$.
[2] $g_0 \in H^{3/2}(\Gamma_0)$, $\quad g_1 \in H^{1/2}(\Gamma_0)$, $\quad g_2 \in H^{-1/2}(\Gamma_0)$, $\quad g_3 \in H^{-3/2}(\Gamma_0)$.

Remark 4.2. The method may be adapted without difficulty to the case of *differential systems* (for example, to the *system of elasticity*). ■

Remark 4.3. We have adapted below the QR method of 2.3.2. To use that of 2.3.1, it suffices to make $\varepsilon_1 = 0$ in (4.4) after multiplication of ε_1^2. ■

4.3 Operators with Singularities

The preceding methods may be adapted to the case of elliptic operators having degeneracies or singularities. We use for this, for example, the methods of H. Morel [1] and the preceding techniques. For example, we can study the problem

$$- \Delta u + \frac{k}{y} \frac{\partial u}{\partial y} = 0 \,,$$

$$\Delta = \frac{\partial^2}{\partial x^2} + \frac{\partial^2}{\partial y^2} \,,$$

$$k \in \mathbf{R} \,,$$

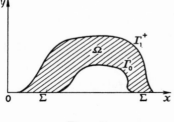

Figure 4

in a domain Ω of the type shown in Figure 4, with the Cauchy data

$$u = g_0 \quad \text{on} \quad \Gamma_0^+ \,,$$

$$\frac{\partial u}{\partial n} = g_1 \quad \text{on} \quad \Gamma_0^+ \,,$$

and

$$\frac{\partial u}{\partial y} = 0 \quad \text{on} \quad \Sigma \,.$$

(For the numerical integration of equations with singularity of the type $\frac{k}{y} \frac{\partial u}{\partial y}$, consult P. Jamet–S. Parter [1], Brelot–Collin [1]).

5. APPLICATION TO HARMONIC PROLONGATION IN AN ANNULUS

5.1 Equations

We propose to solve [in polar coordinates (r, θ)] the following problem (cf. Figure 5):

$$\Delta u = 0 \quad \text{in} \quad \Omega \, ,$$

$$u|_{\Gamma_0} = g_0(\theta) \, ,$$

$$\left. \frac{\partial u}{\partial r} \right|_{\Gamma_0} = g_1(\theta) \, ;$$

(5.1)

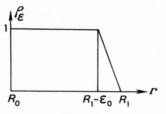

Figure 5

the domain of definition of u, Ω, is an annulus $(R_0 \leqslant r \leqslant R_1)$. ∎

Remark 5.1. This is the problem of 1.2 with $A = -\Delta$ (we can suppose that $\alpha_0 = 0$ in the general theory). ∎

We associate to problem (5.1) (in agreement with 2, 2.3.2) the neighboring problem:

$$\operatorname{div}\left(\rho_{\varepsilon_0}^2(r) \operatorname{grad} u_\varepsilon\right) = \frac{1}{\varepsilon_1^2} \Delta(M_{\varepsilon_0}(r) \Delta u_\varepsilon) \quad \text{in} \quad \Omega \, ,$$

$$u_\varepsilon(R_0, \theta) = g_0(\theta) \, ,$$

(5.2)

$$\left. \frac{\partial u_\varepsilon}{\partial r} \right|_{r = R_0} = g_1(\theta) \, ,$$

where $\rho_{\varepsilon_0}(r)$ and $M_{\varepsilon_0}(r)$ denote respectively the functions: (in agreement with the general theory; cf. Figures 6 and 7).

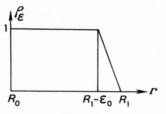

Figure 6

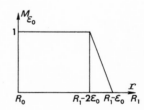

Figure 7

$$\rho_{\varepsilon_0}(r) = \begin{cases} 1 & \text{if} \quad r \in [R_0, R_1 - \varepsilon_0] \\[2mm] \dfrac{R_1 - r}{\varepsilon_0} & \text{if} \quad r \in [R_1 - \varepsilon_0, R_1] \end{cases}$$

$$M_{\varepsilon_0}(r) = \begin{cases} 1 & \text{if} \quad r \in [R_0, R_1 - 2\varepsilon_0] \\[2mm] \dfrac{R_1 - \varepsilon_0 - r}{\varepsilon_0} & \text{if} \quad r \in [R_1 - 2\varepsilon_0, R_1 - \varepsilon_0] \\[2mm] 0 & \text{if} \quad r \in [R_1 - \varepsilon_0, R_1]. \end{cases}$$

The first equation (5.2) is written (suppressing the indices in ε):

$$\frac{\partial}{\partial r}\left[\rho^2\frac{\partial u}{\partial r}\right] + \frac{\rho^2}{r}\frac{\partial u}{\partial r} + \frac{\rho^2}{r^2}\frac{\partial^2 u}{\partial\theta^2} = \frac{\cdot 1}{\varepsilon_1^2}\left[\frac{\partial^2}{\partial r^2}M\left(\frac{\partial^2 u}{\partial r^2} + \frac{1}{r}\frac{\partial u}{\partial r} + \frac{1}{r^2}\frac{\partial^2 u}{\partial\theta^2}\right) + \right.$$
$$\left. + \frac{1}{r}\frac{\partial}{\partial r}M\left(\frac{\partial^2 u}{\partial r^2} + \frac{1}{r}\frac{\partial u}{\partial r} + \frac{1}{r^2}\frac{\partial^2 u}{\partial\theta^2}\right) + \frac{M}{r^2}\frac{\partial^2}{\partial\theta^2}\left(\frac{\partial^2 u}{\partial r^2} + \frac{1}{r}\frac{\partial u}{\partial r} + \frac{1}{r^2}\frac{\partial^2 u}{\partial\theta^2}\right)\right] . \tag{5.3}$$

5.2 Discretization of (5.2)

We decompose the interval $[R_0, R_1]$ into I equal intervals and $[0, 2\pi]$ into J equal intervals. Write:

$$u_{ij} = u(R_0 + \mathrm{i}\,\Delta r, j\,\Delta\theta) ,$$

with

$$\Delta r = \frac{R_1 - R_0}{I} , \quad \Delta\theta = \frac{2\pi}{J} .$$

Remark 5.2. The index j is defined modulo J which corresponds to the transformation $\theta \to \theta + 2J\pi$. ∎

We then discretize (5.3) by means of standard[1] finite differences using the differential equations:

$$\left.\frac{\partial f}{\partial r}\right|_{i,j} \simeq \frac{f_{i+1,j} - f_{i-1,j}}{2\,\Delta r}$$

$$\left.\frac{\partial f}{\partial\theta}\right|_{i,j} \simeq \frac{f_{i,j+1} - f_{i,j-1}}{2\,\Delta\theta}$$

$$\left.\frac{\partial^2 f}{\partial r^2}\right|_{i,j} \simeq \frac{f_{i+1,j} - 2f_{i,j} + f_{i-1,j}}{\Delta r^2}$$

$$\left.\frac{\partial^2 f}{\partial\theta^2}\right|_{i,j} \simeq \frac{f_{i,j+1} - 2f_{i,j} + f_{i,j-1}}{\Delta\theta^2}$$

formulas which naturally permit us to calculate step-by-step all the higher derivatives.

The problem (5.2) may then be put in the form:

$$Au = f \tag{5.4}$$

where A is a 13-diagonal matrix. In effect, (5.3) discretized can be written:

---

[1] We have devoted no particular effort to this stage. We apply usual methods.

$$\frac{\varepsilon_1^2}{4\,\Delta r^2}\left[\rho_{i+1}(u_{i+2,j}-u_{i,j})-\rho_{i-1}(u_{i,j}-u_{i-1,j})\right]+$$

$$+\frac{\varepsilon_1^2\,\rho_i}{r_i^2\,\Delta\theta^2}(u_{i,j+1}+u_{i,j-1}-2\,u_{i,j})+\frac{\varepsilon_1^2\,\rho_i}{2\,r_i\,\Delta r}(u_{i+1,j}-u_{i-1,j})=$$

$$=\frac{1}{\Delta r^4}\left[M_{i+1}(u_{i+1,j}+u_{i,j}-2u_{i+1,j})+M_{i-1}(u_{i,j}+u_{i-2,j}-2u_{i-1,j})-\right.$$

$$\left.-2\,M_i(u_{i+1,j}+u_{i-1,j}-2\,u_{i,j})\right]+\frac{1}{2\,\Delta r^3}\left[\frac{M_{i+1}}{r_{i+1}}(u_{i+2,j}-u_{i,j})\right.$$

$$+\frac{M_{i-1}}{r_{i-1}}(u_{i,j}-u_{i-2,j})-\frac{2\,M_i}{r_i}(u_{i+1,j}-u_{i-1,j})\bigg]$$

$$+\frac{1}{\Delta r^2\,\Delta\theta^2}\left[\frac{M_{i+1}}{r_{i+1}^2}(u_{i+1,j+1}+u_{i+1,j-1}-2\,u_{i+1,j})+\right.$$

$$+\frac{M_{i-1}}{r_{i-1}^2}(u_{i-1,j+1}+u_{i-1,j-1}-2\,u_{i-1,j})$$

$$\left.-\frac{2\,M_i}{r_i^2}(u_{i,j+1}+u_{i,j-1}-2\,u_{i,j})\right]+\frac{1}{2\,r_i\,\Delta r^3}\left[M_{i+1}(u_{i+2,j}\right.$$ (5.5)

$$+u_{i,j}-2\,u_{i+1,j})-M_{i-1}(u_{i,j}+u_{i-2,j}-2\,u_{i-1,j})\big]$$

$$+\frac{1}{4\,r_i\,\Delta r^2}\left[\frac{M_{i+1}}{r_{i+1}}(u_{i+2,j}-u_{i,j})-\frac{M_{i-1}}{r_{i-1}}(u_{i,j}-u_{i-2,j})\right]$$

$$+\frac{1}{2\,r_i\,\Delta r\,\Delta\theta^2}\left[\frac{M_{i+1}}{r_{i+1}^2}(u_{i+1,j+1}+u_{i+1,j-1}-2\,u_{i+1,j})-\right.$$

$$\left.-\frac{M_{i-1}}{r_{i-1}^2}(u_{i-1,j+1}+u_{i-1,j-1}-2\,u_{i-1,j})\right]$$

$$+\frac{M_i}{r_i^2\,\Delta r^2\,\Delta\theta^2}\left[u_{i+1,j+1}+u_{i-1,j+1}-2\,u_{i,j}+u_{i+1,j-1}\right.$$

$$+u_{i-1,j-1}-2\,u_{i,j-1}-2(u_{i+1,j}+u_{i-1,j}-2\,u_{i,j})\big]$$

$$+\frac{M_i}{2\,r_i^3\,\Delta r\,\Delta\theta^2}\left[u_{i+1,j+1}+u_{i+1,j-1}-2\,u_{i+1,j}-(u_{i-1,j+1}\right.$$

$$\left.+u_{i-1,j-1}-2\,u_{i-1,j})\right]+\frac{M_i}{r_i^4\,\Delta\theta^4}\left[u_{i,j+2}-4\,u_{i,j+1}+6\,u_{i,j}-\right.$$

$$-4\,u_{i,j-1}+u_{i,j-2}\big].$$

This discretized equation is valid only for i greater than 2. The index j is calculated modulo J, i.e., the terms in $u_{i,j+2}$ and $u_{i,j+1}$

will be respectively replaced by: $u_{i,j+NJ-2}$ and $u_{i,j+NJ-1}$ when they occur (for $j - 2 \leqslant 0$ and $j - 1 \leqslant 0$). We apply the same procedure to the terms in $u_{i,j+1}$ and $u_{i,j+2}$ when necessary.

We then add to (5.5) the discretized boundary conditions for the system (5.2) which give:

$$u_{0,j} = g_0(j \, \Delta\theta) \tag{5.6}$$

and

$$\frac{u_{-1,j} - u_{1,j}}{2 \, \Delta r} = g_1(j \, \Delta\theta) \tag{5.7}$$

which permits the elimination of the terms in $u_{0,j}$ and $u_{-1,j}$ which appear in (5.5). ∎

Remark 5.3. If we use the first QR method, A (cf. 5.5) remains 13-diagonal. We can assert that the numerical effort is essentially the same. ∎

Remark 5.4. In the neighborhood of $r = R_1$, the shape of the functions ρ and M assures the determination of the linear system associated with the discretized system. In particular, the terms in $u_{I,j} = u(R_1, j \, \Delta\theta)$, $u_{I+1,j}$, $u_{I+2,j}$ which appear in the discretized form (5.5) have zero coefficients. ∎

The system (5.4) is a system of $(I - 1)J$ equations in $(I - 1)J$ unknowns.

5.3 Numerical Examples

We took:

$$R_0 = 1 \qquad R_1 = 1.5 \, .$$

The method was tested by comparison with known analytic solutions. We made the following trials:

EXAMPLE 1. $u(r, t) = \text{Log } r + 1$.

We choose:

$$\Delta r = 0.05 \qquad \Delta\theta = \pi/5$$

$$\varepsilon_0 = 2 \, \Delta r \qquad \varepsilon_1 = 10^{-3} \, .$$

Graph 1 shows the perfect agreement between the calculated function u_ε and the theoretical solution for $r \in [1; 1.5 - 2 \, \Delta r]$ and this is the case despite a quite gross discretization, only ten points in the interval [1; 1.5].

Remark 5.5. At the end of the verification of Remark 2.3, we carried out the calculations for $\varepsilon_1 \to 0$ and $\varepsilon_1 = 0$. We pointed out in that case that the error $|u_\varepsilon - u|$ for fixed ε_0 ($2\,\Delta r$) remained of the order of 2.10^{-4} (this corresponds to truncation and rounding errors). This verification has also been made for the examples which follow. ■

EXAMPLE 2. $u(r, \theta) = \dfrac{\cos 3\,\theta}{r^3}$.

The numerical experiments which were carried out show that the QR solution and the theoretical solution are very close in $R_0 < r \leqslant R_1 - \varepsilon_0$. In the domain where the function ρ_ε decreases from 1 to 0 ($R_1 - \varepsilon_0 \leqslant r \leqslant R_1$), we observed, as expected, an oscillation in the solution.

Graphs 2 and 3 show the exact solution and its QR approximation for:

$$\Delta r = 0.025 \qquad \Delta \theta = \pi/10$$

$$\varepsilon_0 = 2\,\Delta r \qquad \varepsilon_1 = 10^{-3}\,,$$

along the radius $\theta = 0$ and the circle $r = 1.25$ (in fact for $0 \leqslant \theta \leqslant 2\pi/3$, since the function is periodic.

EXAMPLE 3. *Study of a logarithmic singularity.*

We now study the function:

$$u(r, \theta) = \log\left(r^2 + r_0^2 - 2\,rr_0 \cos\theta\right),$$

which is harmonic in any domain not containing $(r_0, 0)$.

Variation of the error with r_0.

We made various numerical trials for different acceptable values of r_0 ($r_0 < 1$ and $r_0 > 1.5$).

Graph 4 shows the variation of the relative error at ($r = 1.25$, $\theta = 0$) for different values of r_0. The line $r_0 = 1$ is asymptatic to this curve, while the line $r_0 = 1.5$ is not. This is due to the shape of the function $\rho_{\varepsilon_0}(r)$ in the neighborhood of $r = R_1$.

Remark 5.6. Thus the approximation fails as $r_0 \to 1$ (i.e., when the singularity approaches the boundary). An adaptation of the QR method permits us, as we will see in Section 6, to overcome this predictable difficulty (cf. Remark 4.1). ∎

The numerical trial for $r_0 = 0$ has already been carried out (it is, apart from a constant, the function $u = \text{Log } r + 1$ of Example 1).

Graphs 5, 6, 7 show the theoretical solution and the solution obtained by the QR method for $r_0 = 0.5$; 1.5; 1.75 and $\theta = 0$, i.e.. on the axis of the point of discontinuity.

We see that, for $r_0 = 0.5$ and $r_0 = 1.75$, the QR approximations are very close to the exact solution in the interval $[R_0, R_1 - \varepsilon_0]$. However, the QR solution worsens proportionately as r increases.

We have studied in more detail the case $r_0 = 0.95$ (Graphs 8, 9, 10). In this case, if we are on the axis of discontinuity $(\theta = 0)$, the error committed by comparison to the theoretical solution is clearly most important; *this had been predicted and made the object of substantial improvement; see* Section 6. ∎

Graph 8 shows two numerical trials carried out, one with a net of (16×16) points; the other with (20×20) points. The gain in precision of the second solution (20×20) over the first solution is minimal for $1 \leqslant r \leqslant 1.3$. For $1.3 \leqslant r \leqslant 1.5$ the second solution worsens even more rapidly than the first.

On Graph 9 we draw the representative curve of the exact solution for $\theta = \pi/2$ and $\theta = \pi$, i.e., on rays not on the axis of discontinuity. We see that the QR solution cannot be distinguished from the theoretical solution for $r \in [R_0, R_1 - \varepsilon_0]$. For $R_1 - \varepsilon_0 \leqslant r \leqslant R_1$, we always observe the same oscillation of the solution.

Finally, we sought to make precise in what sector of the circle the influence of the discontinuity at $(r_0, 0)$ made itself felt (this was of interest in view of the adaptation indicated in Remark 5.6).

Graph 10 shows that, for $r = 1.25$, the QR solution is near the real solution as soon as θ exceed $\pi/5$.

The influence of the discontinuity is felt therefore in the interval $] - \pi/5, + \pi/5[$, about $4 \Delta\theta$.

6. ADAPTATION OF THE QR METHOD IN THE CASE OF A SINGULARITY

All that follows holds for the two methods of 2.3.1 and 2.3.2; we make this explicit below for the second method.[1] In order to im-

---

[1] Here, we have made the same checks as those specified in Remark 5.5.

prove the precision of the results in the case where the function $u(r, \theta)$ possesses a local singularity, we apply Remark 4.1 in the following manner.

6.1 Notation

We denote by (cf. Figure 8):

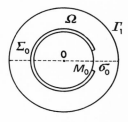

Ω the annulus with boundary Γ_0, Γ_1;

Γ_0 $\Sigma_0 \cup \sigma_0$ the interior boundary (with radius R_0) where σ_0 is a neighborhood on Γ_0 of the point $r = 1$, $\theta = 0$;

Γ_1 the exterior boundary of Ω (with radius R_1).

Figure 8

We suppose that a *logarithmic singularity* exists at M_0, ($r < 1$, $\theta = 0$). Let u be the solution of:

$$\left\{ \begin{array}{ll} \Delta u = 0 & \text{in} \quad \Omega, \\[2mm] u|_{\Gamma_0} = g_0, \\[2mm] \left.\dfrac{\partial u}{\partial n}\right|_{\Gamma_0} = g_1. \end{array} \right.$$

The problem is to determine u in the domain Ω and on Γ_1.

6.2 Associated QR Problem

It is necessary to regularize the functions ρ and M that will be introduced at the points common to σ_0 and Σ_0. We therefore introduce the functions $\rho_{\varepsilon_0}(r, \theta)$ determined by:

$$\rho_{\varepsilon_0}(r, \theta) = \left\{ \begin{array}{ll} 1 & \text{for} \quad R_0 \leqslant r \leqslant R_1 - \varepsilon_0 \\[3mm] \dfrac{R_1 - r}{\varepsilon_0} & \text{for} \quad R_1 - \varepsilon_0 \leqslant r \leqslant R_1 \end{array} \right\} \quad \text{if} \quad \theta \in \Sigma_0 \text{ (cf. Figure 9)}$$

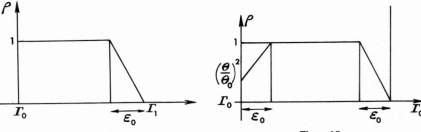

Figure 9   Figure 10

$$\rho_{\varepsilon_0}(r, \theta) = \begin{cases} \dfrac{r - R_0}{\varepsilon_0} + \left(\dfrac{\theta}{\theta_0}\right)^2 \left[1 - \dfrac{2 - R_0}{\varepsilon_0}\right] & \text{for} \quad R_0 \leqslant r \leqslant R_0 + \varepsilon_0 \\ 1 & \text{for} \quad R_0 + \varepsilon_0 \leqslant r \leqslant R_1 - \varepsilon_0 \\ \dfrac{R_1 - r}{\varepsilon_0} & \text{for} \quad R_1 - \varepsilon_0 \leqslant r \leqslant R_1 \end{cases} \begin{array}{l} \text{if} \quad \theta \in \sigma_0 \\ \text{(cf.} \\ \text{Figure 10)} \end{array}$$

Similarly:

$$M_{\varepsilon_0}(r, \theta) = \begin{cases} 1 & \text{for} \quad R_0 \leqslant r \leqslant R_1 - 2\,\varepsilon_0 \\ \dfrac{R_1 - \varepsilon_0 - r}{\varepsilon_0} & \text{for} \quad R_1 - 2\,\varepsilon_0 \leqslant r \leqslant R_1 - \varepsilon_0 \\ 0 & \text{for} \quad R_1 - \varepsilon_0 \leqslant r \leqslant R_1 \end{cases} \begin{array}{l} \text{if} \quad \theta \in \Sigma_0 \\ \text{(cf. Figure 11)} \end{array}$$

and

$$M_{\varepsilon_0}(r, \theta) = \begin{cases} \left(\dfrac{\theta}{\theta_0}\right)^2 & \text{for} \quad R_0 \leqslant r \leqslant R_0 + \varepsilon_0 \\ \dfrac{r - R_0 - \varepsilon_0}{\varepsilon_0} + \left(\dfrac{\theta}{\theta_0}\right)^2 \left[1 - \dfrac{r - R_0 - \varepsilon_0}{\varepsilon_0}\right] & \\ & \text{for} \quad R_0 + \varepsilon_0 \leqslant r \leqslant R_0 + 2\,\varepsilon_0 \\ 1 & \text{for} \quad R_0 + 2\,\varepsilon_0 \leqslant r \leqslant R_1 - 2\,\varepsilon_0 \\ \dfrac{R_1 - \varepsilon_0 - r}{\varepsilon_0} & \text{for} \quad R_1 - 2\,\varepsilon_0 \leqslant r \leqslant R_1 - \varepsilon_0 \\ 0 & \text{for} \quad R_1 - \varepsilon_0 \leqslant r \leqslant R_1 \end{cases} \begin{array}{l} \text{if} \quad \theta \in \sigma_0 \\ \text{(cf. Figure 12)} \end{array}$$

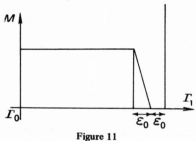

Figure 11   Figure 12

We then wish u_ε the solution of:

$$\left[\begin{array}{l} \operatorname{div}\left[(\rho_{\varepsilon_0}^2(r,\theta)\,\operatorname{grad} u_\varepsilon\right] = \dfrac{1}{\varepsilon_1^2}\,\Delta[M_{\varepsilon_0}^2(r,\theta)\,\Delta u_\varepsilon] \quad \text{in} \quad \Omega, \\[2mm] u_\varepsilon|_{\Sigma_0} = g_0\,, \\[2mm] \left.\dfrac{\partial u_\varepsilon}{\partial n}\right|_{\Sigma_0} = g_1\,. \end{array} \right.$$

The solution u of the original problem is the limit of u_ε as $(\varepsilon_0, \varepsilon_1) \to 0$.

6.3 Numerical Results

Let us again consider the most unfavorable case treated in Section 5, Example 3, namely:

$$u(r, \theta) = \log (r^2 + r_0^2 - 2\,rr_0\,\cos\theta)$$

with

$$r_0 = 0.95\,, \qquad R_0 = 1\,, \qquad R_1 = 1.5\,.$$

The numerical trials were carried out with

$$\left. \begin{array}{ll} \Delta r = 0.025 & \text{i.e., 20 steps in } r\,, \\[1mm] \Delta\theta = \pi/10 & \text{i.e., 20 steps in } \theta\,, \\[1mm] \varepsilon_0 = 5.10^{-2} = 2\,\Delta r\,, \\[1mm] \varepsilon_1 = 10^{-4}\,, \\[1mm] \theta_0 = \pi/10\,; \end{array} \right\} \qquad \text{(I)}$$

$$\Delta r = 0.03125 \qquad \text{i.e., 16 steps in } r,$$
$$\Delta\theta = \pi/8 \qquad \text{i.e., 16 steps in } \theta,$$
$$\varepsilon_0 = 6.25.10^{-2} = 2\,\Delta r,$$
$$\varepsilon_1 = 10^{-4},$$
$$\theta_0 = \pi/10.$$

(II)

On Graph 11, we have drawn the values of exact solution, for (I) and (II) with $\theta = 0$, as a function of i. The improvement by comparison with the former method is quite clear, cf. Graph 8. Naturally, there are various averaging techniques for even further improvement (cf., for example, Chapter 1, 8.4).

Graph 12 represents the exact solutions, calculated for $\theta = \pi/2$ and $\theta = \pi$.

Finally, Graph 13 is to be compared with Graph 10.

Remark 6.1. Other trials carried out for other values of θ_0 show that the solution worsens considerably if θ_0 is chosen too large. ∎

7. CASE OF THE RECTANGLE: OSCILLATORY DATA AND LOGARITHMIC SINGULARITY

7.1 The Problem

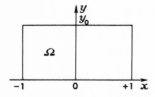

Figure 13

We consider the rectangle (Figure 13)

$$\begin{cases} -1 < x < +1 \\ 0 < y < y_0 \end{cases}$$

and we want $u(x, y)$ satisfying

$$\Delta u = \frac{\partial^2 u}{\partial x^2} + \frac{\partial^2 u}{\partial y^2} = 0 \tag{7.1}$$

$$u(x, 0) = u_0(x) \quad \text{given} \tag{7.2}$$

$$\frac{\partial u}{\partial y}(x, 0) = u_1(x) \quad \text{given} \tag{7.3}$$

We are going to apply the QR method to this "improperly posed" problem, taking in addition, as in the classic example of Hadamard [1], given functions u_0 and u_1 which *oscillate*.

7.2 Associated QR Problem

The application of the general method leads to the problem of solving:

$$\text{div} \left(\rho_{\varepsilon_0} \overrightarrow{\text{grad}} \, u_\varepsilon \right) = \frac{\partial}{\partial x} \left(\rho_{\varepsilon_0} \frac{\partial u_\varepsilon}{\partial x} \right) + \frac{\partial}{\partial y} \left(\rho_{\varepsilon_0} \frac{\partial u_\varepsilon}{\partial y} \right) =$$

$$= \frac{1}{\varepsilon_1} \left(\frac{\partial^2}{\partial x^2} + \frac{\partial^2}{\partial y^2} \right) \left(M_{\varepsilon_0} \frac{\partial^2 u_\varepsilon}{\partial x^2} + M_{\varepsilon_0} \frac{\partial^2 u_\varepsilon}{\partial y} \right), \tag{7.4}$$

$$u_\varepsilon(x, 0) = u_0(x), \tag{7.5}$$

$$\frac{\partial u_\varepsilon}{\partial y}(x, 0) = u_1(x). \tag{7.6}$$

We introduce the following functions $\rho_{\varepsilon_0}(x, y)$ and $M_{\varepsilon_0}(x, y)$:

for $y \leqslant y_0 - \varepsilon_0$,

$$\rho_{\varepsilon_0}(x, y) = \left\{ \begin{array}{ll} \dfrac{1}{\varepsilon_0}(1 + x) & \text{if} \quad -1 < x < -1 + \varepsilon_0 \\[2mm] 1 & \text{if} \quad -1 + \varepsilon_0 < x < 1 - \varepsilon_0 \\[2mm] \dfrac{1}{\varepsilon_0}(1 - x) & \text{if} \quad 1 - \varepsilon_0 < x < 1 \end{array} \right\} \quad \text{cf. Figure 14} \tag{7.7}$$

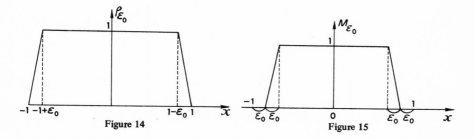

Figure 14   Figure 15

and for $y_0 - \varepsilon_0 \leqslant y \leqslant y_0$,

$$
\rho_{\varepsilon_0}(x, y) = \begin{cases} \dfrac{1}{\varepsilon_0}(y_0 - y)\dfrac{1}{\varepsilon_0}(1 + x) & \text{if} \quad -1 < x < -1 + \varepsilon_0 \\[2mm] \dfrac{1}{\varepsilon_0}(y_0 - y) & \text{if} \quad -1 + \varepsilon_0 < x < 1 - \varepsilon_0 \\[2mm] \dfrac{1}{\varepsilon_0}(y_0 - y)\dfrac{1}{\varepsilon_0}(1 - x) & \text{if} \quad 1 - \varepsilon_0 < x < 1 \; ; \end{cases} \tag{7.8}
$$

for $y \leqslant y_0 - \varepsilon_0$, M_{ε_0} is given by

$$
M_{\varepsilon_0}(x, y) = \begin{cases} 0 & \text{if} \quad -1 < x < -1 + \varepsilon_0 \\[2mm] \dfrac{1}{\varepsilon_0}(1 - \varepsilon_0 + x) & \text{if} \quad -1 + \varepsilon_0 < x < -1 + 2\varepsilon_0 \\[2mm] 1 & \text{if} \quad -1 + 2\varepsilon_0 < x < 1 - 2\varepsilon_0 \\[2mm] \dfrac{1}{\varepsilon_0}(1 - \varepsilon_0 - x) & \text{if} \quad 1 - 2\varepsilon_0 < x < 1 - \varepsilon_0 \\[2mm] 0 & \text{if} \quad 1 - \varepsilon_0 < x < 1 \end{cases} \quad \genfrac{}{}{0pt}{}{\text{cf.}}{\text{Figure 15}} \tag{7.9}
$$

and for $y_0 - \varepsilon_0 \leqslant y \leqslant y_0$, as for ρ_{ε_0} by multiplying the preceding determination of M_{ε_0} by $\dfrac{1}{\varepsilon_0}(y_0 - y)$.

7.3 Discretization

We set:

$$
\Delta x = \frac{1}{I}, \qquad \Delta y = \frac{y_0}{J} \; ;
$$
$$
u_{ij} \simeq u(i\,\Delta x, j\,\Delta y) , \qquad -I \leqslant i \leqslant I, \qquad 0 \leqslant j \leqslant J \; ;
$$
$$
\Delta_{ij} = (\Delta u)_{ij} .
$$

Equation (7.4) is discretized as follows (we suppress ε_0 to simplify the writing):

$$
\text{div}\,[\rho\,\text{grad}\,u]_{ij} \simeq \frac{1}{2\,\Delta x^2}\,[\rho_{i+1,j}(u_{i+1,j} - u_{ij}) - \rho_{i-1,j}(u_{ij} - u_{i-1,j})] + \tag{7.10}
$$
$$
+ \frac{1}{2\,\Delta y^2}\,[\rho_{i,j+1}(u_{i,j+1} - u_{ij}) - \rho_{i,j-1}(u_{ij} - u_{i,j-1})]
$$

$$\Delta[M \, \Delta u]_{ij} \simeq \frac{1}{\Delta x^2} \left[M_{i+1,j} \, \Delta_{i+1,j} - 2 \, M_{ij} \Delta_{ij} + M_{i-1,j} \, \Delta_{i-1,j} \right] +$$

$$+ \frac{1}{\Delta y^2} \left[M_{i,j+1} \, \Delta_{i,j+1} - 2 \, M_{ij} \Delta_{ij} + M_{i,j-1} \, \Delta_{i,j-1} \right],$$

(7.11)

with

$$\Delta_{ij} = \frac{u_{i+1,j} - 2 \, u_{ij} + u_{i-1,j}}{\Delta x^2} + \frac{u_{i,j+1} - 2 \, u_{ij} + u_{i,j-1}}{\Delta y^2} \, .$$

The boundary conditions (7.5) and (7.6) yield, for example:

$$u_{i,0} = u_0(i \, \Delta x)$$

(7.12)

$$\left. \frac{u_{i,1} - u_{i,-1}}{2 \, \Delta y} = u_1(i \, \Delta x) \right\} \quad -I \leqslant i \leqslant +I \, .$$

(7.13)

Thus we have the system of difference equations to be solved.

Remark 7.1. In practice, we take ε_0 so that ρ is zero on the boundary (with the exception of the segment $(-1, +1)$ of the x-axis) and equal to one at all other grid-points, M is zero on the boundary and on the nearest parallel discretization line, and equal to one elsewhere, which is equivalent to taking $\varepsilon_0 = \Delta x = \Delta y$, if $I = J$.

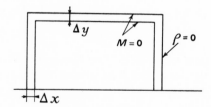

Figure 16

7.4 Numerical Applications: Case of Oscillatory Given Functions

We tried the method with:

$$u_0 = \lambda \sin nx \, ,$$

$$u_1 = 0 \, ,$$

which corresponds to

$$u(x, y) = \lambda \sin nx \operatorname{ch} ny.$$

EXAMPLE 1.

$$n = 1, \qquad \lambda = 1, \qquad y_0 = 1.$$

We choose

$$\varepsilon_1 = 10^{-2}, \qquad I = 10, \qquad J = 8.$$

Graphs 14, 15, 16, respectively, give the results first for two cross sections with y constant and then for two cross sections with x constant.

EXAMPLE 2.

$$n = 4, \qquad \lambda = 4^{-2.5}, \qquad y_0 = 1.$$

We choose

$$\varepsilon_1 = 10^{-3}, \qquad I = 12, \qquad J = 8.$$

The results are given on Graphs 17 and 18.

Influence of ε_1.

A study of the influence of ε_1 is given on Graphs 19, 20, 21 and 22.

Remark 7.2. In agreement with theory, the results remain convincing for $\varepsilon_1 = 0$; we have not drawn the graphs for $\varepsilon_1 = 0$ and $\varepsilon_1 = 10^{-15}$ (which corresponds to the maximum precision on a CDC 3600 in single precision) since they coincide with the foregoing. ∎

7.5 Numerical Applications: Logarithmic Singularity

7.5.1 THE PROBLEM

We now take, with the same geometry as in 7.4, the given functions:

$$u_0(x) = \tfrac{1}{2} \log (x^2 + y_s^2), \qquad y_s \notin [0, 1], \tag{7.14}$$

$$u_1(x) = -\frac{y_S}{x^2 + y_S^2},$$
(7.15)

which correspond to the solution

$$u = \log{(x^2 + (y - y_S)^2)^{1/2}}.$$
(7.16)

which is harmonic in the rectangle *with a singularity at* $S = \{0, y_S\}$. ■

Remark 7.3. The singularity in S requires of course a special treatment in the neighborhood of this point (cf. Remark 4.1 and Section 6). We are nonetheless going to study, at the end of the verification, the results without a special treatment, and then their improvement according to the stated method. ■

7.5.2 BRUTAL METHOD IGNORING THE SINGULARITY

Graph 23 furnishes the relative error at the center of the rectangle as a function of y_S (the approximate solution is calculated using the scheme of 7.4 with $I = J = 10$, $\varepsilon_1 = 10^{-5}$).

We see easily on the graph three predictable results:

— the error increases when the singularity approaches the rectangle;

— the error increases indefinitely *when the singularity moves towards the parts of the boundary where the Cauchy data are given* (it remains finite otherwise);

— at equal distances from the center, the error is greater when the singularity is nearer the part of the boundary where the Cauchy data are given.

It is therefore interesting to study the approximate solution when y_S tends toward zero through negative values; this is done on Graphs 24 to 31. ■

7.5.3 TRUNCATION METHOD TAKING ACCOUNT OF THE SINGULARITY

The method reduces (cf. Remark 4.1 and Section 6) to taking the functions ρ and M zero at the origin and on the nearest lines of discretization (cf. Figure 17).

On Graphs 30 and 31 are given the results obtained for $y_s = -0.25$, *which confirm the essential character of a truncation in the neighborhood of singularities.*

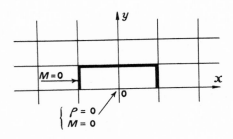

Figure 17

Remark 7.4. Naturally the truncations introduce perturbations *locally* (cf. Graph 31, $0 < y < 0.2$), which can be controlled by a refinement of the truncation of the grid and also by use of grids of variable size. ∎

Graphs 32 to 34 show results for the cross section $x = 0$ as $|y_s| \mapsto 0$. Evidently there is a certain worsening of results as $|y_s| \to 0$. The set of foregoing phenomena is shown on Graph 35 for $y_s = -0.05$ for cross sections in y.

8. SOME SUPPLEMENTARY REMARKS ON STABILIZATION

8.1 The Problem

Let A be an operator, bounded or not, in a Hilbert space $\mathcal{H}$. If A is not bounded, we suppose always that A is *closed* (for these notions, see, for example, Stone [1], Yosida [1]) and with domain $D(A)$ *dense* in $\mathcal{H}$; the domain $D(A)$ has the norm of the graph:

$$\| u \|_{D(A)} = (\| u \|_{\mathcal{H}}^2 + \| Au \|_{\mathcal{H}}^2)^{1/2} \qquad (8.1)$$

which makes it a Hilbert space.

We denote by A^* the *adjoint* of A (with domain $D(A^*)$ if A is not bounded).

We suppose that f is given in H and that the equation

$$Au = f, \qquad u \in \mathcal{H} \qquad \text{if } A \text{ is bounded} \tag{8.2}$$

$$Au = f, \qquad u \in D(A) \qquad \text{if } A \text{ is not bounded} \tag{8.2 bis}$$

possesses a unique solution, without supposing that A is an iso-morphism of $\mathcal{H}$ onto itself [case (8.2)] *or of $D(A)$ onto itself* [case (8.2 bis)] are therefore *unstable*: if in (8.2), for example, we replace f by $f + g$, where $g \in \mathcal{H}$, with $\| a \|_{\mathcal{H}}$ arbitrarily small, the equation

$$Av = f + g, \tag{8.3}$$

need not have a solution, or it can have a solution *which is not near* the solution u of (8.2). ∎

The problem is then to replace (8.2) or (8.2 bis) by a family of "nearby" equations which are "stable." This is therefore, in an abstract situation, a problem analogous to those considered in this chapter. ∎

We will distinguish two cases depending upon whether or not A is bounded.

8.2 Case Where A Is A Bounded Operator in $\mathcal{H}$

We consider the equation (where $\varepsilon > 0$)

$$(A^* A + \varepsilon I) u_\varepsilon = A^* f. \tag{8.4}$$

We have

Theorem 8.1: *For every $\varepsilon > 0$, Eq. (8.5) admits a unique solution*

$$u_\varepsilon = (A^* A + \varepsilon I)^{-1} A^* f. \tag{8.5}$$

In addition, supposing that (8.2) *possesses a unique solution, as $\varepsilon \to 0$, we have*:

$$u_\varepsilon \to u \qquad \text{in} \quad \mathcal{H}. \tag{8.6}$$

Proof. The fact that the operator $A^* A + \varepsilon I$ is invertible for any $\varepsilon > 0$ is well known, whence (8.5).

Taking the scalar product in $\mathcal{H}$ of the two terms of (8.4) with an arbitrary element v of $\mathcal{H}$, the result is (setting $(u, v) = (u, v)_{\mathcal{H}}$):

$$(Au_\varepsilon, Av) + \varepsilon(u_\varepsilon, v) = (f, Av). \tag{8.7}$$

But evidently (8.2) implies

$$(Au, Av) = (f, Av). \tag{8.8}$$

From (8.7) and (8.8), subtracting, we have:

$$(A(u_\varepsilon - u), Av) + \varepsilon(u_\varepsilon, v) = 0 \qquad \forall v \in \mathcal{H}. \tag{8.9}$$

Let us take $v = u_r - u$ in (8.9). We deduce:

$$\| A(u_\varepsilon - u) \|_{\mathcal{H}}^2 + \varepsilon(u_\varepsilon, u_\varepsilon - u) = 0. \tag{8.10}$$

Therefore

$$\varepsilon(u_\varepsilon, u_\varepsilon - u) \leqslant 0$$

and thus (since $\varepsilon > 0$):

$$(u_\varepsilon, u_\varepsilon - u) \leqslant 0. \tag{8.11}$$

But (8.11) implies:

$$\| u_\varepsilon \|_{\mathcal{H}}^2 \leqslant (u_\varepsilon, u) \leqslant \| u_\varepsilon \|_{\mathcal{H}} \| u \|_{\mathcal{H}},$$

whence

$$\| u_\varepsilon \|_{\mathcal{H}} \leqslant \| u \|_{\mathcal{H}} \qquad \forall \varepsilon > 0. \tag{8.12}$$

From every sequence of $\varepsilon \to 0$ we can extract a subsequence η such that $u_\eta \to w$ *weakly* in $\mathcal{H}$. But, from (8.10) and (8.12) we have

$$A(u_\varepsilon - u) \to 0 \quad \text{in} \quad \mathcal{H}$$

therefore

$$Au_\eta \to Au = f.$$

Since A is closed, it is weakly closed; thus

$$w \in D(A) \quad \text{and} \quad Aw = Au = f.$$

From uniqueness (by hypothesis) of the solution of (8.2), we have

$$w = u .$$

Therefore the extraction of a subsequence is unnecessary and $u_\varepsilon \to u$ *weakly* in $\mathscr{H}$. But

$$\| u_\varepsilon - u \|_{\mathscr{H}}^2 = (u_\varepsilon, u_\varepsilon - u) - (u, u_\varepsilon - u)$$

and, from (8.11), we therefore have

$$\| u_\varepsilon - u \|_{\mathscr{H}}^2 \leqslant - (u, u_\varepsilon - u) \to 0 ;$$

whence (8.6). ∎

8.3 Case Where A is A Unbounded Operator in $\mathscr{H}$

We return to (8.4) but it is now necessary to interpret $A^* f$ suitably (in the dual space of $D(A)$; cf. Lions–Magenes [1]). We can present the solution in the following way:

Proposition 8.1: *For every $\varepsilon > 0$, there exists a unique $u_\varepsilon \in D(A)$ such that*

$$(Au_\varepsilon, Av) + \varepsilon(u_\varepsilon, v) = (f, Av) \qquad \forall v \in D(A) . \tag{8.14}$$

The mapping $f \to u_\varepsilon$ is continuous from $\mathscr{H} \to D(A)$ (in other words: (8.14) is stable).

Proof. We use the same principle as in Lemma 3.3; for $u, v \in D(A)$, we put

$$Q_\varepsilon(u, v) = (Au, Av) + \varepsilon(u, v) \tag{8.15}$$

and it suffices to show the *coerciveness* of Q_ε:

$$Q_\varepsilon(v, v) = \| Av \|_{\mathscr{H}}^2 + \varepsilon \| v \|_{\mathscr{H}}^2 \geqslant \min(1, \varepsilon) \| v \|_{D(A)}^2 . \quad ∎$$

We now have (as in Theorem 8.1)

Theorem 8.2: *We suppose that (8.2 bis) possesses a unique solution in $D(A)$ (f is fixed). Then if u_ε is the solution of (8.14), we have*

$$u_\varepsilon \to u \quad \text{in} \quad D(A) \quad \text{when} \quad \varepsilon \to 0 . \tag{8.16}$$

Proof. We again have (8.8) where now $v \in D(A)$ and by subtraction we deduce [as in (8.9)]:

$$(A(u_\varepsilon - u), Av) + \varepsilon(u_\varepsilon, v) = 0 \qquad \forall v \in D(A). \tag{8.17}$$

Taking, as is permissible, $v = u_\varepsilon - u$ in (8.17), we deduce:

$$\| A(u_\varepsilon - u) \|_{\mathscr{H}}^2 + \varepsilon(u_\varepsilon, u_\varepsilon - u) = 0, \tag{8.18}$$

whence (8.11) and (8.12). We thus deduce, from (8.18):

$$\| A(u_\varepsilon - u) \|_{\mathscr{H}}^2 \leqslant 2\,\varepsilon \, \| u \|_{\mathscr{H}}^2. \tag{8.19}$$

We again extract $u_\eta \to w$ weakly in $\mathscr{H}$ and have from (8.19) $Au_\eta \to Au$. Since A is *closed*, we deduce that $w \in D(A)$ and that

$$Au = Aw.$$

Therefore $Aw = f$.

From the uniqueness (by hypothesis) of the solution of (8.2 *bis*) we have $w = u$, and it is not necessary to extract a subsequence. Therefore

$$u_\varepsilon \to u \text{ weakly in } \mathscr{H}$$

$$Au_\varepsilon \to Au \text{ strongly in } \mathscr{H}$$

But we show then that $u_\varepsilon \to u$ *strongly* in $\mathscr{H}$ as at the end of the proof of Theorem 8.1, whence the theorem. ∎

Remark 8.1. The preceding methods are evidently related to the methods of Tychonoff [1], [2], [3]; cf. also B. L. Phillips [1], R. Bellman–Kalaba–Lockett [1].

8.4 Example

We again consider the problem (1.5), (1.6), (1.7) in the present context. We take:

$$\mathscr{H} = L_2(\Omega). \tag{8.20}$$

Consider a function $\Phi \in H^2(\Omega)$ such that

$$\Phi|_{\Gamma_0} = g_0, \qquad \frac{\partial \Phi}{\partial v_A}\bigg|_{\Gamma_0} = g_1. \tag{8.21}$$

(There exist an infinity of such functions.)

If u satisfies (1.5), (1.6), (1.7), then

$$w = u - \Phi \tag{8.22}$$

satisfies:

$$Aw = f \quad \text{(where } f = - A\Phi \text{ is } known\text{)}, \tag{8.23}$$

$$w|_{\Gamma_0} = 0, \tag{8.24}$$

and

$$\frac{\partial w}{\partial v_A}\bigg|_{\Gamma_0} = 0. \tag{8.25}$$

We will apply the theory of 8.3 to the problem (8.23), (8.24), (8.25) (we change u into w). ∎

The operator A

The domain $D(A)$ is given (by definitioni) by:

$$D(A) = \left\{ v \mid v \in L_2(\Omega), Av \in L_2(\Omega), v|_{\Gamma_0} = 0, \frac{\partial v}{\partial v_A}\bigg|_{\Gamma_0} = 0 \right\}. \tag{8.26}$$

Remark 8.2. In (8.26), Av is calculated in the sense of $D'(\Omega)$.

If $v \in L^2(\Omega)$, $Av \in L^2(\Omega)$, then $v|_{\Gamma_0}$ and $\dfrac{\partial v}{\partial v_A}\bigg|_{\Gamma_0}$ have a *meaning* according to Lions–Magenes [1], Chapter 2 (in the spaces $H^{-1/2}(\Gamma_0)$ and $H^{-3/2}(\Gamma_0)$ respectively). Therefore (8.26) has a meaning. ∎

Let us verify that the conditions of application of the general theory are satisfied. This is a consequence of the two lemmas which follow.

Lemma 8.1: *The operator A, defined by (8.26), is closed.*

Proof. Let $v_n \to v$ in $L_2(\Omega)$, $Av_n \to \chi$ in $L_2(\Omega)$. From the continuity of the derivative in the sense of distributions, we have:

$$Av_n \to Av \text{ in } D'(\Omega) \text{ and therefore } \chi = Av.$$

It remains to show that $v \in D(A)$.

From Lions–Magenes, *loc*. *cit*., Chapter 2, we have:

$$v_n|_{\Gamma_0} \to v|_{\Gamma_0} \qquad \text{in} \quad H^{-1/2}(\Gamma_0),$$

$$\left.\frac{\partial v_n}{\partial v_A}\right|_{\Gamma_0} \to \left.\frac{\partial v}{\partial v_A}\right|_{\Gamma_0} \quad \text{in} \quad H^{-3/2}(\Gamma_0)$$

and therefore $v|_{\Gamma_0} = 0$, $\left.\dfrac{\partial v}{\partial v_A}\right|_{\Gamma_0} = 0$, whence the result. ∎

Remark 8.3. The space $A(D(A))$ *is dense in* $\mathscr{H} = L_2(\Omega)$.

Proof. Let $f \in L_2(\Omega)$ with

$$(Av, f) = 0 \qquad \forall v \in D(A). \tag{8.27}$$

Then, to begin with

$$A^* f = 0 \quad \text{in} \quad \Omega. \tag{8.28}$$

Then we have Green's formula, valid according to Lions–Magenes, *loc*. *cit*., Chapter 2, if $v \in H^2(\Omega)$ and $f \in L_2(\Omega)$, with $A^* f \in L_2(\Omega)$:

$$(Av, f) - (v, A^* f) = \int_{\Gamma_0} \frac{\partial v}{\partial v_A} f \, d\Gamma_0 + \int_{\Gamma_1} \frac{\partial v}{\partial v_A} f \, d\Gamma_1 - \\ - \int_{\Gamma_0} v \frac{\partial f}{\partial v_{A^*}} d\Gamma_0 - \int_{\Gamma_1} v \frac{\partial f}{\partial v_{A^*}} d\Gamma_1 \tag{8.29}$$

where, in the second term, the first and second (respectively third and fourth) indicate the duality between $H^{1/2}(\Gamma_i)$, $H^{-1/2}(\Gamma_i)$, $i = 0$ (respectively $H^{3/2}(\Gamma_i)$, $H^{-3/2}(\Gamma_i)$, $i = 0, 1$).

If now we take:

$$v \in H^2(\Omega), \qquad v|_{\Gamma_0} = \left.\frac{\partial v}{\partial v_A}\right|_{\Gamma_0} = 0 \tag{8.30}$$

then $v \in D(A)$ and taking account of (8.27), (8.28), we deduce from (8.29):

$$\int_{\Gamma_1} \frac{\partial v}{\partial v_A} f \, d\Gamma_1 - \int_{\Gamma_1} v \frac{\partial f}{\partial v_{A^*}} d\Gamma_1 = 0 \tag{8.31}$$

and this $\forall v$ with (8.30). That implies

$$f|_{\Gamma_1} = 0, \qquad \left.\frac{\partial f}{\partial \nu_{A^*}}\right|_{\Gamma_1} = 0 \tag{8.32}$$

which, with (8.28) implies $f = 0$, whence the result. ∎

Consequence: by application of Theorem 8.2, we have:

> *to approach the solution w of* (8.23), (8.24), (8.25) (which
> exists for *the particular f* corresponding to $f = -A\Phi$
> where Φ satisfies (8.21) for $\{g_0, g_1\}$ *admissible*), *we de-*
> *fine u_ε as the unique solution in* $D(A)$ of
>
> $$(Au_\varepsilon, Av) + \varepsilon(u_\varepsilon, v) = (f, Av) \qquad \forall v \in D(A).$$
>
> *Then $u_\varepsilon \to w$ in $D(A)$ when $\varepsilon \to 0$.* ∎ \tag{8.33}

Remark 8.4. We have therefore

$$\begin{aligned}
u_\varepsilon + \Phi \quad &\to u & \text{in} \quad &L_2(\Omega), \\
A(u_\varepsilon + \Phi) &\to Au = 0 & \text{in} \quad &L_2(\Omega). \blacksquare
\end{aligned} \tag{8.34}$$

Let us now interpret *the variational equation* (8.14).

To begin with, taking $v = \varphi \in D(\Omega)$, we deduce that

$$A^* Au_\varepsilon + \varepsilon u_\varepsilon = A^* f. \tag{8.35}$$

Since $u_\varepsilon \in D(A)$ we have:

$$u_\varepsilon|_{\Gamma_0} = \left.\frac{\partial u_\varepsilon}{\partial \nu_A}\right|_{\Gamma_0} = 0. \tag{8.36}$$

Multiplying (8.35) by v and integrating over Ω, with $v \in D(A)$, we have:

$$\int_{\Gamma_0} \left(\frac{\partial}{\partial \nu_{A^*}} Au_\varepsilon\right) v \, d\Gamma_0 + \int_{\Gamma_1} \left(\frac{\partial}{\partial \nu_{A^*}} Au_\varepsilon\right) v \, d\Gamma_1 -$$

$$- \int_{\Gamma_0} Au_\varepsilon \frac{\partial v}{\partial \nu_A} d\Gamma_0 - \int_{\Gamma_1} Au_\varepsilon \frac{\partial v}{\partial \nu_A} d\Gamma_1 +$$

$$+ (Au_\varepsilon, Av) + \varepsilon(u_\varepsilon, v) = \int_{\Gamma_0} \frac{\partial f}{\partial \nu_A} v \, d\Gamma_0 + \int_{\Gamma_1} \frac{\partial f}{\partial \nu_{A^*}} v \, d\Gamma_1 -$$

$$- \int_{\Gamma_0} f \frac{\partial v}{\partial \nu_A} d\Gamma_0 - \int_{\Gamma_1} f \frac{\partial u}{\partial \nu_A} d\Gamma_1 + (f, Av).$$

Taking account of (8.14) and of $v = \partial v/\partial v_A = 0$ on Γ_0, we have:

$$\int_{\Gamma_1} \left(\frac{\partial}{\partial v_{A^*}} Au_\varepsilon\right) v \, d\Gamma_1 - \int_{\Gamma_1} Au_\varepsilon \frac{\partial v}{\partial v_A} \, d\Gamma_1 = \int_{\Gamma_1} \frac{\partial f}{\partial v_{A^*}} v \, d\Gamma_1 - \int_{\Gamma_1} f \frac{\partial v}{\partial v_A} \, d\Gamma_1$$

whence

$$Au_\varepsilon = f \quad \text{on} \quad \Gamma_1,$$

$$\frac{\partial}{\partial v_{A^*}} Au_\varepsilon = \frac{\partial f}{\partial v_{A^*}} \quad \text{on} \quad \Gamma_1. \tag{8.37}$$

We have thus demonstrated

Theorem 8.3: *Let A be an elliptic operator of the second order* (as in Section 1) *and let u be the solution of* (1.5), (1.6), (1.7),

1) *We choose Φ satisfying* (8.21), *and we put $f = -A\Phi$.*
2) *We define u_ε as the solution of the* (properly posed) *problem* (8.35), (8.36), (8.37). *Then, as $\varepsilon \to 0$, we have* (8.34). ∎

Remark 8.5. We can also (this is particular to the example considered and is not always valid) *make $\varepsilon = 0$ in the process indicated in Theorem 8.3.*

The justification of this remark requires some efforts. To begin with let us use the *norm* $\|Au\|_{L_2(\Omega)}$ in $D(A)$ and $\hat{D}(A)$ be the completion of $D(A)$ with this norm. From what we have seen in 3.5, $\hat{D}(A) \subset H^2(\Omega_\varepsilon)$ $\forall \varepsilon > 0$.

We observe that there *exists a function, one and only one function u_0 in $D(A)$ such that*

$$(Au_0, Av) = (f, Av) \qquad \forall v \in D(A) \tag{8.38}$$

(this is (8.14) with $\varepsilon = 0$).

We then have

$$u_0 + \Phi = u. \tag{8.39}$$

From (8.38) we deduce that

$$A^*(Au_0 - f) = 0, \tag{8.40}$$

and that [cf. (8.37)]:

$$Au_0 - f = 0 \quad \text{on} \quad \Gamma_1,$$

$$\frac{\partial}{\partial v_{A^*}} (Au_0 - f) = 0 \quad \text{on} \quad \Gamma_1. \tag{8.41}$$

Therefore (uniqueness of the solution of the Cauchy problem):

$$Au_0 - f = 0 \quad \text{in} \quad \Omega$$

i.e.,

$$A(u_0 + \Phi) = 0 \quad \text{in} \quad \Omega . \tag{8.42}$$

But then

$$A(u_0 + \Phi - u) = 0 \quad \text{in} \quad \Omega \tag{8.43}$$

and

$$(u_0 + \Phi - u)|_{\Gamma_0} = (\Phi - u)|_{\Gamma_0} = 0 ,$$
$$\frac{\partial}{\partial v_A} (u_0 + \Phi - u)|_{\Gamma_0} = \frac{\partial}{\partial v_A} (\Phi - u)|_{\Gamma_0} = 0 ,$$

therefore (relying always on the uniqueness of the solution of the Cauchy problem)

$$u_0 + \Phi - u = 0 \quad \text{in} \quad \Omega ,$$

therefore (8.39) holds. ∎

Remark 8.6. We have therefore obtained a third QR method (the two first were presented in Section 2) for the solution of the problem (1.5), (1.6), (1.7). ∎

Remark 8.7. As opposed to the two QR methods presented in Section 2, there is *no deterioration of the approximation in the neighborhood of Γ_1*. This is an advantage of the present method, an advantage moreover obtained at the price of the following two difficulties:

    i) *practical* construction of the function Φ;
    ii) a discretization of (8.37) which leads to slightly heavier calculation than that of the first two methods where the boundary conditions on Γ_1 were "suppressed" by the introduction of the weight functions ρ and M, a difficulty which is indeed confirmed in numerical results. ∎

Remark 8.8. Apart from the difficulty noted in (ii), Remark 8.7, the discretization of (8.35), (8.36), (8.37) (or the analogue with $\varepsilon = 0$), leads to calculations (and to a matrix) of the same order as in the first two methods. ∎

Remark 8.9. All the variants noted for the two first methods are valid here: different geometry, operators of order superior to 2, etc. ∎

Remark 8.10. We have replaced the equation $Au = f$ by the equation

$$A^* Au + \varepsilon u = A^* f$$

with suitable boundary conditions.

We could, more generally, introduce the equation

$$A^* Au + \varepsilon Bu = A^* f,$$

where B is chosen in a way to *improve the conditioning* of the matrix $A^* A + \varepsilon B$ (after having been discretized). For the case of finite dimension, cf. J. D. Riley [1], G. Ribiere [1] and the bibliography of this last work.

8.5 Numerical Results

We again considered the example of 7.5. The results are shown on Graph 36 where in addition we presented the results obtained using the penalization method (Section 7); we see that this last method is better than the present method, the principal difficulty of which resides in taking account of the boundary conditions. We remark finally that decreasing ε does not improve results, in accord with Remark 8.10. ∎

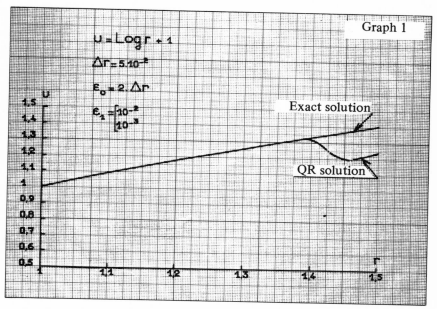

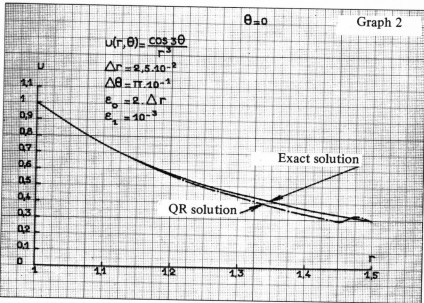

NOTE: In all graphs commas represent decimal points.

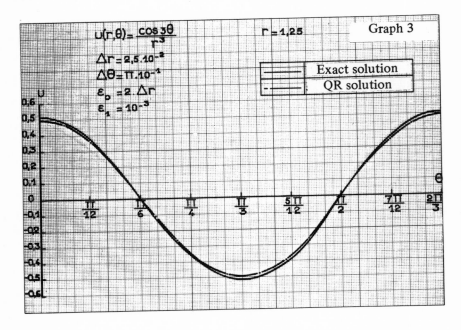

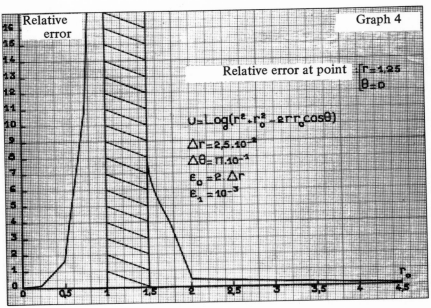

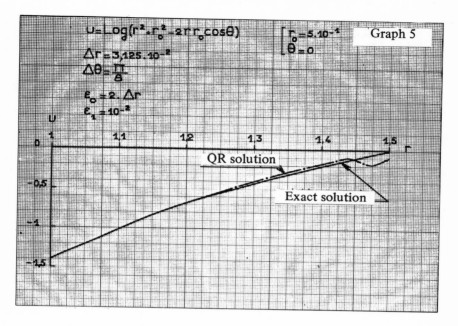

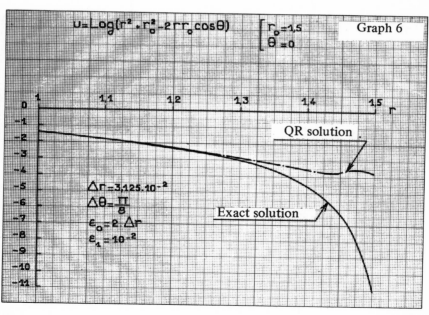

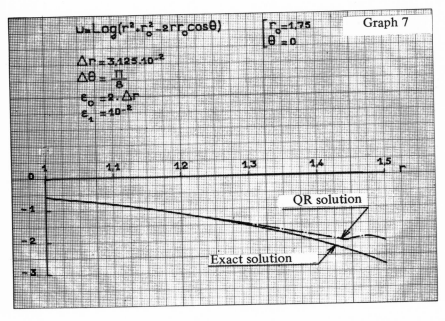

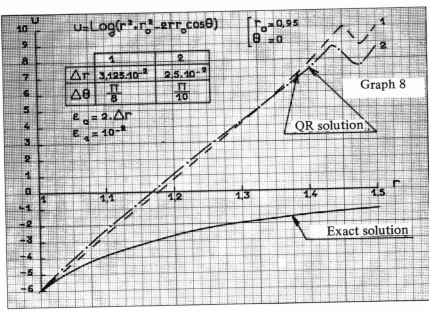

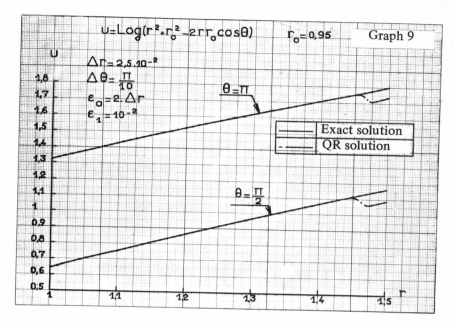

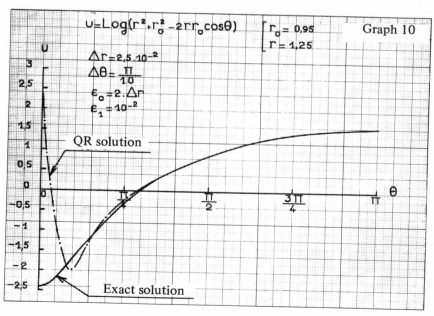

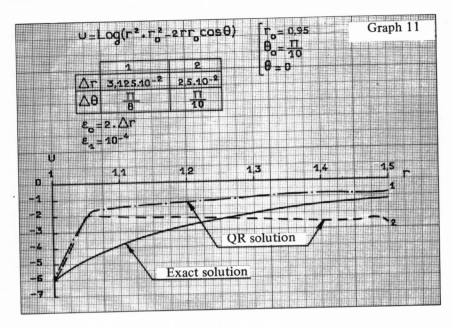

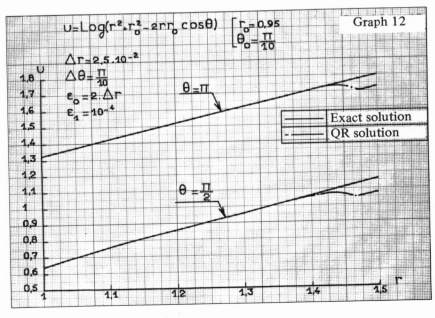

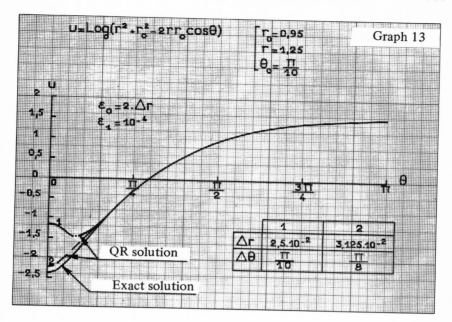

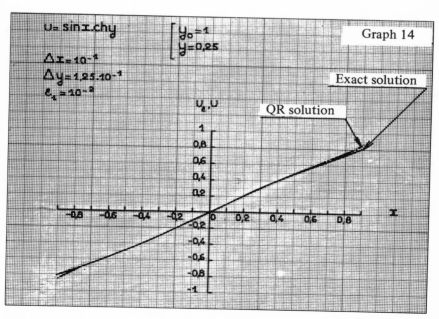

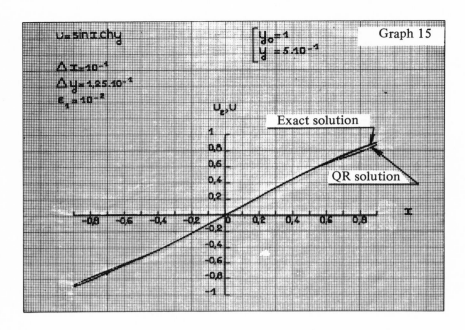

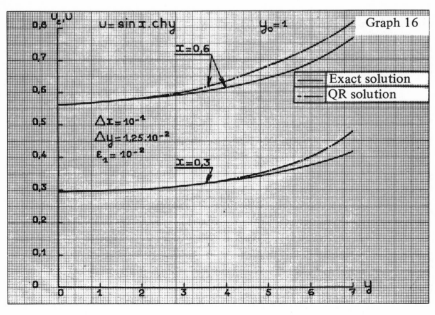

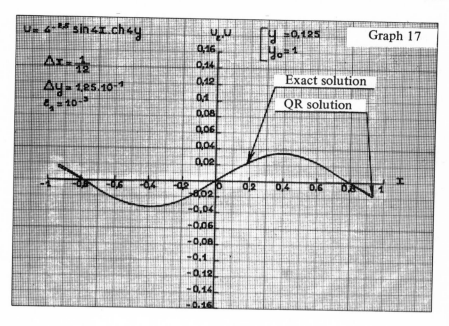

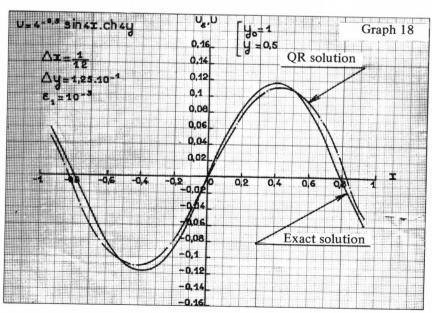

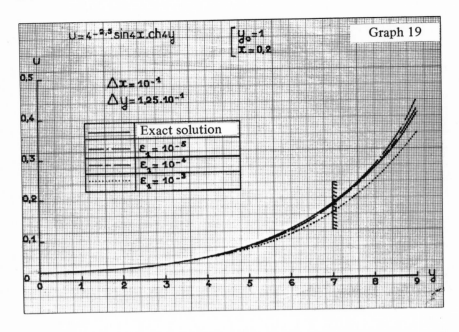

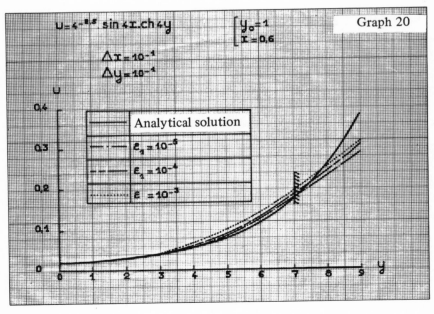

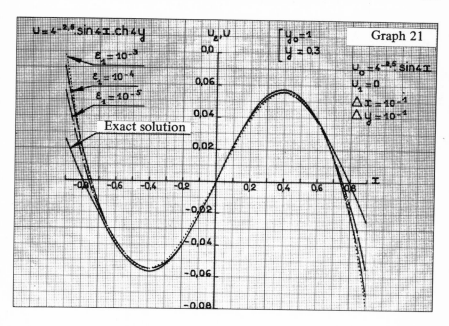

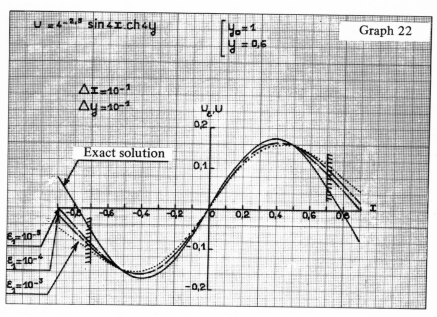

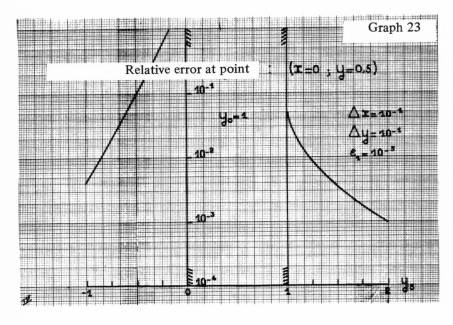

Graph 23

Relative error at point $(x=0\;;\;y=0.5)$

y_o-1

$\triangle x = 10^{-1}$
$\triangle y = 10^{-1}$
$\varepsilon_1 = 10^{-5}$

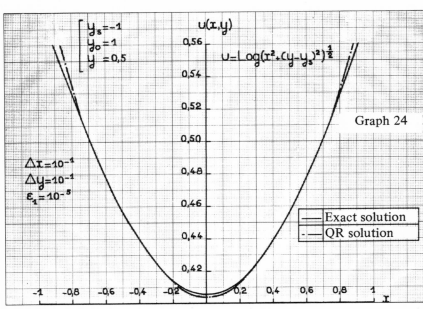

$\left[\begin{array}{l} y_s = -1 \\ y_o = 1 \\ y = 0.5 \end{array}\right]$

$u(x,y)$

$u = \log(x^2 + (y - y_s)^2)^{\frac{1}{2}}$

Graph 24

$\triangle x = 10^{-1}$
$\triangle y = 10^{-1}$
$\varepsilon_1 = 10^{-5}$

——— Exact solution
-·— QR solution

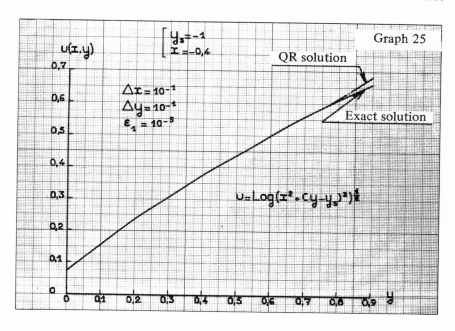

Graph 25

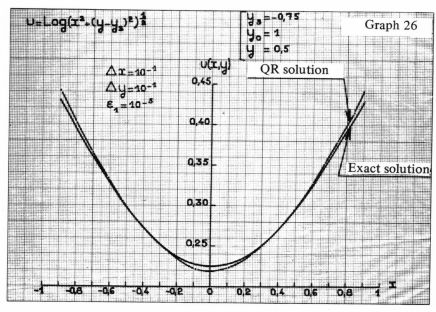

Graph 26

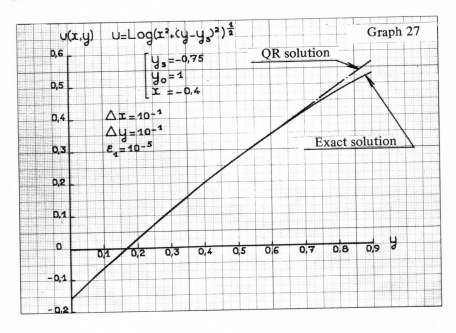

Graph 27

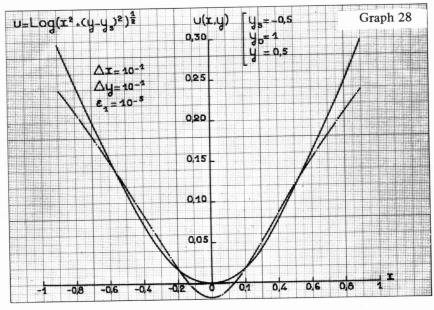

Graph 28

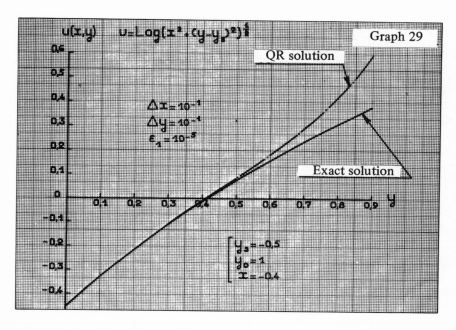

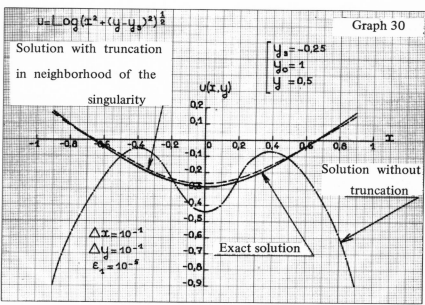

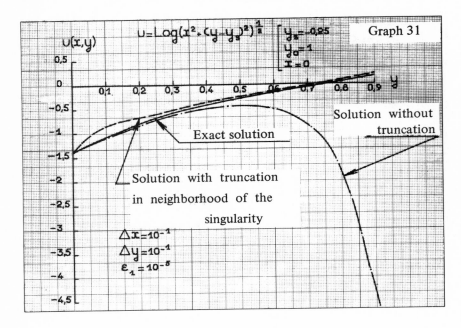

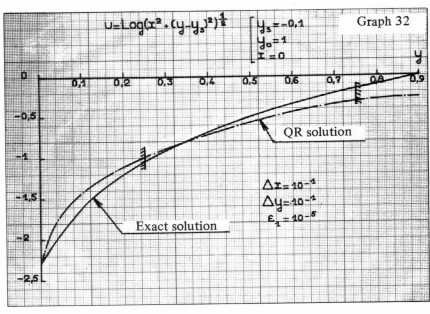

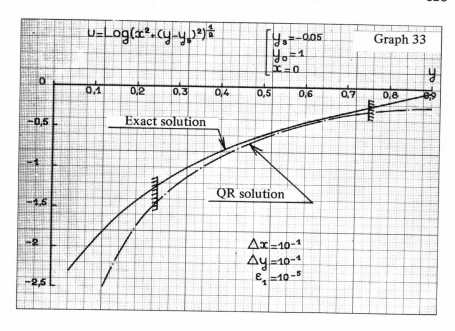

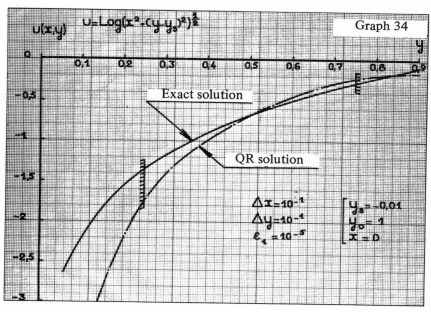

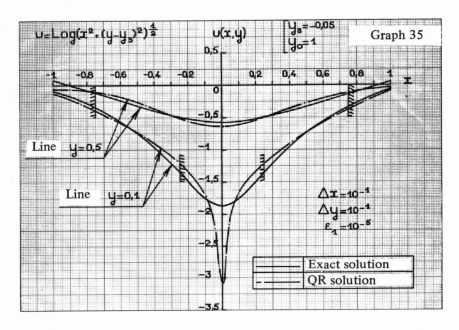

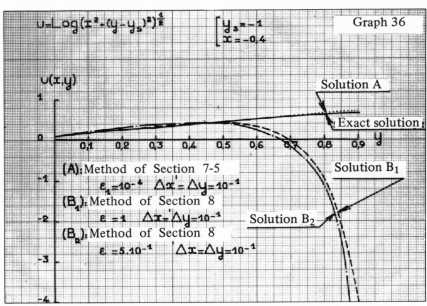

Chapter 5

QUASI-REVERSIBILITY AND CONTINUATION OF SOLUTIONS OF PARABOLIC EQUATIONS

DETAILED OUTLINE

1. FORMULATION OF THE PROBLEM

1.1 Notation

Let Ω be an open region in $\mathbf{R}^n$ with the boundary $\Gamma_0 \cup \Gamma_1$ (cf. Figure 1) (as in Chapter 4, Section 1, 1.2). Let A be an *elliptic differential operator of the second order* in Ω:

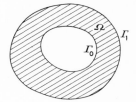

$$Au = -\sum \frac{\partial}{\partial x_i}\left(a_{ij}(x)\frac{\partial u}{\partial x_j}\right) \qquad (1.1)$$

Figure 1

$$a_{ij} \in C^3(\overline{\Omega})\,, \qquad a_{ij} \text{ real valued to simplify,} \qquad (1.2)$$

$$\sum_{i,j=1}^{n} a_{ij}(x)\,\xi_i\,\xi_j \geq \alpha(\xi_1^2 + \cdots + \xi_n^2)\,, \quad \alpha > 0\,. \qquad (1.3)$$

We then consider the *finite cylinder*:

$$Q = \Omega \times \,]0, T[\,. \qquad (1.4)$$

We denote by t the *time variable*, $t \in [0, T]$. The notation will be:

$$\Sigma_i = \Gamma_i \times [0, T]\,, \quad i = 0, 1 \qquad (1.5)$$

(therefore Σ_0 (respectively Σ_1) is the interior (exterior) side boundary of Q). ∎

In this cylinder Q, we consider the *parabolic operator of the second order*

$$P = \frac{\partial}{\partial t} + A\,. \qquad (1.6)$$

1.2 The Problems

Let $u = u(x, t)$, a function in Q, be the solution of

$$Pu = 0\,; \qquad (1.7)$$

the precise hypotheses concerning u will be given further on; let us formally *set* for the moment

$$u\,|_{\Sigma_0} = g_0 \quad \text{(value (or trace) of } u \text{ on } \Sigma_0\text{)},\qquad\qquad (1.8)$$

$$\left.\frac{\partial u}{\partial \nu_A}\right|_{\Sigma_0} = g_1 \qquad\qquad\qquad\qquad (1.9)$$

(where $\dfrac{\partial u}{\partial \nu_A} = \sum\limits_{i,j=1}^{n} a_{ij}(x)\,\dfrac{\partial u}{\partial x_j}\cos(\nu, x_i)$, ν = normal to Γ_0 exterior to Ω); let us likewise set

$$u(x, 0) = u_0(x) . \blacksquare \qquad\qquad\qquad (1.10)$$

Problem (I) *is now the following*: *calculate u satisfying* (1.7) *from the given functions* g_0, g_1 *and* u_0.

Remark 1.1. (This is analogous to Remark 1.1 of Chapter 4, Section 1). It g_0, g_1 and u_0 are given, the problem (1.7), . . . , (1.10) does not in general possess a solution. If it does possess a solution, there is *instability*: the (actual) solution does not depend continuously on the initial data, since it even ceases to exist if the given functions undergo "arbitrarily small" variations starting from a compatible solution. $\blacksquare$

Remark 1.2. (This is analogous to Remark 1.3 of Chapter 4). The function u (if it exists) is *defined in a unique way* by (1.7), . . . , (1.10). If u_1 and u_2 are two actual solutions of the problem, the difference

$$w = u_1 - u_2$$

satisfies

$$Pw = 0$$
$$w\,|_{\Sigma_0} = 0 \qquad \left.\frac{\partial w}{\partial \nu_A}\right|_{\Sigma_0} = 0 \qquad\qquad (1.11)$$

$$w(x, 0) = 0 \qquad\qquad\qquad\qquad (1.12)$$

whence $w = 0$; cf. S. Mizohata [1]. $\blacksquare$

In fact, condition (1.12) has not been used here. Thus we can equally well consider *problem* (II): *calculate u satisfying* (1.7) *given the functions* g_0 *and* g_1.

In the following sections we will present the solutions of these problems using the QR method.

Remark 1.3. Changing, as is classical, u into $\exp(\lambda t)u$ we can always replace A by $A + \lambda$, where λ is a positive constant. We suppose that this has been done in what follows; this introduces certain technical simplifications in the *proofs*. The theory is valid *without* this change which may or may not be effected in the numerical examples. ∎

Remark 1.4. In a large number of practical problems, we actually wish to determine u on Σ_1. However, the QR method in Section 2 yields only an approximation to u at a certain distance from Σ_1. We therefore determine u on Σ_1 by extrapolation.

This is not the case for the QR method given in Section 5, 5.3; however, the discretization is then more complicated in the neighborhood of Σ_1; there is therefore in the long run almost an equivalence of the methods. ∎

2. QR PROBLEM ASSOCIATED WITH PROBLEM (I)

2.1 Notation

We use notation similar to that of Chapter 4, Section 2. Therefore:

$$d(x, \Gamma_1) = \text{distance from } x(\in \Omega) \text{ to } \Gamma_1 \; ; \tag{2.1}$$

$$M_{\varepsilon_0}, \rho_{\varepsilon_0} \text{ defined by (2.2) (2.3), Chapter 4}. \tag{2.2}$$

We set:

$$A_{\varepsilon_0} = - \sum_{i,j=1}^{n} \frac{\partial}{\partial x_i} \left(\rho_{\varepsilon_0}^2 \, a_{ij} \frac{\partial}{\partial x_j} \right) + \lambda \tag{2.3}$$

and

$$P_{\varepsilon_0} = \frac{\partial}{\partial t} + A_{\varepsilon_0}. \tag{2.4}$$

Let A^* be the *adjoint* of A and P^* the adjoint of P:

$$P^* = -\frac{\partial}{\partial t} + A^* .$$

(2.5)

2.2 The QR Problem

We introduce three parameters $\varepsilon_0, \varepsilon_1, \varepsilon_2 > 0$ and we denote by $u_\varepsilon = u_{\varepsilon_0, \varepsilon_1, \varepsilon_2}$ "the"[1] solution of:

$$\frac{1}{\varepsilon_1} P^*(M_{\varepsilon_0}^2 P u_\varepsilon) + P_\varepsilon u_\varepsilon - \varepsilon_2 \frac{\partial^2 u_\varepsilon}{\partial t^2} = 0$$

(2.6)

with the boundary conditions

$$u_\varepsilon |_{\Sigma_0} = g_0 ,$$

(2.7)

$$\left. \frac{\partial u_\varepsilon}{\partial v_A} \right|_{\Sigma_0} = g_1 ,$$

(2.8)

$$u_\varepsilon(x, 0) - \varepsilon_2 \frac{\partial u_\varepsilon}{\partial t}(x, 0) - \frac{1}{\varepsilon_1} M_{\varepsilon_0}^2 P u_\varepsilon(x, 0) = u_0(x)$$

(2.9)

and

$$\varepsilon_2 \frac{\partial u_\varepsilon}{\partial t}(x, T) + \frac{1}{\varepsilon_1} M_{\varepsilon_0}^2 (P u_\varepsilon)(x, T) = 0 . \quad \blacksquare$$

(2.10)

We shall return to the *interpretation* of this problem. We are going to show first that, in a suitable sense, the problem (2.6), . . . , (2.10) is *properly posed and that $u_\varepsilon \to u$ when $\varepsilon \to 0$.*

Remark 2.1. We *always* introduce here a *penalty term* [first term of (2.6)]. This was not so in the elliptic case (cf. Chapter 4, 2.3.1 and 2.3.2); a method of the type of 2.3.1 would consist here of trying to solve

$$P^*(M_{\varepsilon_0}^2 P u_\varepsilon) = 0$$

(2.11)

[1] In fact, it is necessary *to show* that *the* solution of the QR problem exists and is unique.

with **(2.7)**, **(2.8)** and

$$(M_{\varepsilon_0} \, Pu_\varepsilon) \, (x, 0) = 0 \,, \qquad (M_{\varepsilon_0} \, Pu_\varepsilon) \, (x, T) = 0 \,. \tag{2.12}$$

But this problem is probably not *well-posed*.[1]

We could equally well think of replacing **(2.11)** by

$$P^*(N_\varepsilon^2 \, Pu_\varepsilon) = 0 \tag{2.13}$$

where N_ε is a function of x *and* t (then M_{ε_0} in **(2.11)** depends only on x) which is zero not only (like M_{ε_0}) in the neighborhood of $\Gamma_1 \times (0, T)$ but also in the neighborhood of $t = 0$ and $t = T$.

We seem thus to run into additional technical difficulties. We can, on the other hand, use the method (without penalization) of Chapter 4, Section 8; this will be pointed out in Sections 5, 5.3. ∎

3. CONVERGENCE THEOREM

3.1 Notation

To begin with we set

$$U_{\varepsilon_0} = \left\{ v \mid v \in L_2(\Omega) \,, \qquad \rho_{\varepsilon_0} \frac{\partial v}{\partial x_i} \in L_2(\Omega) \,, \right.$$
$$\left. i = 1, ..., n \quad \text{and} \quad v = 0 \text{ on } \Gamma_0 \right\} \,; \tag{3.1}$$

U_{ε_0} is a Hilbert space with the norm

$$\| v \|_{U_{\varepsilon_0}} = \left\{ | v |^2 + \sum_{i=1}^n \left| \rho_{\varepsilon_0} \frac{\partial v}{\partial x_i} \right|^2 \right\}^{1/2} \,,$$

where, for example,

$$| v |^2 = \left[\int_\Omega (v(x))^2 \, dx \right]^{1/2} \,.$$

---

[1] An argument analogous to that of 3.5, Chapter 4, is not valid here.

We now use the notation of Chapter 1; we set:

$$L_2(U_\varepsilon) = L_2(0, T; U_\varepsilon) \tag{3.2}$$

space of (classes of) functions which are square summable on $]0, T[$ with values in U_ε; therefore, if $u \in L_2(U_\varepsilon)$, we have:

$$u \in L_2(Q), \qquad \frac{\partial u}{\partial x_i} \in L_2(Q), \qquad i = 1, ..., n$$

$$u|_{\Sigma_0} = 0 \tag{3.3}$$

and reciprocally.

We introduce

$$X_\varepsilon = \left\{ v \mid v \in L_2(U_\varepsilon), \qquad M_{\varepsilon_0} \, Pv \in L_2(Q), \qquad \frac{\partial v}{\partial t} \in L_2(Q), \qquad \frac{\partial v}{\partial \nu_A}\bigg|_{\Sigma_0} = 0 \right\}. \tag{3.4}$$

We admit here (cf. Lions–Magenes [1]) that if $v \in L_2(U_\varepsilon)$, with $M_{\varepsilon_0} \, Pv \in L_2(Q)$, then $\dfrac{\partial v}{\partial \nu_A}\bigg|_{\Sigma_0}$ *has meaning*, such that the definition (3.4) for X_ε is correct. The space X_ε is a Hilbert space with the norm:

$$\| v \|_{X_\varepsilon} = \left[\| v \|^2_{L_2(U_\varepsilon)} + \| M_{\varepsilon_0} \, Pv \|^2_{L_2(Q)} + \left\| \frac{\partial v}{\partial t} \right\|^2_{L_2(Q)} \right]^{1/2}$$

3.2 Bilinear Forms

For $u, v \in X_\varepsilon$, we set

$$\pi_\varepsilon(u, v) = \int_0^T a_{\varepsilon_0}(u, v) \, dt - \int_0^T (u, v') \, dt + \varepsilon_2 \int_0^T (u', v') \, dt +$$

$$+ \frac{1}{\varepsilon_1} (M_{\varepsilon_0} \, Pu, M_{\varepsilon_0} \, Pv)_{L_2(Q)} + (u(T), v(T)) \tag{3.6}$$

where the notation is the following:

$$a_{\varepsilon_0}(\varphi, \psi) = \sum_{i,j=1}^n \int_\Omega a_{ij} \, \rho_{\varepsilon_0}^2 \, \frac{\partial \varphi}{\partial x_j} \frac{\partial \psi}{\partial x_i} \, dx + \lambda \int_\Omega \varphi \psi \, dx \, ;$$

$$a_{\varepsilon_0}(u, v) = a_{\varepsilon_0}(u(t), v(t)),$$

where

$$u(t) = « x \to u(x, t) » ;$$

therefore

$$a_{\varepsilon_0}(u, v) = \sum_{i,j=1}^{n} \int_{\Omega} a_{ij}(x) \, \rho_{\varepsilon_0}^2(x) \, \frac{\partial u}{\partial x_j}(x, t) \, \frac{\partial v}{\partial x_i}(x, t) \, dx \; + \lambda \int_{\Omega} u(x, t) \, v(x, t) \, dx \; ;$$

$$(f, g) = \int_{\Omega} f(x) \, g(x) \, dx \; ,$$

thus

$$(u, v') = \int_{\Omega} u(x, t) \, v'(x, t) \, dx \; ,$$

where

$$v' = \frac{\partial v}{\partial t} \; ;$$

$$(u(T), v(T)) = \int_{\Omega} u(x, T) \, v(x, T) \, dx \; .$$

We readily verify:

Lemma 3.1: *The bilinear form* $u, v \to \pi_\varepsilon(u, v)$ *is continuous on* X_ε.
In addition, noting that

$$-\int_{0}^{T} (v, v') \, dt = -\tfrac{1}{2} | v(T) |^2 + \tfrac{1}{2} | v(0) |^2 \quad (^1)$$

we obtain

$$\pi_\varepsilon(v, v) = \int_{0}^{T} a_{\varepsilon_0}(v, v) \, dt \; + \varepsilon_2 \int_{0}^{T} | v'(t) |^2 \, dt \; + \frac{1}{\varepsilon_1} \| M_{\varepsilon_0} \, Pv \|_{L_2(Q)}^2$$

$$+ \tfrac{1}{2} | v(0) |^2 + \tfrac{1}{2} | v(T) |^2 \; .$$

<hr>

(¹) **For example** $| v(T) |^2 = \int_{\Omega} v(x, T)^2 \, dx \; .$

We deduce from this:

Lemma 3.2: *For every* $\{\varepsilon_0, \varepsilon_1, \varepsilon_2\}, \varepsilon_i > 0,$ *there exists a positive constant* $\alpha(\varepsilon)$ *such that*

$$\pi_\varepsilon(v, v) \geqslant \alpha(\varepsilon) \| v \|_{X_\varepsilon}^2 \qquad \forall v \in X_\varepsilon.$$

3.3 Hypotheses On u; the Solution w_ε

We now make the *following hypotheses* concerning u:

$$u \in L_2(Q), \quad \frac{\partial u}{\partial x_i} \in L_2(Q), \quad \frac{\partial^2 u}{\partial x_i\, \partial x_j} \in L_2(Q), \quad i, j = 1, ..., n$$

$$\frac{\partial u}{\partial t}, \quad \frac{\partial^2 u}{\partial t^2} \in L_2(Q). \tag{3.9}$$

We then have:

Lemma 3.3: *For a given u satisfying* (3.9), *the linear form $v \to L_\varepsilon(v)$ defined by*

$$L(v) = + \int_0^T a_{\varepsilon_0}(u, v)\, dt + \int_0^T (u', v)\, dt - \varepsilon_2 \int_0^T (u''. v)\, dt +$$

$$+ \varepsilon_2(u'(T), v(T)) - \varepsilon_2(u'(0), v(0))$$

is continuous on X_ε.

From Lemma 3.1, 3.2 and 3.3 results:

Lemma 3.4: *For every $\varepsilon_0, \varepsilon_1, \varepsilon_2, \varepsilon_i > 0$, there exists a unique w_ε in X_ε satisfying:*

$$\pi_\varepsilon(w_\varepsilon, v) = L_\varepsilon(v), \qquad \forall v \in X_\varepsilon. \tag{3.11}$$

In addition, using (3.7) we obtain:

$$w_\varepsilon \; ; \rho_{\varepsilon_0} \frac{\partial w_\varepsilon}{\partial x_i}, \quad i = 1, ..., n \; ; \quad \frac{1}{\sqrt{\varepsilon_1}} M_{\varepsilon_0} P w_\varepsilon \; ;$$

$$\sqrt{\varepsilon_2}\, w_\varepsilon' \text{ stays in a bounded region of } L_2(Q) \qquad \text{as} \qquad \varepsilon \to 0, \tag{4.12}$$

and

$$w_\varepsilon(x, T) \text{ stays in a bounded region of } L_2(\Omega) \quad \text{as} \quad \varepsilon \to 0 . \qquad (3.13)$$

3.4 First Result Concerning Convergence

Proposition 3.1: *Under hypotheses* (1.2), (1.3), (3.9) *we have, as* $\varepsilon \to 0$,

$$w_\varepsilon \to 0 \quad \text{in} \quad L_2(Q) ; \qquad (3.14)$$

$$\rho_{\varepsilon 0} \frac{\partial w_\varepsilon}{\partial x_i} \to 0 \quad \text{in} \quad L_2(Q) , \qquad i = 1, ..., n . \qquad (3.15)$$

Proof. 1) From (3.12) we can extract a sequence from w_ε which we still denote by w_ε such that:

$$\left. \begin{array}{l} w_\varepsilon \to w \quad \text{in} \quad L_2(Q) \quad \text{weakly,} \\[2mm] \rho_{\varepsilon 0} \dfrac{\partial w_\varepsilon}{\partial x_i} \to w_i \quad \text{in} \quad L_2(Q) \quad \text{weakly,} \\[2mm] \dfrac{1}{\sqrt{\varepsilon_1}} M_{\varepsilon 0} Pw \to \chi_1 \quad \text{in} \quad L_2(Q) \quad \text{weakly,} \\[2mm] \sqrt{\varepsilon_2}\, w_\varepsilon' \to \chi_2 \quad \text{in} \quad L_2(Q) \quad \text{weakly.} \end{array} \right\} \qquad (3.16)$$

Necessarily, however (cf. Chapter 4 for remarks of the same type),

$$w_i = \frac{\partial w}{\partial x_i} ,$$
$$\chi_1 = 0 , \qquad \chi_2 = 0 .$$

But $M_{\varepsilon 0} Pw_\varepsilon \to Pw$ in the sense of distributions in Q, therefore

$$Pw = 0 . \qquad (3.17)$$

2) If Ω_η is the set of x in Ω at distance $\Gamma_1 > \eta$, we see that for every fixed $\eta > 0$, we have:

$$w_\varepsilon \to w \quad \text{in} \quad L_2(0, T; L_2(\Omega_\eta)) \text{ weakly,}$$

$$i = 1, ..., n \qquad (3.18)$$

$$\frac{\partial w_\varepsilon}{\partial x_i} \to \frac{\partial w}{\partial x_i} \quad \text{in} \quad L_2(0, T; L_2(\Omega_\eta)) \text{ weakly,}$$

whence

$$w_\varepsilon|_{\Sigma_0} \to w|_{\Sigma_0} \quad \text{in particular weakly in } L_2(\Sigma_0) .$$

Since $w_\varepsilon|_{\Sigma_0} = 0$ (because $w_\varepsilon \in X_\varepsilon \subset L_2(U_\varepsilon)$), we deduce that

$$w|_{\Sigma_0} = 0 . \qquad (3.19)$$

But we also have

$$Pw_\varepsilon \to 0 = Pw \quad \text{in} \quad L_2(0, T; L_2(\Omega)) \text{ weakly,} \qquad (3.20)$$

and therefore (by Lions–Magenes [1]):

$$\frac{\partial w_\varepsilon}{\partial v_A} \to \frac{\partial w}{\partial v_A} \quad \text{on} \quad \Sigma_0$$

in the sense, for example, of distributions of order 1 of Σ_0. But since $w_\varepsilon \in X_\varepsilon$, we have

$$\left.\frac{\partial w_\varepsilon}{\partial v_A}\right|_{\Sigma_0} = 0 ,$$

whence

$$\left.\frac{\partial w}{\partial v_A}\right|_{\Sigma_0} = 0 .$$

However, (3.17), (3.19), (3.21) imply that $w = 0$.

Since the limit of w_ε is unique, it is meaningful to extract a subsequence of w_ε. We therefore already have (3.14), (3.15) with *weak* convergence.

It remains to show the *strong* convergence. Now:

$$L_\varepsilon(w_\varepsilon) = + \int_0^T a_{\varepsilon_0}(u, w_\varepsilon)\, dt + \int_0^T (u', w_\varepsilon)\, dt + \varepsilon_2 \int_0^T (u'', w_\varepsilon)\, dt +$$

$$+ \varepsilon_2(u'(T), w_\varepsilon(T)) - \varepsilon_2(u'(0), w_\varepsilon(0)) \to 0 .$$

(For example,

$$\int_0^T a_{ij}\, \rho_{\varepsilon_0} \frac{\partial u}{\partial x_j}\, \rho_{\varepsilon_0} \frac{\partial w_\varepsilon}{\partial x_i}\, dx\, dt \to 0\,,$$

since $a_{ij}\, \rho_{\varepsilon_0} \dfrac{\partial u}{\partial x_j} \to a_{ij} \dfrac{\partial u}{\partial x_j}$ in $L_2(Q)$ strongly and $\rho_{\varepsilon_0} \dfrac{\partial w_\varepsilon}{\partial x_i} \to 0$ in $L_2(Q)$ weakly.)

Therefore $\pi_\varepsilon(w_\varepsilon, w_\varepsilon) \to 0$, which, in view of (3.7), leads to the result. ∎

Remark 3.1. We deduce from the fact that $\pi_\varepsilon(w_\varepsilon, w_\varepsilon) \to 0$ that

$$\| M_{\varepsilon_0} P w_\varepsilon \|_{L_2(Q)} = o(\sqrt{\varepsilon_1}) \tag{3.22}$$

where $o(\lambda)$ is a quantity such that $o(\lambda)/\lambda \to 0$ if $\lambda \to 0$.

This can be useful for a practical choice of ε_1; this estimate illustrates the *dominant* role of ε_1 by comparison with the other parameters ε_0, ε_2. This will be confirmed by the subsequent numerical results. ∎

3.5 Convergence of the QR Method

Theorem 3.1: *We suppose that the hypotheses* (1.2), (1.3) *hold. Let* g_0, g_1, u_0 *be given on* Σ_0 *and* Ω, *such that there exists a* u *(necessarily unique) satisfying* (1.7), (1.8), (1.9), (1.10) *and* (3.9). *Then:*

(i) *there exists a unique* u_ε *such that*

$$\begin{cases} u_\varepsilon,\ \rho_{\varepsilon_0} \dfrac{\partial u_\varepsilon}{\partial x_i} \in L_2(Q), \qquad i = 1, \ldots, n\,, \\[2mm] M_{\varepsilon_0} P u_\varepsilon \in L_2(Q)\,, \qquad \dfrac{\partial u_\varepsilon}{\partial t} \in L_2(Q) \end{cases} \tag{3.23}$$

and satisfying (2.6), (2.7), (2.8), (2.9);

(ii) *when* $\varepsilon = \{\varepsilon_0, \varepsilon_1, \varepsilon_2\} \to 0$, *we have:*

$$u_\varepsilon \to u\,, \quad \rho_{\varepsilon_0} \frac{\partial u_\varepsilon}{\partial x_i} \to \frac{\partial u}{\partial x_i} \quad \text{in} \quad L_2(Q) \text{ strongly,} \tag{3.24}$$

(iii) $\| M_{\varepsilon_0} P u_\varepsilon \|_{L_2(Q)} = o(\sqrt{\varepsilon_1})$.

Proof. 1) If w_ε is defined as in Lemma 3.5, let us set (for a moment; we will see that $\varphi_\varepsilon = u_\varepsilon$):

$$\varphi_\varepsilon = u - w_\varepsilon.$$

Taking first $v \in \mathscr{D}(Q)$ in (3.11), we obtain:

$$A_{\varepsilon_0} w_\varepsilon + w'_\varepsilon - \varepsilon_2 w''_\varepsilon + \frac{1}{\varepsilon_1} P^*(M_{\varepsilon_0}^2 Pw_\varepsilon) = A_{\varepsilon_0} u + u' - \varepsilon_2 u'' \qquad (3.25)$$

and since $Pu = 0$, we have:

$$P_{\varepsilon_0} \varphi_\varepsilon - \varepsilon_2 \varphi''_\varepsilon + \frac{1}{\varepsilon_1} P^*(M_{\varepsilon_0}^2 P\varphi_\varepsilon) = 0. \qquad (3.26)$$

Then, since $w_\varepsilon, \rho_{\varepsilon_0} \dfrac{\partial w_\varepsilon}{\partial x_i} \in L_2(Q)$, we have the same properties for φ_ε and since $w_\varepsilon|_{\Sigma_0} = 0$, we have:

$$\varphi_\varepsilon|_{\Sigma_0} = u|_{\Sigma_0} = g_0. \qquad (3.27)$$

Since $M_{\varepsilon_0} Pw_\varepsilon \in L_2(Q)$ and since (3.9) holds, we can define $\dfrac{\partial u}{\partial v_A}\Big|_{\Sigma_0}$ and therefore also $\dfrac{\partial \varphi_\varepsilon}{\partial v_A}\Big|_{\Sigma_0}$ and:

$$\frac{\partial \varphi_\varepsilon}{\partial v_A}\Big|_{\Sigma_0} = \frac{\partial u}{\partial v_A}\Big|_{\Sigma_0} = g_1. \qquad (3.28)$$

Now multiply (3.25) by $v \in X_\varepsilon$ and integrate by parts. The result is:

$$\int_0^T a_{\varepsilon_0}(w_\varepsilon, v)\, dt - \int_0^T (w_\varepsilon, v')\, dt + (w_\varepsilon(T), v(T)) - (w_\varepsilon(0), v(0)) +$$

$$+ \varepsilon_2 \int_0^T (w'_\varepsilon, v')\, dt - \varepsilon_2(w'_\varepsilon(T), v(T)) + \varepsilon_2(w'_\varepsilon(0), v(0))$$

$$+ \frac{1}{\varepsilon}(M_{\varepsilon_0} Pw_\varepsilon, M_{\varepsilon_0} Pv)_{L_2(Q)} - \frac{1}{\varepsilon_1}(M_{\varepsilon_0}^2 Pw_\varepsilon(T), v(T))$$

$$+ \frac{1}{\varepsilon_1}(M_{\varepsilon_0}^2 Pw_\varepsilon(0), v(0)) =$$

$$= \int_0^T a_{\varepsilon_0}(u, v)\, dt + \int_0^T (u', v)\, dt - \varepsilon_2 \int_0^T (u'', v)\, dt,$$

namely:

$$\pi_\varepsilon(w_\varepsilon, v) - (w_\varepsilon(0), v(0)) - \varepsilon_2(w'_\varepsilon(T), v(T)) + \varepsilon_2(w'_\varepsilon(0), v(0))$$

$$- \frac{1}{\varepsilon_1} (M^2_{\varepsilon_0} Pw_\varepsilon(T), v(T)) + \frac{1}{\varepsilon_1} (M^2_{\varepsilon_0} Pw_\varepsilon(0), v(0)) =$$

$$= L_\varepsilon(v) + \varepsilon_2(u'(0), v(0)) - \varepsilon_2(u'(T), v(T))$$

and since $\pi_\varepsilon(w_\varepsilon, v) = L_\varepsilon(v)$, we deduce that:

$$- w_\varepsilon(0) + \varepsilon_2 w'_\varepsilon(0) + \frac{1}{\varepsilon_1} M^2_{\varepsilon_0} Pw_\varepsilon(0) = \varepsilon_2 u'(0) +$$

$$+ \varepsilon_2 w'_\varepsilon(T) + \frac{1}{\varepsilon_1} M^2_{\varepsilon_0} Pw_\varepsilon(T) = \varepsilon_2 u'(T).$$

Then φ_ε satisfies (since $Pu = 0$, we have $Pu(0) = 0$):

$$\varphi_\varepsilon(0) - \varepsilon_2 \varphi'_\varepsilon(0) - \frac{1}{\varepsilon_1} M^2_{\varepsilon_0} P\varphi_\varepsilon(0) = u_0 - \varepsilon_2 u'(0) + \varepsilon_2 u'(0) = u_0 \quad \textbf{(3.29)}$$

$$\varepsilon_2 \varphi'_\varepsilon(T) + \frac{1}{\varepsilon_1} M^2_{\varepsilon_0} P\varphi_\varepsilon(0) = 0. \quad \textbf{(3.30)}$$

Reciprocally, if φ_ε satisfies these relations, $w_\varepsilon = u - \varphi_\varepsilon$ satisfies (3.11). This property (i) holds with $\varphi_\varepsilon = u_\varepsilon$.

2) The parts (ii), (iii) of the theorem are now a consequent of Proposition 3.1 and of Remark 3.1. ∎

4. VARIOUS REMARKS

4.1 Coefficients Depending On t

The preceding methods and results extend to the case where the coefficients a_{ij} dependent on x and t:

$$Au = - \sum_{i,j=1}^{n} \frac{\partial}{\partial x_i} \left(a_{ij}(x, t) \frac{\partial u}{\partial x_j} \right).$$

4.2 Different Geometry for Γ_0, Γ_1

The method and results may be adapted—as in Chapter 4—to the case where Γ_0 is a portion of $\Gamma_0 \cup \Gamma_1$, as in Figure 2.

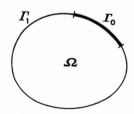

Figure 2

4.3 Order of Magnitude of g_0 and g_1

As soon as g_0 and (or) g_1 become ''too'' large, it suffices to ''neglect'' the portions of the boundary where this situation holds (cf. Chapter 4, Sections 4, 4.3).

4.4 Higher Order in x

The preceding method may be adapted to the case where A is an *elliptic operator of arbitrary order* (cf. Chapter 4, Section 4).[1]

4.5 Higher Orders in t

It seems probable (but the problem is open) that the preceding method can be extended to general parabolic operators (cf. Agrano-vich–Visik [1], R. Arima [1]). In the same vein, 5.3 below very probably extends to the cases studied by those authors.

5. QR PROBLEM ASSOCIATED WITH PROBLEM (II)

5.1 Preliminary Remarks

Analyzing the demonstration of Proposition 3.1, we observe the following: let $\tilde{w}_\varepsilon$ be *the* solution in X_ε of:

$$\pi_\varepsilon(\tilde{w}_\varepsilon, v) = L_\varepsilon(v) + (u_0^*, v(0)) \qquad \forall v \in X_\varepsilon \tag{5.1}$$

where u_0^* is chosen in *whatever* fashion in $L_2(\Omega)$; then:

$$\tilde{w}_\varepsilon \to 0 \quad \text{in} \quad L_2(Q) \text{ weakly,}$$

$$\rho_{\varepsilon_0} \frac{\partial \tilde{w}_\varepsilon}{\partial x_i} \to 0 \quad \text{in} \quad L_2(Q) \text{ weakly,} \tag{5.2}$$

in other words, *we have the same conclusions as in Proposition* 3.1.

We verify that $\tilde{w}_\varepsilon$ satisfies the *same* conditions as w_ε *except* for the boundary conditions at $t = 0$ which become:

$$\tilde{w}_\varepsilon(0) - \varepsilon_2 \tilde{w}_\varepsilon'(0) - \frac{1}{\varepsilon_1} M_{\varepsilon_0}^2 P\tilde{w}_\varepsilon(0) = u_0^* - \varepsilon_2 u'(0). \tag{5.3}$$

[1] Provided that uniqueness of the Cauchy problem holds.

Then, if we introduce:

$$\tilde{u}_\varepsilon = u - \tilde{w}_\varepsilon, \tag{5.4}$$

we obtain:

$$\frac{1}{\varepsilon_1} P^*(M_{\varepsilon_0}^2 P\tilde{u}_\varepsilon) + P_\varepsilon \tilde{u}_\varepsilon - \varepsilon_2 \frac{\partial^2 \tilde{u}_\varepsilon}{\partial t^2} = 0, \tag{5.5}$$

$$\tilde{u}_\varepsilon |_{\Sigma_0} = g_0, \tag{5.6}$$

$$\left.\frac{\partial \tilde{u}_\varepsilon}{\partial \nu_A}\right|_{\Sigma_0} = g_1, \tag{5.7}$$

$$\tilde{u}_\varepsilon(x, 0) - \varepsilon_2 \frac{\partial \tilde{u}_\varepsilon}{\partial t}(x, 0) - \frac{1}{\varepsilon_1} M_{\varepsilon_0}^2 P\tilde{u}_\varepsilon(x, 0) = u_0(x) - u_0^*(x), \tag{5.8}$$

$$\varepsilon_2 \frac{\partial u_\varepsilon}{\partial t}(x, T) + \frac{1}{\varepsilon_1} M_{\varepsilon_0}^2 (Pu_\varepsilon)(x, T) = 0. \tag{5.9}$$

The conclusion is: for any $u_0^* \in L_2(\Omega)$, $\tilde{u}_\varepsilon$ *the solution* of (5.5), ... , (5.9) *furnishes an approximation to u.* ∎

Naturally, the reason for this result is evidently that *being given g_0 and g_1 it is sufficient to determine u in a unique fashion.*

We can consider the prescription of u_0 as a *supplementary condition* which must be taken account of in "the best way" in the QR solution proposed in Section 2.

5.2 Problem (II)

In problem (II) (cf. 1.2) u_0 is not given.

One QR method which seems reasonable—and which is *convergent* as follows from the preceding remarks—consists in taking $u_0^* = u_0$, so as *to eliminate all unknowns* from (5.5), . . . , (5.9).

We are finally led to choose *as an approximation to u in the problem* (II) *the* solution U_ε of:

$$\frac{1}{\varepsilon_1} P^*(M_{\varepsilon_0}^2 PU_\varepsilon) + P_\varepsilon U_\varepsilon - \varepsilon_2 \frac{\partial^2 U_\varepsilon}{\partial t^2} = 0 \tag{5.10}$$

$$U_\varepsilon |_{\Sigma_0} = g_0 \tag{5.11}$$

$$\left.\frac{\partial U_\varepsilon}{\partial v_A}\right|_{\Sigma_0} = g_1 \qquad\qquad (5.12)$$

$$U_\varepsilon(x, 0) - \varepsilon_2 \frac{\partial U_\varepsilon}{\partial t}(x, 0) - \frac{1}{\varepsilon_1} M_{\varepsilon_0}^2 P U_\varepsilon(x, 0) = 0 , \qquad (5.13)$$

$$\varepsilon_2 \frac{\partial U_\varepsilon}{\partial t}(x, T) + \frac{1}{\varepsilon_1} M_{\varepsilon_0}^2 P U_\varepsilon(x, T) = 0 . \ \blacksquare \qquad (5.14)$$

Remark 5.1. We have, for one example, made a number of trials for various u_0^* (cf. Section 8). $\blacksquare$

5.3 Another Method

We will now show how to adapt to problem (II) the QR method of Chapter 4, Section 8 (we could also perform a corresponding adaptation for problem (I)). $\blacksquare$

We take (with the notation of Chapter 4, Section 8):

$$\mathcal{H} = L_2(Q) \qquad\qquad (5.15)$$

and, with P defined by (1.6), we take:

$$D(P) = \left\{ v \mid v \in L_2(Q), Pv \in L_2(Q) ; \ v|_{\Sigma_0} = \left.\frac{\partial v}{\partial v_A}\right|_{\Sigma_0} = 0 \right\} . \qquad (5.16)$$

The space $D(P)$ is *dense* in $L_2(Q)$. Definition (5.16) is meaningful (for this use Lions–Magenes [1], Chapter 5).

We show that:

$$P \text{ is closed} \qquad\qquad (5.17)$$

$$P(D(P)) \text{ is dense in } L_2(Q) \qquad\qquad (5.18)$$

(this last fact is not indispensable; it can be obtained using the uniqueness of the solution of the Cauchy problem). $\blacksquare$

We introduce (compare Chapter 4, Sections 8, 8.4) $\Phi = \Phi(x, t)$ a function "sufficiently regular" in $\bar{Q}$ satisfying:

$$\Phi|_{\Sigma_0} = g_0 , \qquad\qquad (5.19)$$

$$\left.\frac{\partial \Phi}{\partial v_A}\right|_{\Sigma_0} = g_1 . \qquad\qquad (5.20)$$

Then, if u satisfies (1.7), (1.8), (1.9), the function

$$w = u - \Phi \tag{5.21}$$

satisfies:

$$Pw = f \quad \text{(where} \quad f = - P\Phi \quad \text{is known)} \tag{5.22}$$

$$w\,|_{\Sigma_0} = 0\,, \tag{5.23}$$

$$\frac{\partial w}{\partial v_A}\bigg|_{\Sigma_0} = 0\,. \tag{5.24}$$

We now apply the general method of Theorem 8.2, Chapter 4 [with A replaced by P given by (1.6) and (5.16)]. ∎

We therefore determine u_ε as *the* solution in $D(P)$ of:

$$(Pu_\varepsilon, Pv) + \varepsilon(u_\varepsilon, v) = (f, Pv) \quad (^1) \qquad \forall v \in D(P)\,. \tag{5.25}$$

From Theorem 8.2, Chapter 4, we have:

$$u_\varepsilon + \Phi \to u \quad \text{in} \quad L_2(Q) \quad \text{when} \quad \varepsilon(> 0) \to 0, P(u_\varepsilon + \Phi) \to 0 \quad \text{in} \quad L_2(Q). \tag{5.26}$$

It remains to *interpret* the (variational) problem (5.25). We have to begin with (taking $v = \varphi \in \mathscr{D}(Q)$):

$$P^* \, Pu_\varepsilon + \varepsilon u_\varepsilon = P^* f\,. \tag{5.27}$$

To obtain the boundary conditions, multiply (5.27) by v [in $D(P)$] and integrate over Q. The result is:

$$-\int_\Omega (Pu_\varepsilon)\,(x, T)\, v(x, T)\, dx + \int_\Omega (Pu_\varepsilon)\,(x,\,0)\, v(x, 0)\, dx + \int_{\Sigma_0 \cup \Sigma_1} \frac{\partial Pu_\varepsilon}{\partial v_{A^*}}\, v\, d\Sigma -$$

$$-\int_{\Sigma_0 \cup \Sigma_1} Pu_\varepsilon \cdot \frac{\partial v}{\partial v_A}\, d\Sigma + (Pu_\varepsilon, Pv) + \varepsilon(u_\varepsilon, v) = -\int_\Omega f(x, T)\, v(x, T)\, dx +$$

$$+\int_\Omega f(x, 0)\, v(x, 0)\, dx + \int_{\Sigma_0 \cup \Sigma_1} \frac{\partial f}{\partial v_{A^*}}\, v\, d\Sigma - \int_{\Sigma_0 \cup \Sigma_1} f\, \frac{\partial v}{\partial v_A}\, d\Sigma + (f, Pv)\,.$$

$(^1)$ **Where** $(\varphi, \psi) = \displaystyle\int_Q \varphi(x, t)\, \psi(x, t)\, dx\, dt\,.$

Taking account of (5.25) and of the fact that $v|_{\Sigma_0} = \dfrac{\partial v}{\partial v_A}\Big|_{\Sigma_0} = 0$, this reduces to:

$$
-\int_\Omega ((Pu_\varepsilon)(x, T) - f(x, T))\, v(x, T)\, dx + \int_\Omega (Pu_\varepsilon(x, 0) - f(x, 0))\, v(x, 0)\, dx +
$$

$$
+ \int_{\Sigma_1} \left(\frac{\partial(Pu_\varepsilon)}{\partial v_{A^*}} - \frac{\partial f}{\partial v_{A^*}} \right) v\, d\Sigma_1 - \int_{\Sigma_1} (Pu_\varepsilon - f)\frac{\partial v}{\partial v_A}\, d\Sigma_1 = 0 \qquad \forall v \in D(P),
$$

which implies:

$$
Pu_\varepsilon = f \quad \text{on} \quad \Sigma_1, \tag{5.28}
$$

$$
\frac{\partial(Pu_\varepsilon)}{\partial v_{A^*}} = \frac{\partial f}{\partial v_{A^*}} \quad \text{on} \quad \Sigma_1, \tag{5.29}
$$

and

$$
Pu_\varepsilon(x, 0) = f(x, 0), \tag{5.30}
$$

$$
Pu_\varepsilon(x, T) = f(x, T). \tag{5.31}
$$

We thus have finally:

Theorem 5.1: *We choose a function Φ satisfying (5.19), (5.20). We determine u_ε, the solution of (5.27) with the boundary conditions:*

$$
u_\varepsilon|_{\Sigma_0} = 0, \qquad \frac{\partial u_\varepsilon}{\partial v_A}\Big|_{\Sigma_0} = 0
$$

and (5.28), (5.29), (5.30), (5.31).
    Then we have (5.26). ∎

Remark 5.2. We can show that it is permissible to take $\varepsilon = 0$ in the preceding method (cf. Chapter 4, Remark 8.4). ∎

Remark 5.3. We can make remarks here analogous to Remarks 8.5 to 8.8 of Chapter 4. ∎

Remark 5.4. A disadvantage of the preceding method (as also that of Section 8, Chapter 4) is the apparent impossibility of extending it to *nonlinear* problems (cf. Chapter 6). ∎

6. FURTHER EXAMPLES

6.1 Noncylindrical Domain (cf. also Chapter 1, 13.1)

Let $u = u(x, t)$ be a function which we know satisfies:

$$\frac{\partial u}{\partial t} - \frac{\partial^2 u}{\partial x^2} = 0 \quad \text{in the domain} \quad 0 \leqslant x \leqslant S(t) \quad \text{for} \quad t > 0, \text{(cf. Figure 3)}, \qquad (6.1)$$

$$u = g \quad \text{given on} \quad \Sigma \qquad (6.2)$$

$$\frac{\partial u}{\partial x} = -S'(t) \quad \text{on} \quad \Sigma, \qquad (6.3)$$

$$u(x, 0) = u_0(x) \quad \text{given.} \qquad (6.4)$$

The problem is to determine u, when g, S and u_0 are given.

Remark 6.1. Naturally (6.3) can be replaced by giving the normal derivative of u on (Σ). ∎

Remark 6.2. The condition of (6.4) is not indispensable, for reasons analogous to that of Section 5. ∎

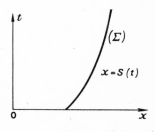

Figure 3

QR Method

We take:

$$\rho_{\varepsilon_0}(x, t) = \begin{cases} \dfrac{1}{\varepsilon_0} x & \text{for} \quad 0 \leqslant x \leqslant \varepsilon_0, \\[2mm] 1 & \text{elsewhere;} \end{cases} \qquad (6.5)$$

$$M_{\varepsilon_0}(x, t) = \begin{cases} 0 & \text{for} \quad 0 \leqslant x \leqslant \varepsilon_0, \\[2mm] \dfrac{1}{\varepsilon_0}(x - \varepsilon_0) & \text{for} \quad \varepsilon_0 \leqslant x \leqslant 2\varepsilon_0, \\[2mm] 1 & \text{elsewhere.} \end{cases} \qquad (6.6)$$

We adapt the notation of Section 2 in the following fashion; we set:

$$A_{\varepsilon_0} = -\frac{\partial}{\partial x}\left(\rho_{\varepsilon_0}^2 \frac{\partial}{\partial x}\right),$$

$$P_{\varepsilon_0} = \frac{\partial}{\partial t} + A_{\varepsilon_0}. \qquad (6.7)$$

We denote by $u_\varepsilon = u_{\varepsilon_0, \varepsilon_1, \varepsilon_2}$ *the* solution of:

$$\frac{1}{\varepsilon_1} P^*(M_{\varepsilon_0}^2 \, Pu_\varepsilon) + P_{\varepsilon_0} \, u_\varepsilon - \varepsilon_2 \frac{\partial^2 u_\varepsilon}{\partial t^2} = 0 \,, \tag{6.8}$$

with the boundary conditions:

$$u_\varepsilon \big|_\Sigma = g \,, \tag{6.9}$$

$$\frac{\partial u_\varepsilon}{\partial x}\bigg|_\Sigma = -S' \,, \tag{6.10}$$

1) $u_\varepsilon(x, 0) - \varepsilon_2 \dfrac{\partial u_\varepsilon}{\partial t}(x, 0) - \dfrac{1}{\varepsilon_1} M_{\varepsilon_0}^2 \, Pu_\varepsilon(x, 0) = u_0(x) \,, \qquad 0 \leqslant x \leqslant S(0) \,, \tag{6.11}$

(6.12) $\varepsilon_2 \dfrac{\partial u_\varepsilon}{\partial t}(x, T) + \dfrac{1}{\varepsilon_1} M_{\varepsilon_0}^2 (Pu_\varepsilon)(x, T) = 0 \,, \qquad 0 \leqslant x \leqslant S(T) \,, \tag{6.12}$

where T has been chosen taking account of the given data g and S and of practical considerations. ∎

We show that the problem (6.8), . . . , (6.12) is properly posed and that as $\varepsilon = \{\varepsilon_0, \varepsilon_1, \varepsilon_2\} \to 0$, u_ε converges towards u in a topology corresponding to that of Theorem 3.1. ∎

6.2   A Hyperbolic Case

The problems which precede have many variants. *For example*, let us suppose, with the geometry of Figure 1, that we have:

$$\frac{\partial^2 u}{\partial t^2} + Au = 0 \quad \text{in} \quad \Omega \times \,]0, T[\,, \tag{6.13}$$

with:

$$u\big|_{\Sigma_0} = g_0 \,, \tag{6.14}$$

$$\frac{\partial u}{\partial \nu_A}\bigg|_{\Sigma_0} = g_1 \,; \tag{6.15}$$

(it is not necessary to know $u(x, 0)$, $\dfrac{\partial u}{\partial t}(x, 0)$, cf. Section 5).

The problem is then to determine u in $\Omega \times \,]0, T[$.

We can extend the QR method, using the preceding ideas to-gether with "elliptic regularization" for hyperbolic problems (cf. W. Strauss [1]). We could also utilize "parabolic regularization" introduced in Lions–Magenes [1].

7. NUMERICAL APPLICATIONS (I)

7.1 The Problem

We consider to begin with an example of problem (I), Section 1 and Section 2.

Let $u(x, t)$ be a function satisfying:

$$\frac{\partial u}{\partial t} - \frac{\partial^2 u}{\partial x^2} = 0 \qquad \begin{array}{l} 0 < x < x_0, \\ 0 < t < T, \end{array} \tag{7.1}$$

$$u(x, 0) = u_0(x) \quad \text{given}, \tag{7.2}$$

$$u(x_0, t) = g_0(t), \tag{7.3}$$

$$\frac{\partial u}{\partial x}(x_0, t) = g_1(t), \tag{7.4}$$

where g_0 and g_1 are known functions of t on $(0, T)$. *The problem is to determine $u(0, t)$.*

7.2 The Associated QR Problem

We introduce, in agreement with the general theory (Section 2),

$$\rho_{\varepsilon_0}(x) = \begin{cases} 1 & \text{if } x \geqslant \varepsilon_0, \\ \dfrac{x}{\varepsilon_0} & \text{if } 0 \leqslant x < \varepsilon_0 \end{cases} \qquad \text{(cf. Figure 4)} \tag{7.5}$$

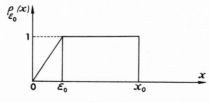

Figure 4

$$M_{\varepsilon_0}(x) = \begin{cases} 1 & \text{if } x \geqslant 2\,\varepsilon_0\,, \\ 0 & \text{if } 0 < x < \varepsilon_0\,, \\ \text{continuous, bounded between 0 and 1 for } x \in (\varepsilon_0,\, 2\,\varepsilon_0) \end{cases}$$

(cf. Figure 5).   (7.6)

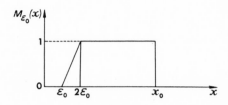

Figure 5

Let us write:

$$P \;\;= \frac{\partial}{\partial t} - \frac{\partial^2}{\partial x^2}\,, \tag{7.7}$$

$$P^* = -\frac{\partial}{\partial t} - \frac{\partial^2}{\partial x^2}\,, \tag{7.8}$$

$$P_{\varepsilon_0} = \frac{\partial}{\partial t} - \frac{\partial}{\partial x}\left[\rho_{\varepsilon_0}^2\,\frac{\partial}{\partial x}\right]. \tag{7.9}$$

Let ε_1 and ε_2 be two small numbers, eventually tending towards zero. We write $\varepsilon = (\varepsilon_0,\, \varepsilon_1,\, \varepsilon_2)$.

The QR method leads to solving:

$$\frac{1}{\varepsilon_1}\,P^*(M_{\varepsilon_0}^2\,Pu_\varepsilon) + P_{\varepsilon_0}\,u_\varepsilon - \varepsilon_2\,\frac{\partial^2 u_\varepsilon}{\partial t^2} = 0 \qquad \begin{array}{l} 0 < x < x_0\,, \\ 0 < t < T\,, \end{array} \tag{7.10}$$

$$u_\varepsilon(x,\,0) - \varepsilon_2\,\frac{\partial u_\varepsilon}{\partial t}\,(x,\,0) - \frac{1}{\varepsilon_1}\,[M_{\varepsilon_0}^2\,Pu_\varepsilon(x,\,0)] = u_0(x)\,, \tag{7.11}$$

$$u_\varepsilon(x_0,\,t) = g_0(t)\,, \tag{7.12}$$

$$\frac{\partial u_\varepsilon}{\partial x}\,(x_0,\,t) = g_1(t)\,, \tag{7.13}$$

$$\varepsilon_2\,\frac{\partial u_\varepsilon}{\partial t}\,(x,\,T) + \frac{1}{\varepsilon_1}\,M_{\varepsilon_0}^2\,Pu_\varepsilon(x,\,T) = 0\,. \tag{7.14}$$

7.3 Finite Difference Equations

We set:

$$\Delta x = \frac{x_0}{J}, \quad \Delta t = \frac{T}{N},$$
$$u_j^n \simeq u(j\,\Delta x, n\,\Delta t).$$

Developing Eq. (7.10), we obtain (replacing $M_{\varepsilon_0}^2$ by M^2 and $\rho_{\varepsilon_0}^2$ by ρ^2):

$$-\left[\frac{M^2}{\varepsilon_1} + \varepsilon_2\right]\frac{\partial^2 u}{\partial t^2} + \frac{\partial u}{\partial t} + \frac{M^2}{\varepsilon_1}\frac{\partial}{\partial t}\left(\frac{\partial^2 u}{\partial x^2}\right) - \frac{1}{\varepsilon_1}\frac{\partial^2}{\partial x^2}\left(M^2\frac{\partial u}{\partial t}\right) + $$
$$+ \frac{1}{\varepsilon_1}\frac{\partial^2}{\partial x^2}\left(M^2\frac{\partial^2 u}{\partial x^2}\right) - \frac{\partial}{\partial x}\left(\rho^2\frac{\partial u}{\partial x}\right) = 0. \tag{7.15}$$

Equation (7.15) is approximated to by a finite difference equation, estimating the derivatives by differential quotients:

$$\left(\frac{\partial f}{\partial t}\right)_j^n \simeq \frac{f_j^{n+1} - f_j^{n-1}}{2\,\Delta t}, \tag{7.16}$$

$$\left(\frac{\partial^2 f}{\partial t^2}\right)_j^n \simeq \frac{f_j^{n+1} - 2f_j^n + f_j^{n-1}}{\Delta t^2}, \tag{7.17}$$

$$\left(\frac{\partial^2 f}{\partial x^2}\right)_j^n \simeq \frac{f_{j+1}^n - 2f_j^n + f_{j-1}^n}{\Delta x^2}. \tag{7.18}$$

The term $\dfrac{\partial}{\partial x}\left(\rho^2\dfrac{\partial u}{\partial x}\right)$ is discretized in the following way:

$$\left(\frac{\partial}{\partial x}\left(\rho^2\frac{\partial u}{\partial x}\right)\right)_j^n \simeq \frac{1}{2\,\Delta x}(\rho_{j+1}^2(u_{j+1}^n - u_j^n) - \rho_{j-1}^2(u_j^n - u_{j-1}^n)). \tag{7.19}$$

We thus obtain a system of finite difference equations of the form:

$$A_j u_{j+1}^{n+1} + B_j u_j^{n+1} + C_j u_{j-1}^{n+1} + D_j u_{j+2}^n + E_j u_{j+1}^n + F_j u_j^n + G_j u_{j-1}^n + $$
$$+ H_j u_{j-2}^n + I_j u_{j+1}^{n-1} + J_j u^{n-1} + K_j u_{j-1}^{n-1} = 0,$$

where the set of points associated with the point (j, n) is represented in Figure 6. The system of Eqs. (7.10) is written for $1 \leqslant j \leqslant J - 1$ and $1 \leqslant n \leqslant N - 1$.

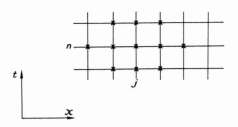

The determination of the solution of this system is possible because of the behavior of the functions ρ and M in the neighborhood of $x = 0$ and the boundary conditions at $x = x_0$, $t = 0$ and $t = T$. These are taken into account by use of the finite difference equations:

$$u_j^0 = u_0(j\,\Delta x) \qquad\qquad \text{for} \quad (7.11),\ (^1) \qquad\qquad (7.21)$$

$$u_J^n = g_0(n\,\Delta t) \qquad\qquad \text{for} \quad (7.12), \qquad\qquad (7.22)$$

$$u_{J+1}^n - u_{J-1}^n = 2\,\Delta x\, g_1(n\,\Delta t) \quad \text{for} \quad (7.13). \qquad\qquad (7.23)$$

The condition (7.14) at time T can be written in the form:

$$\left(\varepsilon_2 + \frac{M^2}{\varepsilon_1}\right)\frac{\partial u_\varepsilon}{\partial t}(x,\,T) - \frac{M^2}{\varepsilon_1}\frac{\partial^2 u_\varepsilon}{\partial x^2}(x,\,T) = 0. \qquad (7.24)$$

We can associate with this condition various relations for finite differences according to the estimate of the derivatives which appear. In fact, having used:

$$\left(\frac{\partial u}{\partial t}\right)_j^N \simeq \frac{u_j^N - u_j^{N-1}}{\Delta t}, \qquad\qquad (7.25)$$

the three following possibilities have been studied (cf. Figure 7):

(i)

$$\left(\frac{\partial^2 u}{\partial x^2}\right)_j^N \simeq \frac{u_{j+1}^N - 2\,u_j^N + u_{j-1}^N}{\Delta x^2}, \qquad\qquad (7.26)$$

$(^1)$ This is done at the end of a certain simplification and to verify the weak influence of certain terms at $t = 0$ (cf. also Section 8).

(ii)

$$\left(\frac{\partial^2 u}{\partial x^2}\right)_j^N \simeq \frac{1}{2} \frac{u_{j+1}^N - 2u_j^N + u_{j-1}^N}{\Delta x^2} + \frac{u_{j+1}^{N-1} - 2u_j^{N-1} + u_{j-1}^{N-1}}{\Delta x^2}. \tag{7.27}$$

(iii)

$$\left(\frac{\partial^2 u}{\partial x^2}\right)_j^N \simeq \frac{1}{2} \frac{-u_{j+2}^N + 16u_{j+1}^N - 30u_j^N + 16u_{j-1}^N - u_{j-2}^N}{12\,\Delta x^2} +$$
$$+ \frac{-u_{j+2}^{N-1} + 16u_{j+1}^{N-1} - 30u_j^{N-1} - 16u_{j-1}^{N-1} - u_{j-2}^{N-1}}{12\,\Delta x^2}. \tag{7.28}$$

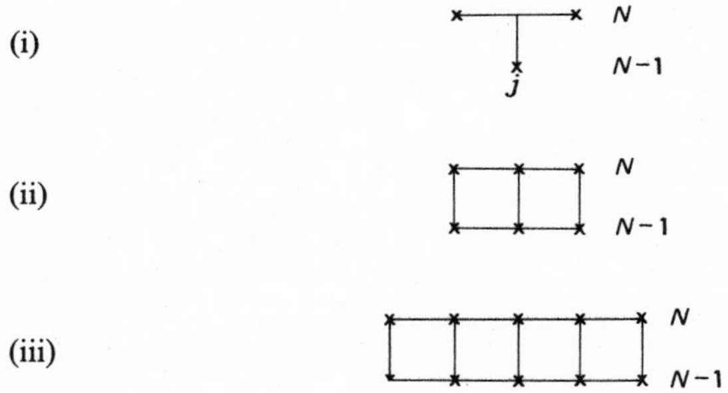

(i)

(ii)

(iii)

Figure 7

The system of Eqs. (7.20) combined with relations (7.21), (7.22), (7.23) and with one of the equations derived from (7.26), (7.27) and (7.28) can be written in the matrix form:

$$AU = F, \tag{7.21}$$

where A is an 11-diagonal matrice in case (7.26) (cf. Figure 8) or (7.27) (cf. Figure 9) and 13-diagonal in the Case (7.28) (cf. Figure 10).

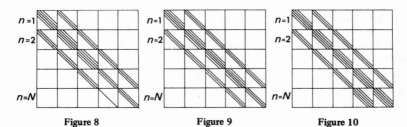

Figure 8   Figure 9   Figure 10

7.4 Numerical Results

The method just described has been applied to a simple analytic example in order to obtain a basis for comparison of results. In particular we have studied the influence of the following parameters: ε_0, ε_1, ε_2, N, T and thus the influence of the discretization used for (7.14).

7.4.1 ANALYTIC EXPRESSION

We choose

$$u(x, t) = \exp\left(-\frac{\pi^2}{4}t\right)\left(\sin\frac{\pi x}{2} + 3\cos\frac{\pi x}{2}\right),$$

which is the solution of the problem:

$$\frac{\partial u}{\partial t} - \frac{\partial^2 u}{\partial x^2} = 0 \qquad \begin{array}{c} t > 0 \\ 0 < x < 1 \end{array};$$

$$u(x, 0) = \sin\frac{\pi x}{2} + 3\cos\frac{\pi x}{2};$$

$$u(1, t) = \exp\left(-\frac{\pi^2}{4}t\right);$$

$$\frac{\partial u}{\partial x}(1, t) = -3\exp\left(-\frac{\pi^2}{4}t\right).$$

All the numerical trials were made with:

$$\Delta x = \frac{1}{10} \quad \text{or} \quad \Delta x = \frac{1}{25}.$$

7.4.2 INFLUENCE OF ε_0

Graph 1 represents the solutions obtained at $t = 0.5$ for various values of ε_0.

We see then that the value of the parameter ε_0 practically has no influence on the accuracy of the results.

For these calculations we had to vary from 0 to $T = 0.6$ by steps of 10^{-2}. The results obtained at the 50th step ($t = 0.5$) are very satisfactory; the curve obtained for $\varepsilon_0 = 0.25$ is nonetheless a little closer than the others to the actual solution.

7.4.3 INFLUENCE OF ε_1 (GRAPH 2)

We had ε_1 vary from 10 to 10^{-5}. On Graph 2 we reproduce the profile of the solution after 10 time steps ($t = 0.1$; $\Delta t = 10^{-2}$). The curves all correspond to a final instant $T = 0.2$.

The results, agreeing with the theory (cf. Remark 3.1), are extremely sensitive to variations in ε_1. We see on Graph 2 that when $\varepsilon_1 \to 0$, $u_\varepsilon \to u$.

7.4.4 INFLUENCE OF ε_2

The calculations carried out do not allow us to exhibit any evidence of an influence of the parameter ε_2.

7.4.5 INFLUENCE OF NUMBER N OF STEPS IN TIME WITH T FIXED (GRAPH 3)

Letting t vary in the interval $(0, 0.3)$, we varied the couple $(N, \Delta t)$. We drew the profiles for $t = 0.15$.

Graph 3 shows that a finer partition in time does not appreciably improve the results.

7.4.6 INFLUENCE OF CHOICE OF T

7.4.6.1 Disturbances introduced by condition (7.14) at $t = T$

The solution $u_\varepsilon(x, t)$ undergoes important disturbances in the neighborhood of $t = T$, due, as is to be expected, to the introduction of condition (7.14) at $t = T$.

Graph 4 illustrates the distortion of the solution as T is approached, here equal to 0.2.

Graph 4 is to be compared with Graph 5 showing the exact solution.

7.4.6.2 Influence of T on the solution at a given time t (Graph 6)

Graph 6 shows that to obtain good accuracy at $t = 0.3$, it is necessary to use the scheme (7.20) for a time T greater than $t = 0.3$.

7.4.7 INFLUENCE OF THE DISCRETIZATION METHOD USED FOR CONDITION (7.14)

All the preceding calculations have been carried out with the discretization scheme of (7.27) for six points.

In (7.3), we indicated two other possible schemes: (7.26) and (7.28). To compare them, we carried out calculations with the following parameters: $\Delta x = 0.04$, $\Delta t = 0.01$ $(T = 0.2)$.

Remark 7.1. Schemes (7.26) and (7.27) require the conditions:

$$\left.\begin{array}{l} \rho(0) \ = 0 \\ M_\varepsilon(0) \ = M_\varepsilon(\Delta x) = 0 \end{array}\right\} \forall t \in (0, T),$$

while (7.28) requires the supplementary condition:

$$M_\varepsilon(2 \, \Delta x) = 0 \qquad \forall t \in (0, T). \ \blacksquare$$

We compared the results obtained by the three schemes at two values of t (cf. Graphs 7 and 8).

Remark 7.2. The improvement of the discretization scheme at the point $t = T$ leads to a very important gain in accuracy.

At $x = 2 \, \Delta x = 0.08$, at $t = 0.1$, the accuracy goes from 5% (for the scheme at 4 points) to 1.5% for the scheme at 10 points.

At $t = T$, it goes from 36% (for the scheme with 4 points) to 1% for the scheme with 10 points (a natural improvement). $\blacksquare$

Remark 7.3. With schemes (7.26) and (7.27), there do not appear to be disturbances due to the regularization functions $\rho_\varepsilon(x)$ and $M_\varepsilon(x)$.

On the contrary, with the 10 point scheme (7.28), the results are sensitive to the influence of the functions ρ_ε and M_ε: on Graphs 7 and 8 we observe a discontinuity in the neighborhood of $x = 0$; a disturbance which affects only the extremity of the curve and therefore not an actual extrapolation out to $x = 0$. $\blacksquare$

8. NUMERICAL APPLICATIONS (II)

8.1 The Problem

We consider a situation analogous to that of Section 7, but *without an initial condition this time* (cf. the general theory, problem II, Sections 1 and 5).

8.2 Associated QR Problem

We introduce the same functions ρ_ε and M_ε as in Section 7; in agreement with the general theory of 5.2, the problem to solve is:

$$\frac{1}{\varepsilon_1} P^*(M_{\varepsilon_0}^2 \, Pu_\varepsilon) + P_{\varepsilon_0} \, u_\varepsilon - \varepsilon_2 \frac{\partial^2 u_\varepsilon}{\partial t^2} = 0, \qquad \begin{array}{c} 0 < x < x_0, \\ 0 < t < T \end{array} \tag{8.1}$$

$$u_\varepsilon(x_0, t) = g_0(t), \tag{8.2}$$

$$\frac{\partial u_\varepsilon}{\partial x}(x_0, t) = g_1(t), \tag{8.3}$$

$$\varepsilon_2 \frac{\partial u_\varepsilon}{\partial t}(x, T) + \frac{1}{\varepsilon_1} M_{\varepsilon_0}^2 \, Pu(x, T) = 0, \tag{8.4}$$

$$u_\varepsilon(x, 0) - \varepsilon_2 \frac{\partial u_\varepsilon}{\partial t}(x, 0) - \frac{1}{\varepsilon_1} [M_{\varepsilon_0}^2 \, Pu_\varepsilon(x, 0)] = u_0(x) - u^*(x), \tag{8.5}$$

where P, P^* and P_{ε_0} have the same definitions as in (7.7), (7.8), (7.9).

Remark 8.1. Following the method of 7.2, we replaced (7.11) by (8.5), leaving the other equations unchanged. ∎

Remark 8.2. The function $u^*(x)$ of (8.5) can be arbitrary (cf. Section 5) and we will verify with examples that, except naturally in the neighborhood of $t = 0$, the numerical results are essentially independent of the choice of $u^*(x)$. ∎

8.3 Difference Equations

$$\Delta x = \frac{x_0}{J} \qquad \Delta t = \frac{T}{N}.$$

We discretize as in Section 7, the only change now being due to relation (8.5).

8.4 Numerical Results

8.4.1 ANALYTIC EXPRESSION

As in 7.4.1, we choose:

$$\Delta x = 10^{-1}, \qquad \Delta t = 10^{-2}, \qquad T = 0.5.$$

8.4.2 CHOICE OF $u^*(x)$

We carried out the calculations with four functions:

$$u^*(x) = 0 \,, \qquad\qquad \text{(identical to Section 7)} \qquad\qquad \textbf{(1)}$$

$$u^*(x) = x^4 \,, \qquad\qquad \textbf{(2)}$$

$$u^*(x) = u_0(x) = \sin\frac{\pi x}{2} + 3\cos\frac{\pi x}{2} \,, \qquad\qquad \textbf{(3)}$$

$$u^*(x) = 1 \,. \qquad\qquad \textbf{(4)}$$

8.4.3 RESULTS

Graphs 9, 10, 11, respectively, represent the solutions obtained at three times $t = 0.2$, $t = T = 0.5$ and finally $t = 0$.

Remark 8.3. At time $t = 0$, the solutions are naturally very different (cf. Remark 8.2). Then, as soon as t increases, the solution no longer depends on $u^*(x)$.

9. PROBLEMS WITH SINGULARITIES

9.1 The Problem

The notation is the same as in Section 7 and Section 8. We choose as functions $u(x, t)$ the solutions of the heat equation, with singularities:

$$u(x, t) = \frac{1}{\sqrt{\beta + t}} \exp\left[\frac{-x^2}{4(\beta + t)}\right] \,, \qquad\qquad \textbf{(9.1)}$$

$$u(x, t) = \frac{1}{\sqrt{\beta + t}} \exp\left[\frac{-(1 - x)^2}{4(\beta + t)}\right] \,, \qquad\qquad \textbf{(9.2)}$$

where β is positive and "small"; the singularities then occur at $x = 0$ or at $x = 1$, in the neighborhood of $t = 0$.

9.2 The Associated QR Problem

We have two possibilities: either the method of Section 7, or a particular treatment of the singularities (as done, for example, in

Chapter 4, Section 6). We began with the method of Section 7, to which we adhered, because the results obtained were satisfactory.

9.3 Numerical Results

These are given in Graphs 12 to 15 for two values of t.

10. PARABOLIC PROBLEM IN A NONCYLINDRICAL DOMAIN

10.1 The Initial Value Problem and the Associated QR Problem

We consider the problem of 6.1 with:

$$S(t) = \lambda(1 + t), \qquad \lambda > 0. \tag{10.1}$$

To test the calculations, we choose:

$$u(x, t) = \exp(-\lambda x + \lambda^2(1 + t)), \tag{10.2}$$

whence, with the notation of 6.1:

$$g = 1, \tag{10.3}$$

$$S' = \lambda, \tag{10.4}$$

$$u_0(x) = \exp(-\lambda x + \lambda^2). \tag{10.5}$$

The QR method is that of 6.1.

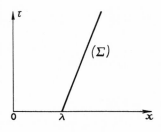

Figure 11

10.2 Discretization

We use the same method as in Section 7, apart evidently from the neighborhood of the boundary (Σ).

Remark 10.1. We know (Mignot [1]) that taking account of boundary conditions on (Σ)—since the problem is not cylindrical—leads to numerical difficulties; we can overcome them (Mignot, *loc. cit.*) with the aid of elaborate methods,[1] something we have not tried to do here.[2] ∎

We choose the steps Δx and Δt so that, for all the q steps in time (q an arbitrary integer), a gridpoint is located on (Σ).

Finally, with the help of the data on (Σ), we calculated, using interpolations and linear extrapolation, the value of $u(x, t)$ in the gridpoints situated, for all t, on one part or the other of (Σ). Thus it was possible to use the same program as for Section 7, for the sake of simplification, but at the expense of accuracy of results.

10.3 Numerical Results

We took $\lambda = 1$, $q = 10$, $T = 0.5$.

Graphs 16 and 17, respectively, give the calculated solution and the exact solution as functions of x for five values of t.

Remark 10.2. The results are only fairly satisfactory, especially for $t = T$, which is natural in view of the remarks of 10.2. ∎

Graphs 18, 19 and 20 show the influences of the parameters ε_i at the time $t = 0.3$.

---

[1] In particular elliptic regularization, and especially *the method of "auxiliary domains."*

[2] It is very interesting to apply the method of 5.3 here. Naturally, in the present case, we could make a change of variable leading to the cylindrical case!

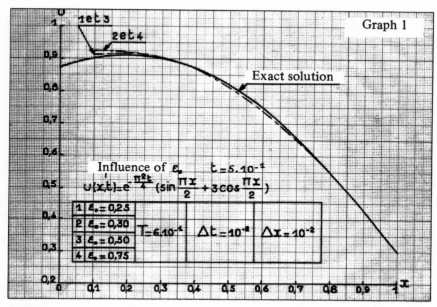

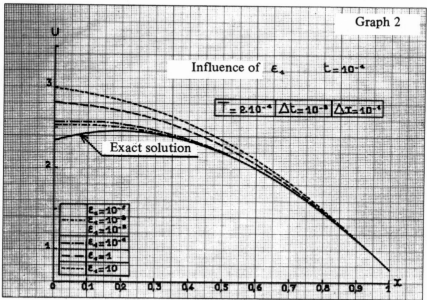

NOTE: In all graphs commas represent decimal points.

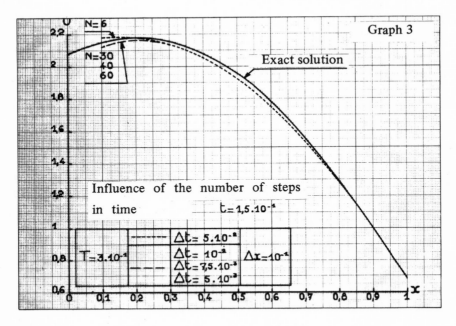

Graph 3

Exact solution

Influence of the number of steps in time $\quad t = 1,5.10^{-1}$

$T = 3.10^{-1}$

	$\Delta t = 5.10^{-2}$	$\Delta x = 10^{-1}$
	$\Delta t = 10^{-2}$	
	$\Delta t = 7,5.10^{-3}$	
	$\Delta t = 5.10^{-3}$	

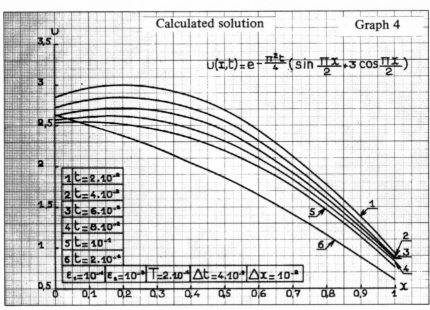

Calculated solution   Graph 4

$$U(x,t) = e^{-\frac{\pi^2 t}{4}}\left(\sin\frac{\pi x}{2} + 3\cos\frac{\pi x}{2}\right)$$

1	$t = 2.10^{-2}$
2	$t = 4.10^{-2}$
3	$t = 6.10^{-2}$
4	$t = 8.10^{-2}$
5	$t = 10^{-1}$
6	$t = 2.10^{-1}$

$\varepsilon_1 = 10^{-4}$ $\varepsilon_2 = 10^{-5}$ $T = 2.10^{-1}$ $\Delta t = 4.10^{-3}$ $\Delta x = 10^{-2}$

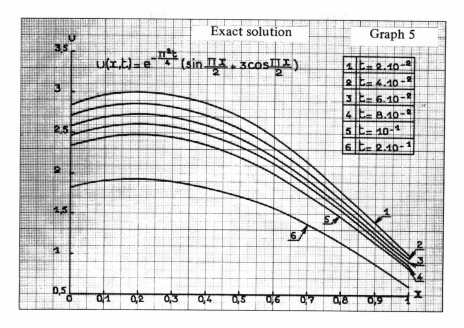

Exact solution   Graph 5

$$U(x,t) = e^{-\frac{\pi^2 t}{4}} \left(\sin \frac{\pi x}{2} + 3 \cos \frac{\pi x}{2} \right)$$

1	$t = 2.10^{-2}$
2	$t = 4.10^{-2}$
3	$t = 6.10^{-2}$
4	$t = 8.10^{-2}$
5	$t = 10^{-1}$
6	$t = 2.10^{-1}$

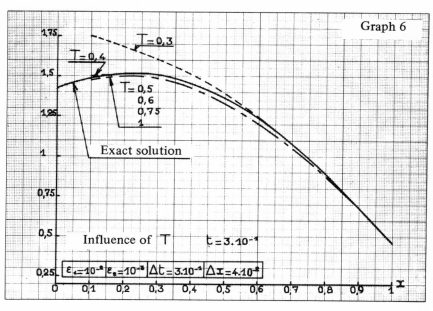

Graph 6

$T = 0,3$

$T = 0,4$

$T = 0,5$
$0,6$
$0,75$
1

Exact solution

Influence of T $t = 3.10^{-1}$

| $\varepsilon_1 = 10^{-2}$ | $\varepsilon_2 = 10^{-3}$ | $\Delta t = 3.10^{-1}$ | $\Delta x = 4.10^{-2}$ |

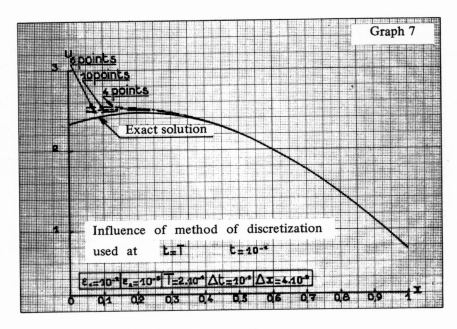

Graph 7

Influence of method of discretization used at $t=T$ $t=10^{-4}$

Exact solution

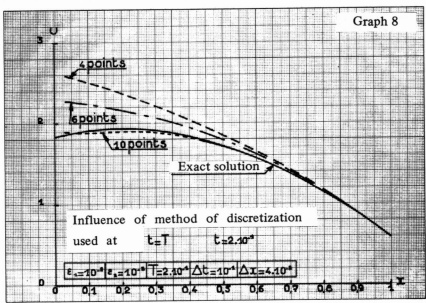

Graph 8

Influence of method of discretization used at $t=T$ $t=2.10^{-3}$

Exact solution

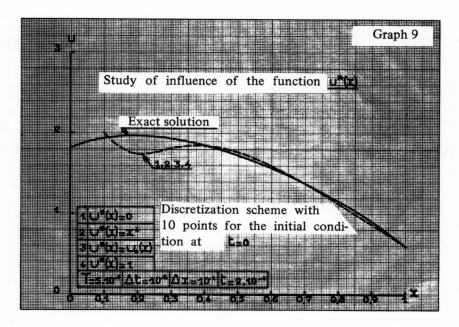

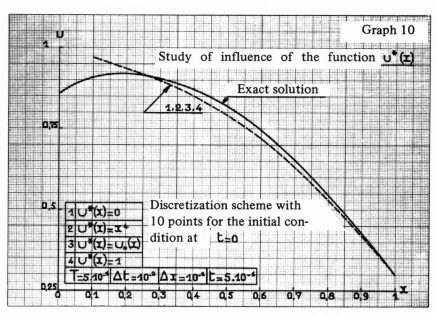

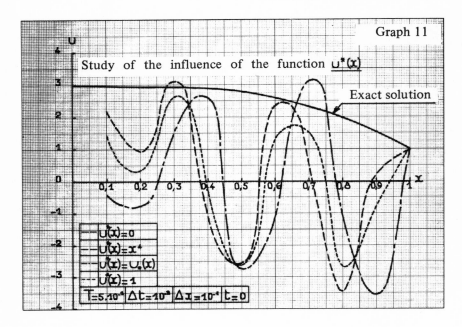

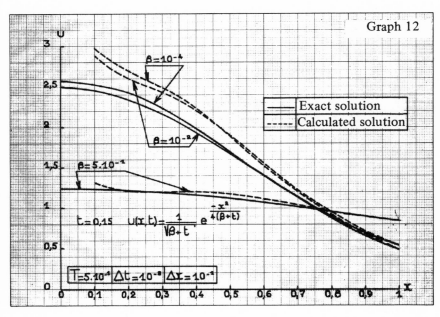

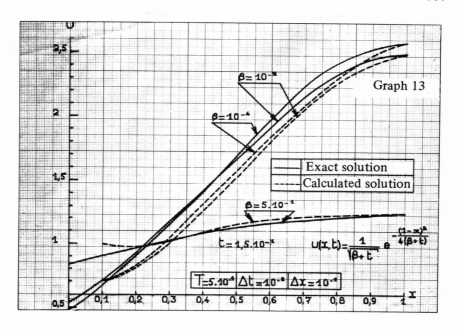

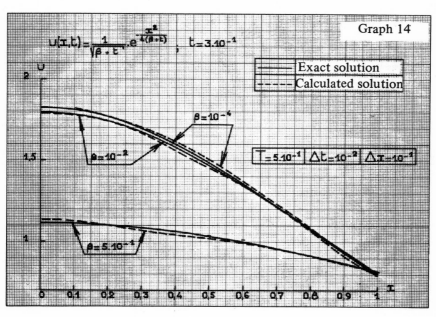

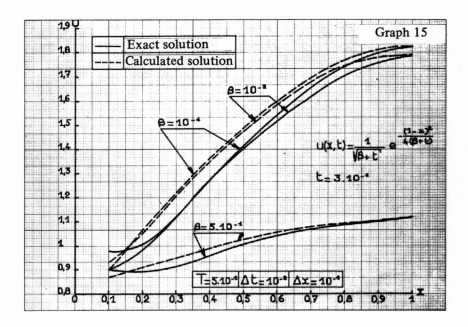

Graph 15

Exact solution
---- Calculated solution

$\beta = 10^{-2}$

$\beta = 10^{-4}$

$u(x,t) = \dfrac{1}{\sqrt{\beta + t}}\, e^{-\frac{(1-x)^2}{4(\beta + t)}}$

$t = 3.10^{-1}$

$\beta = 5.10^{-1}$

$\boxed{T = 5.10^{-1}\quad \Delta t = 10^{-5}\quad \Delta x = 10^{-1}}$

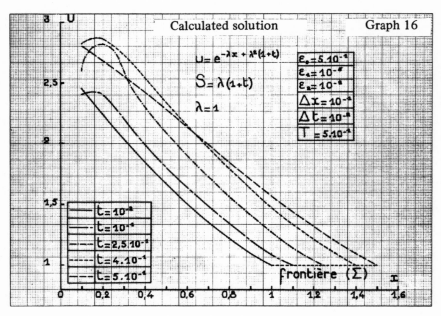

Calculated solution

Graph 16

$u = e^{-\lambda x + \lambda^2(1+t)}$

$S = \lambda(1+t)$

$\lambda = 1$

$\varepsilon_0 = 5.10^{-1}$
$\varepsilon_1 = 10^{-5}$
$\varepsilon_2 = 10^{-5}$
$\Delta x = 10^{-1}$
$\Delta t = 10^{-5}$
$T = 5.10^{-1}$

$t = 10^{-2}$
--- $t = 10^{-1}$
--- $t = 2,5.10^{-1}$
--- $t = 4.10^{-1}$
--- $t = 5.10^{-1}$

frontière (Σ)

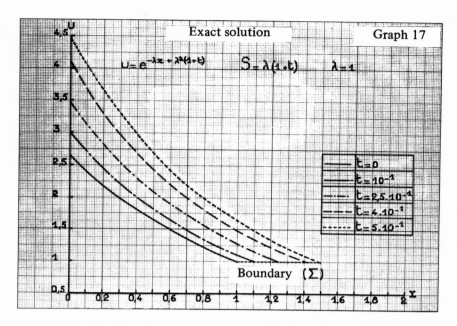

Graph 17

Exact solution

$U = e^{-\lambda x + \lambda^2(1+t)}$ $S = \lambda(1+t)$ $\lambda = 1$

- — — $t = 0$
- — · — $t = 10^{-1}$
- — · · — $t = 2,5.10^{-1}$
- — — — $t = 4.10^{-1}$
- · · · · · $t = 5.10^{-1}$

Boundary (Σ)

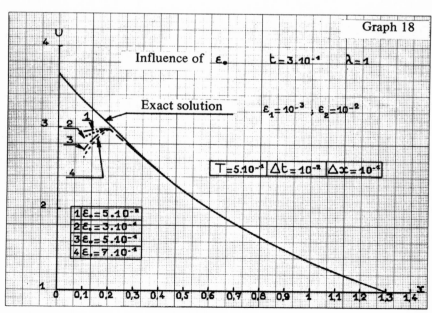

Graph 18

Influence of ε_0 $t = 3.10^{-1}$ $\lambda = 1$

Exact solution $\varepsilon_1 = 10^{-3}$; $\varepsilon_2 = 10^{-2}$

$T = 5.10^{-1}$ $\Delta t = 10^{-3}$ $\Delta x = 10^{-1}$

1	$\varepsilon_0 = 5.10^{-3}$
2	$\varepsilon_0 = 3.10^{-4}$
3	$\varepsilon_0 = 5.10^{-4}$
4	$\varepsilon_0 = 7.10^{-4}$

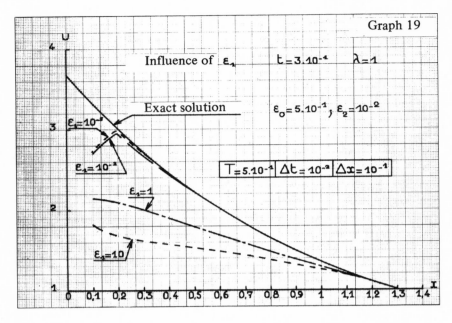

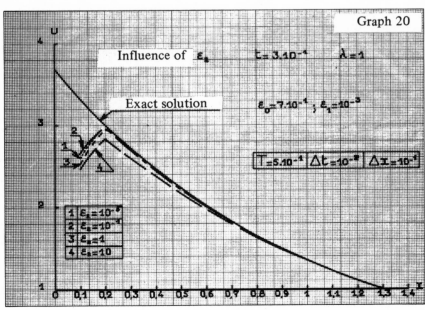

Chapter 6

SOME EXTENSIONS

This short chapter has a little different character from the preceding chapters; we limit ourselves to giving here—most often in a heuristic fashion—some extensive and new applications of the QR method, only some of which have been tried numerically.

DETAILED OUTLINE

1. QR METHOD AND NONLINEAR PROBLEMS OF EVOLUTION

1.1 Orientation

In Chapters 1 and 3, we considered various minimization problems (leading to improperly posed problems) associated with *linear*
equations of evolution:

$$\frac{\partial u}{\partial t} + Au = 0, \tag{1.1}$$

$$\frac{\partial^2 u}{\partial t^2} + Au = 0, \tag{1.2}$$

etc. (where A is a linear elliptic operator).

We can formulate analogous problems for equations of evolution
of the form (1.1) (or (1.2), etc.) but where now *the operator A is
not linear*.

A very large variety of problems (apparently non trivial!) present themselves in these directions. We limit ourselves to two
simple examples.

1.2 Examples (I)

1.2.1 FORMULATION OF PROBLEM

We consider the (nonlinear) equation:

$$\frac{\partial u}{\partial t} - \frac{\partial^2 u}{\partial x^2} + u \frac{\partial u}{\partial x} = 0, \quad 0 < x < 1, \quad t > 0, \tag{1.3}$$

with the *boundary conditions*:

$$u(0, t) = u(1, t) = 0. \tag{1.4}$$

If the *initial condition* is fixed:

$$u(x, 0) = \xi(x), \quad 0 < x < 1, \tag{1.5}$$

then we can show[1] that u exists and is unique:

$$u = u(x, t) = u(x, t \; ; \xi) . \blacksquare \qquad (1.6)$$

If χ is a given function in $(0, 1)$, square summable and if $T > 0$ is given, we consider:

$$\operatorname*{Inf}_{\xi} J(\xi), \qquad J(\xi) = \int_0^1 |u(x, t \; ; \xi) - \chi(x)|^2 \, dx . \qquad (1.7)$$

Remark 1.1. Problems of the same type may be posed for the same equations with:

- control in the boundary conditions (cf. Chapter 3),
- thick functionals (cf. Chapter 1),
- etc.

We can then (formally) adapt the QR solution proposed in Chapters 1 and 3, as is indicated below for the functional (1.7). $\blacksquare$

Remark 1.2. We conjecture that:

$$\operatorname{Inf} J(\xi) = 0 . \qquad (1.8)$$

The proofs we have given in the linear case (Chapter 1) are not valid here. It is evidently necessary to take care that the set traversed in $L_2(0, 1)$ by $u(x, T)$ *is not* a vector space. $\blacksquare$

If we accept (1.8), *the problem is then*:

for given $\eta > 0$, find ξ such that:

$$J(\xi) \leqslant \eta . \qquad (1.9)$$

Remark 1.3. (Analogous to Chapter 1, Remark 1.2.) There *will be no* uniqueness of ξ satisfying (1.9), but on the contrary *an infinite number* of such ξ. We observe then *an instability* in the controls ξ (cf. Chapter 1, Graphs 1 *bis*, 2 *bis*, 3 *bis*, etc.). $\blacksquare$

---

[1] We do not wish to give complete proofs here. Formally, we note that there are "the same" energy inequalities for (1.3), (1.4) as for the heat equation (i.e., without the term $\partial u/\partial x$); in effect, when we multiply by u and integrate over x, we have

$$\int_0^1 \left(u \frac{\partial u}{\partial x} \right) u \, dx = 0.$$

1.2.2 THE QR METHOD

By formal analogy with Chapter 1, we consider for every $\varepsilon > 0$ *the QR problem associated with* (1.3), (1.4), (1.7):

$$\frac{\partial u_\varepsilon}{\partial t} - \frac{\partial^2 u_\varepsilon}{\partial x^2} + u_\varepsilon \frac{\partial u_\varepsilon}{\partial t} - \varepsilon \frac{\partial^4 u_\varepsilon}{\partial x^4} = 0, \quad 0 < x < 1, \quad 0 < t < T, \tag{1.10}$$

$$u_\varepsilon(0, t) = u_\varepsilon(1, t) = 0$$

$$\frac{\partial^2 u_\varepsilon}{\partial x^2}(0, t) = \frac{\partial^2 u_\varepsilon}{\partial x^2}(1, t) = 0 , \tag{1.11}$$

$$u_\varepsilon(x, T) = \chi(x) . \tag{1.12}$$

We then take

$$\xi(x) = u_\varepsilon(x, 0) . \tag{1.13}$$

1.2.3 JUSTIFICATION. CHOICE OF ε

Naturally, the method proposed in 1.2.2 supposes that *we know the existence and uniqueness of u_ε the solution of* (1.10), (1.11), (1.12). We show that this is indeed the case: we obtain [as in (1)] "the same" a priori estimates as for the QR problem in the linear case, since:

$$\int_0^1 \left(u_\varepsilon(x, t) \frac{\partial u_\varepsilon}{\partial x}(x, t) \right) u_\varepsilon(x, t) \, dx = 0 \qquad \forall t . \tag{1.14}$$

Therefore the choice (1.13) is well defined without ambiguity.

As a consequence of (1.14) we can choose ε *as in the linear case* (cf. Chapter 1, Sections 5, 5.5).

1.2.4 GENERALIZATIONS. NAVIER–STOKES EQUATIONS

In place of Eq. (1.3), consider now the Navier–Stokes equations in an open domain Ω of R^n (for these equations, cf. J. Leray [1, 2], O. A. Ladyzenskaya [1], J. L. Lions–G. Prodi [1]):

($u = u_1, ..., u_n$ is the velocity vector, p the pressure)

$$\frac{\partial u}{\partial t} - \nu \, \Delta u + \sum_{i=1}^{n} u_i \frac{\partial u}{\partial x_i} = \text{grad } p, \quad x \in \Omega, \quad t > 0 \; ; \tag{1.15}$$

$$\text{div } u = 0, \tag{1.16}$$

$$u(x, t) = 0 \quad \text{if} \quad x \in \partial\Omega = \Gamma \quad (\text{boundary of } \Omega), \tag{1.17}$$

$$u(x, 0) = \xi(x) \; (= \{ \xi_1, ..., \xi_n \}). \tag{1.18}$$

The question of existence and uniqueness of solution is not resolved if $n = 3;$[1] but it is when $n = 2$ (cf. the cited works).

We suppose implicitly that $n = 2$ below. Then:

$$u = u(x, t) = u(x, t \; ; \xi)$$

and we can again consider

$$J(\xi) = \int_{\Omega} | \, u(x, T \; ; \xi) - \chi(x) \, |^2 \, dx \, ,$$

or, explicitly:

$$J(\xi) = \sum_{i=1}^{n} \int_{\Omega} | \, u_i(x, T) - \chi_i(x) \, |^2 \, dx \, . \tag{1.19}$$

We again conjecture that:

$$\text{Inf}_{\xi} J(\xi) = 0 \, . \tag{1.20}$$

The problem is then stated as in 1.2.1.

The QR *method* consists in solving[2]

$$\left. \begin{aligned} &\frac{\partial u_\varepsilon}{\partial t} - \nu \, \Delta u_\varepsilon + \sum_{i=1}^{n} u_\varepsilon \frac{\partial u_\varepsilon}{\partial x_i} - \varepsilon \, \Delta^2 u_\varepsilon = \text{grad } p_\varepsilon \, , \\ &\text{div } u_\varepsilon = 0 \, , \\ &u_\varepsilon \, \Delta u_\varepsilon = 0 \quad \text{if} \quad x \in \Gamma \, , \\ &u_\varepsilon(x, T) = \chi(x) \, , \end{aligned} \right\} \tag{1.21}$$

then in choosing ξ by the formula corresponding to (1.13). ∎

---

[1] Thus if $n = 3$ the problems posed above are, for the moment, out of reach, except for numerical trials.
[2] We can show that this problem is properly posed.

Remark 1.4. We can, from the numerical point of view, "suppress" the constraints div $u = 0$ and div $u_\varepsilon = 0$; cf. R. Teman [3]. ∎

Remark 1.5. As in Chapter 1, in the linear case, there are an infinite number of possible ways of applying the QR method; for example, we could add, in place of the term $- \varepsilon \Delta^2 u_\varepsilon$, the term $\varepsilon \Delta^3 u_\varepsilon$ (thus adding a boundary condition that $\Delta^2 u_\varepsilon = 0$ if $x \in \Gamma$). In the present case, we made the simplest choice; in applications, we can often be guided by physical considerations in a choice. ∎

1.3 Examples (II)

1.3.1 FORMULATION OF THE PROBLEM

We now consider the equation (always nonlinear):

$$\frac{\partial u}{\partial t} - \frac{\partial^2 u}{\partial x^2} + u^3 = 0, \quad 0 < x < 1, \quad t > 0, \tag{1.22}$$

with the boundary conditions:

$$u(0, t) = u(1, t) = 0 \tag{1.23}$$

and the initial condition:

$$u(x, 0) = \xi(x) . \blacksquare \tag{1.24}$$

We can show existence and uniqueness of a solution[1] of the system (1.22), (1.23), (1.24); the fundamental inequality for the proof (not given here) is obtained, as usual, by multiplying by u and integrating over $(0, 1)$; after integration by parts, we have:

$$\frac{1}{2} \frac{1}{dt} \int_\Omega u(x, t)^2 \, dx + \int_\Omega \left(\frac{\partial u}{\partial x}(x, t) \right)^2 \, dx + \int_\Omega u(x, t)^4 \, dx = 0 .$$

We thus obtain a priori estimates analogous to the linear case of the heat equation and *in addition* an a priori estimate for:

$$\int_0^T \int_\Omega u(x, t)^4 \, dx \, dt . \blacksquare \quad ..$$

---

[1] For given ξ in $L_2(0, 1)$, "solution" is understood in a suitable weak sense.

As in the preceding section, we conjecture that (with χ given in $L_2(0, 1)$):

$$\underset{\xi \in L_2(0,1)}{\text{Inf}} \quad J(\xi) = 0 \,, \tag{1.25}$$

where

$$J(\xi) = \int_\Omega | u(x, T) - \chi(x) |^2 \, dx \tag{1.26}$$

[with u the solution of (1.22), (1.23), (1.24)]. ∎

The problem is then (as in the preceding section): for given $\eta > 0$, find ξ (not unique, cf. Remark 1.3) such that:

$$J(\xi) \leqslant \eta \,. \tag{1.27}$$

1.3.2 THE QR METHOD

By analogy with 1.2 above, it is natural to consider the following problem for fixed $\varepsilon > 0$:

$$\left.\begin{aligned}
&\frac{\partial u_\varepsilon}{\partial t} - \frac{\partial^2 u_\varepsilon}{\partial x^2} + u_\varepsilon^3 - \varepsilon \frac{\partial^4 u_\varepsilon}{\partial x^4} = 0 \,, \quad 0 < x < 1 \,, \quad t \in (0, T) \,, \\[2mm]
&u_\varepsilon(0, t) = u_\varepsilon(1, t) = 0 \,, \quad \frac{\partial^2 u_\varepsilon}{\partial x^2}(0, t) = \frac{\partial^2 u_\varepsilon}{\partial x^2}(1, t) = 0 \,, \\[2mm]
&u_\varepsilon(x, T) = \chi(x) \,.
\end{aligned}\right\} \tag{1.28}$$

This problem, however, has no reason to be properly posed;[1] in effect, multiplying by u_ε, integrating by parts in x, then integrating from t to T, we have:

$$\left.\begin{aligned}
\frac{1}{2} \int_0^1 u_\varepsilon(x, t)^2 \, dx &+ \varepsilon \int_t^T \int_0^1 \left(\frac{\partial^2 u_\varepsilon}{\partial x^2}(x, \sigma) \right)^2 dx \, d\sigma = \\[2mm]
&= \frac{1}{2} \int_0^1 \chi(x)^2 \, dx + \int_t^T \int_0^1 \left(\frac{\partial u_\varepsilon}{\partial x}(x, \sigma) \right)^2 dx \, d\sigma + \\[2mm]
&\qquad\qquad + \int_t^T \int_0^1 u_\varepsilon(x, \sigma)^4 \, dx \, d\sigma
\end{aligned}\right\} \tag{1.29}$$

---

[1] We think, without proof, that (1.28) is improperly posed. *Examples* where we can *demonstrate* that a nonlinear problem is *improperly posed* are rare; see the interesting cases of S. Kaplan [1] and H. Fujita [1].

and we do not know how to majorize the last integral (involving u^4) with the help of the quantities appearing in the first part of (1.29). ∎

Here is how we could modify the QR method. We introduce two parameters ε_1, ε_2 and modify the "irreversible" operator:

$$u \to \frac{\partial u}{\partial t} - \frac{\partial^2 u}{\partial x^2} + u^3 \,,$$

not only by a linear term $(-\varepsilon_1 \, \partial^4 u/\partial x^4)$ as before, *but in addition with a nonlinear term* $(-\varepsilon_2 \, u^5)$.[1] We are thus led to solve, for ε_1 and $\varepsilon_2 > 0$:

$$\left. \begin{array}{l} \dfrac{\partial u_\varepsilon}{\partial t} - \dfrac{\partial^2 u_\varepsilon}{\partial x^2} + u_\varepsilon^3 - \varepsilon_1 \dfrac{\partial^4 u_\varepsilon}{\partial x^4} - \varepsilon_2 \, u_\varepsilon^5 = 0 \,, \\[2ex] u_\varepsilon(0, t) = u_\varepsilon(1, t) = 0 \,, \quad \dfrac{\partial^2 u_\varepsilon}{\partial x^2}(0, t) = \dfrac{\partial^2 u_\varepsilon}{\partial x^2}(1, t) = 0 \,, \\[2ex] u_\varepsilon(x, T) = \chi(x) \,. \end{array} \right\} \quad (1.30)$$

We can show that (1.30) *possesses a unique solution*; without giving the details of the proof, let us indicate the essential point: multiplying by u_ε and integrating by parts, we obtain [compare (1.29)]:

$$\left. \begin{array}{l} \dfrac{1}{2} \displaystyle\int_0^1 u_\varepsilon(x, T)^2 \, dx + \varepsilon_1 \int_t^T \int_0^1 \left(\dfrac{\partial^2 u_\varepsilon}{\partial x^2}(x, \sigma) \right)^2 dx \, d\sigma + \\[3ex] \qquad + \varepsilon_2 \displaystyle\int_t^T \int_0^1 u_\varepsilon(x, \sigma)^6 \, dx \, d\sigma = \\[3ex] = \dfrac{1}{2} \displaystyle\int_0^1 \chi(x)^2 \, dx + \int_t^T \int_0^1 \left(\dfrac{\partial u_\varepsilon}{\partial x}(x, \sigma) \right)^2 dx \, d\sigma + \int_t^T \int_0^1 u_\varepsilon^4 \, dx \, d\sigma \,. \end{array} \right\} \quad (1.31)$$

But

$$\int_t^T \int_0^1 \left(\frac{\partial u_\varepsilon}{\partial x} \right)^2 dx \, d\sigma = \int_t^T \int_0^1 \left(-\frac{\partial^2 u_\varepsilon}{\partial x^2} \right) u_\varepsilon \, dx \, d\sigma \leqslant$$

$$\leqslant \frac{\varepsilon_1}{2} \int_t^T \int_0^1 \left(\frac{\partial^2 u_\varepsilon}{\partial x^2} \right)^2 dx \, d\sigma + \frac{1}{2\,\varepsilon_1} \int_t^T \int_0^1 u_\varepsilon^2 \, dx \, d\sigma$$

---

[1] Note that the "pseudo-viscosity" term of Von Neumann–Richtmyer [1] is *a nonlinear term*.

and

$$\int_t^T \int_0^1 u_\varepsilon^4 \, dx \, d\sigma = \int_t^T \int_0^1 u_\varepsilon^3 \, u_\varepsilon \, dx \, d\sigma \leqslant$$

$$\leqslant \frac{\varepsilon_2}{2} \int_t^T \int_0^1 u_\varepsilon^6 \, dx \, d\sigma + \frac{1}{2\,\varepsilon_2} \int_t^T \int_0^1 u_\varepsilon^2 \, dx \, d\sigma$$

and (1.31) then gives:

$$\left. \begin{array}{l} \displaystyle\int_0^1 u_\varepsilon(x, t)^2 \, dx + \varepsilon_1 \int_t^T \int_0^1 \left(\frac{\partial^2 u_\varepsilon}{\partial x^2}\right)^2 dx \, d\sigma + \varepsilon_2 \int_t^T \int_0^1 u_\varepsilon^6 \, dx \, d\sigma \leqslant \\[4mm] \displaystyle\leqslant \int_0^1 \chi(x)^2 \, dx + \left(\frac{1}{\varepsilon_1} + \frac{1}{\varepsilon_2}\right) \int_t^T \int_0^1 u_\varepsilon(x, \sigma)^2 \, dx \, d\sigma \, , \end{array} \right\} \qquad (1.32)$$

whence we deduce the a priori estimates on u_ε^2, u_ε^6, and $(\partial^2 u_\varepsilon/\partial x^2)^2$ by an application of the Bellman–Gronwall lemma. ∎

Remark 1.6. The term $\varepsilon_2 u_\varepsilon^5$ is the simplest term—but not the only— permitting us to obtain an inequality of the type of (1.32); as in the linear case (and "even more so"!) *there is no uniqueness in the choice of the QR method.* ∎

Remark 1.7. If we consider, in place of (1.22) an equation of the form (1.33):

$$\frac{\partial u}{\partial t} - \frac{\partial^2 u}{\partial x^2} + |u|^\rho u = 0, \qquad \rho \text{ real} > 0, \qquad (1.33)$$

we introduce the equation:

$$\frac{\partial u_\varepsilon}{\partial t} - \frac{\partial^2 u_\varepsilon}{\partial x^2} + |u_\varepsilon|^\rho u_\varepsilon - \varepsilon_1 \frac{\partial^4 u}{\partial x^4} - \varepsilon_2 |u_\varepsilon|^{2\rho} u_\varepsilon = 0 \, . \blacksquare \qquad (1.34)$$

1.3.3 VARIANTS

1.3.3.1

In the preceding examples, the nonlinearity of the operator A did not enter into the term of maximum derivative. Consider now the solution u of the following problem[1]:

$$\left.\begin{array}{l} \dfrac{\partial u}{\partial t} - \dfrac{\partial}{\partial x}\left(\left|\dfrac{\partial u}{\partial x}\right|^{p}\dfrac{\partial u}{\partial x}\right) = 0\,, \qquad 0 < x < 1\,, \qquad t > 0\,, \\[2mm] u(0,t) = u(1,t) = 0\,, \\[2mm] u(x,0) = \xi(x)\,. \end{array}\right\} \qquad (1.35)$$

We can again *formulate* problems of the same type. But everything seems quite open in this direction.

1.3.3.2 Nonlinear boundary conditions

We can also consider problems of the same type for operators, linear or not, with *nonlinear* boundary conditions. The QR method consists then of introducing "regularizing" terms or "penalization" terms, linear or *not*, not only in the equations, but also in the boundary conditions (cf. Cicurel [1] and, in particular, Chapter 5, 2.2).

2. NUMERICAL EXAMPLES

2.1 Discretization

We limited ourselves to a numerical treatment of 1.2.1 and 1.2.2. We discretized (1.10) with the help of an implicit scheme of the Crank–Nicholson type; setting $v(x,t) = u_\varepsilon(x, T - t)$, we discretize the nonlinear term $v\,\partial v/\partial x$ by:

$$v_j^{n-1} \times \frac{1}{4\,\Delta x}\left(v_{j+1}^n - v_{j-1}^n + v_{j+1}^{n-1} - v_{j-1}^{n-1}\right).$$

The initial conditions and boundary conditions are taken care in a standard fashion as has already been done many times.

---

[1] A very particular case of a "monotone" problem.

We deduce from Eq. (1.10) in approximate fashion:

$$\xi_\varepsilon(x) = v(x, T) = u_\varepsilon(x, 0) \tag{2.2}$$

For the sake of verification, we then integrate the system (1.3), (1.4), (1.5) (with $\xi = \xi_\varepsilon$), directly which yields:

$$\chi_\varepsilon^*(x) = u(x, T ; \xi_\varepsilon) . \tag{2.3}$$

We set:

$$e = \int_0^1 (\chi - \chi^*)^2 \, dx . \frac{1}{\displaystyle\int_0^1 \chi(x)^2 \, dx} , \tag{2.4}$$

which we calculated simply using:

$$\sum_j \left(\chi(j \, \Delta x) - \chi^*(j \, \Delta x)\right)^2 \frac{1}{\displaystyle\sum_j \chi(j \, \Delta x)^2} .$$

Remark 2.1. Naturally, all that precedes holds in the case where, in place of the homogeneous boundary conditions (1.4) we use the inhomogeneous conditions:

$$u(0, t) = g_0(t), \quad u(1, t) = g_1(t) . \tag{2.5}$$

2.2 Numerical Results

We considered four examples:

EXAMPLE 1

$$\chi(x) = x , \qquad g_0(t) = 0 , \qquad g_1(t) = 1 .$$

The calculations were made for various values of ε and in particular:

$$\varepsilon = 5.10^{-2} , \qquad e = 2.4.10^{-4} ,$$

$$\varepsilon = 10^{-2} \quad , \qquad e = 5.2.10^{-5} ,$$

$$\varepsilon = 5 \; 10^{-3} , \qquad e = 2.4.10^{-5} .$$

The functions χ^* and $\tilde{\xi}$ are given on Graphs 1 and 1 *bis*.

EXAMPLE 2 (cf. Graphs 2 and 2 *bis*)

$$\chi(x) = \begin{cases} 0 & \text{if} \quad 0 \leqslant x \leqslant 0.4 \; ; \\ \tfrac{1}{2}(1 + \sin 5\,\pi(x - 0.5)) & \text{if} \quad 0.4 \leqslant x \leqslant 0.6 \; ; \\ 1 & \text{if} \quad 0.6 \leqslant x \leqslant 1 \quad ; \end{cases}$$

$$g_0(t) = 0 \, , \qquad g_1(t) = 1 \, .$$

EXAMPLE 3 (cf. Graphs 3 and 3 *bis*)

$$\chi(x) = \frac{1 + x}{1 + T} \, , \quad g_0(t) = \frac{1}{1 + t} \, , \quad g_1(t) = \frac{2}{1 + t} \, .$$

We have in this case the analytic solution of the direct problem:

$$u(x, t) = \frac{x + 1}{t + 1} \, .$$

In this quite special case, for $\varepsilon = 5.10^{-3}$, we have: $e = 6.19^{-9}$.

EXAMPLE 4 (cf. Graphs 4 and 4 *bis*)

$$\chi(x) = \sin \pi x \; ; \quad g_0 = g_1 = 0 \, .$$

3. REMARKS ON "ELLIPTIC CONTINUATION"

3.1 Linear Case

(Refer to Chapter 4 for the notation and the hypotheses).
Let u be a solution of

$$Au = 0 \; (^1) \tag{3.1}$$

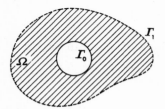

Figure 1

satisfying

$$u\,|_{\Gamma_0} = g_0 \, , \qquad \frac{\partial u}{\partial v_A}\bigg|_{\Gamma_0} = g_1 \, .$$

in a domain Ω, bounded by (Figure 1).

(^{1}) The coefficients of A are known in the entire space.

- Γ_0 known,
- Γ_1 *unknown.*

When Γ_1 is *known*, we have given in Chapter 4 many stable procedures for the approximate calculation of u valid in Ω, using the QR method.

Suppose now that Γ_1 is not given a priori; we then look for u in its "maximum domain" of definition (the unknown part of the boundary will then be Γ_1): this is a problem of "analytic continuation" type. ∎

Remark 3.1. Taking in particular $A = -\Delta$ and separating an analytic function into its real and imaginary part, the problem of analytic continuation reduces essentially to the foregoing.[1] ∎

The QR method may be adapted to this situation, but leads to long calculations. The principal is the following[2]: if $\Gamma_1^{(1)}$ is a chosen curve (or a surface)—at first in arbitrary fashion—we apply a QR method as in Chapter 4, for the open region $\Omega^{(1)}$ bounded by Γ_0 and Γ_1; let $u_\varepsilon^{(1)}$ be the corresponding solution; then $u_\varepsilon^{(1)} \to u$ as $\varepsilon \to 0$ (for the topology given in Ω Chapter 4) if and only if u exists in $\Omega^{(1)}$. Therefore:

i) if $u_\varepsilon^{(1)}$ does not converge as $\varepsilon \to 0$,[3] $\Omega^{(1)}$ is "too large," i.e., more precisely: $\Omega^{(1)}$ contains a "set" not contained in Ω;

ii) if $u_\varepsilon^{(1)}$ converges as $\varepsilon \to 0$, then (evidently with the—miraculous—possibility that $\Omega^{(1)} = \Omega$!); in the first case it is necessary "to diminish" $\Omega^{(1)}$, in the second case it is necessary "to enlarge" it and to go from $\Omega^{(1)}$ to $\Omega^{(2)}$—and we begin again with $\Omega^{(2)}$.

We have not tried to detail an algorithm for the systematic passage from $\Omega^{(n)}$ to $\Omega^{(n+1)}$. ∎

3.2 Nonlinear Problems

Let us suppose that Ω is known, bounded by Γ_0 and Γ_1 (Figure 2).

---

[1] We indicate these properties after discussions with L. Amerio. Consult also the work (in a different spirit) of this author: L. Amerio [1].

[2] There is evidently a serious numerical difficulty as far as appreciating this fact is concerned!

[3] For *analytic* functions and the location of singularities, see K. Miller [1].

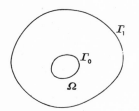

Figure 2

Let u satisfy (we can obviously considerably generalize this):

$$- \Delta u + u^3 = 0 \quad \text{in} \quad \Omega \qquad (3.3)$$

and

$$u \mid_{\Gamma_0} = g_0 , \qquad \frac{\partial u}{\partial n} \bigg|_{\Gamma_0} = g_1 .$$

We wish to calculate u in the neighborhood of Γ_1. This type of problem is basically open (cf. Chapters 4 and 5) because of the essential role played by the adjoint operators.

4. PROBLEMS WITH CONSTRAINTS

4.1 Orientation

Let us consider, as in Chapter 1, an equation—or a system—of linear evolution[1]:

$$\frac{\mathrm{d}u}{\partial t} + Au = 0 , \qquad (4.1)$$

$$u(t) \text{ satisfying (3.10) and (3.11), Chapter 1,} \qquad (4.2)$$

$$u(0) = \xi . \qquad (4.3)$$

For a given χ in H, let

$$J(\xi) = | u(T) - \chi |^2 . \qquad (4.4)$$

We have up until now looked for $J(\xi)$ *without a constraint on* ξ—i.e., ξ spans the *entire* H space.

Suppose now that ξ belongs to a subset E of H (a case said to be "with constraints"). We then wish:

$$\text{Inf } J(\xi) , \qquad \xi \in E . \ \blacksquare \qquad (4.5)$$

---

[1] Naturally, we have the same type of *problem* with nonlinear operators.

Our object here is *not at all* a systematic study of this problem of control. We indicate only very briefly in the following section *one* possible approach to this problem, using the QR method.

4.2 The QR Method

In the case where ξ spans the entire space H, the function $u(x, T; \xi)$ spans, in the linear case, a space *dense* in H (cf. Chapter 1). If ξ spans a set E belonging to H, the image set

$$F = \{ u(x, T; \xi) \mid \xi \in E \}$$

will not in general be dense in H. We say that there is *controllability* if χ is in the closure $\bar{F}$ of F; this is equivalent to

$$J_0 = \inf_{\xi \in E} J(\xi) = 0 .$$

In the general case where controllability does not hold, we have

$$J_0 = \operatorname{Inf} J(\xi) > 0 .$$

We can apply the QR method as follows; we solve

$$\frac{du_\varepsilon}{dt} + Au_\varepsilon - \varepsilon A^* A u_\varepsilon = 0 , \tag{4.6}$$

$$u_\varepsilon \text{ satisfying conditions (5.7) of Chapter 1}, \tag{4.7}$$

$$u_\varepsilon(T) = \chi . \tag{4.8}$$

Let then

$$\varphi = u_\varepsilon(0) . \tag{4.9}$$

In general $\varphi \notin E$.

We then chose N functions in H

$$w_1, w_2, ..., w_N \tag{4.10}$$

(chosen as functions of the given χ): for each function w_i, we apply the QR method; let $u_{\varepsilon,i}$ be the solution of

$$\frac{du_{\varepsilon,i}}{dt} + Au_{\varepsilon,i} - \varepsilon A^* Au_{\varepsilon,i} = 0,$$

$$u_{\varepsilon,i}, \quad Au_{\varepsilon,i} \in D(A), \qquad (1 \leqslant i \leqslant N) \tag{4.11}$$

$$u_{\varepsilon,i}(T) = w_i,$$

and let

$$\psi_i = u_{\varepsilon,i}(0), \qquad 1 \leqslant i \leqslant N. \tag{4.12}$$

We then look for a ξ of the form

$$\xi = \varphi + \sum_{i=1}^{N} \lambda_i \psi_i, \tag{4.13}$$

where the $\lambda_i \in \mathbf{R}$ *are determined "optimally"*; there are two objectives to satisfy:

1) ξ must be "optimal" in E;
2) $J(\varphi + \sum_{i=1}^{N} \lambda_i \psi_i) = \tilde{J}(\lambda_1, ..., \lambda_N)$ must be a minimum.

Let us suppose, as is the case in practice, that E is determined by inequalities of the type

$$E = \{f \mid f \in H, \quad (f, \theta_j) \leqslant \alpha_j, \quad j = 1, 2, ... \}. \tag{4.14}$$

Then we consider the set Λ in $\mathbf{R}^N$ of $\lambda = \{\lambda_1, ..., \lambda_N\}$ such that

$$\left(\varphi + \sum_{i=1}^{N} \lambda_i \psi_i, \theta_j \right) \leqslant \alpha_j \tag{4.15}$$

for $j \in J$, set of suitable indices, and we determine

$$\text{Inf } \tilde{J}(\lambda_1, ..., \lambda_N), \quad \lambda \in \Lambda. \tag{4.16}$$

This problem—one of nonlinear programming—to which we are led is itself nontrivial. *But using the QR method we can hope to obtain some useful result for small N* (if the index is 3 or 4). ∎

Remark 4.1. A supplementary difficulty resides in the evaluation of the result obtained—since $J_0 = \operatorname*{Inf}_E J(\xi)$ is not linear.[1] ∎

Remark 4.2. The foregoing remarks may be adapted to the case where we impose *constraints on the state of the system*.

For example, in many physical problems, a fundamental constraint would be

$$u(x, t) \geqslant 0 \qquad \forall x \text{ and } t \geqslant 0 . \blacksquare$$

Remark 4.3. All of the preceding considerations apply naturally to control problems with constraints on the boundary conditions (cf. Chapter 3).

5. INVERSE PROBLEMS[2]

5.1 Generalities

In the class of "inverse problems," we have those where the unknowns are the *coefficients* of differential operators or partial differential operators, where the general form is otherwise known.

These unknown coefficients can be:

1) either coefficients of operators *"inside the domain,"*
2) or coefficients of the *boundary operator*. ∎

In case 1), many methods have been proposed:

— method of "linear forms" G. Cahen [1], J. Loeb—G. Cahen [1], M. G. Marchouk [2];
— method of differential approximation, Bellman—Gluss—Roth [1] accompanied by various iterative methods (cf. Kagiwada [1]);
— reduction to an integral equation (cf. Douglas, Jones [1]).

We wish here only to indicate that the QR method (in particular that of Chapters 4, 5) is also a method of solution of inverse problems in *case* 2.

---

[1] We can always use the duality method of Legendre.
[2] Or identification of systems.

5.2 An Example

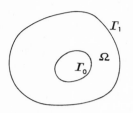

Figure 3

Let u be the solution in Ω (Figure 3) of:

$$- \Delta u + au = 0 \qquad (5.1)$$

satisfying

$$u\,|_{\Gamma_0} = g_0 , \qquad (5.2)$$

$$\frac{\partial u}{\partial n} + \alpha u = 0 \qquad \text{on} \qquad \Gamma_1 , \qquad (5.3)$$

where α is a function (continuous, for example) bounded and $\geqslant 0$ on Γ_1, *a function which is not known a priori* (but which is known to exist).

We suppose *in addition* that

$$\left.\frac{\partial u}{\partial n}\right|_{\Gamma_0} = g_1 \qquad (5.4)$$

is known (i.e., in practice *measured* at a set of discrete points!). *The problem is to determine the coefficient α "optimally"* from the data (which would be superfluous if α were known). ∎

Remark 5.1. Although the operator (5.1) is linear *in u*, the problem is really *nonlinear*. ∎

The QR method of Chapter 4 immediately furnished a solution to this problem: forgetting (5.3) we determine by the QR method of Chapter 4, u and $\partial u/\partial n$ in Ω (from which we actually delete a small band in the region of Γ_1); we thus deduce u and $\partial u/\partial n$ on Γ_1 (by extrapolation if necessary) then we determine α by (5.3) solved for α:

$$\alpha = -\left.\left(\frac{1}{u}\frac{\partial u}{\partial n}\right)\right|_{\Gamma_1}$$

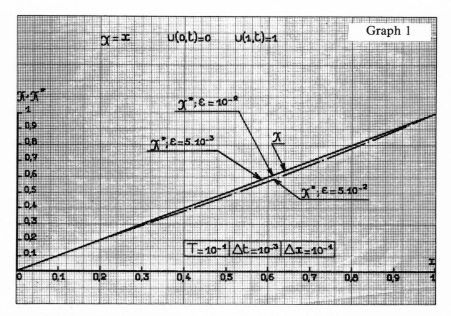

Graph 1

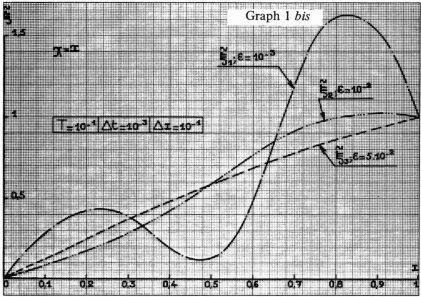

Graph 1 *bis*

NOTE: In all graphs commas represent decimal points.

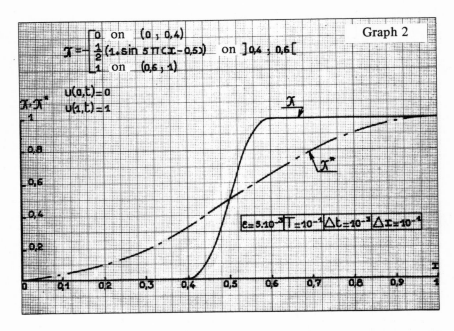

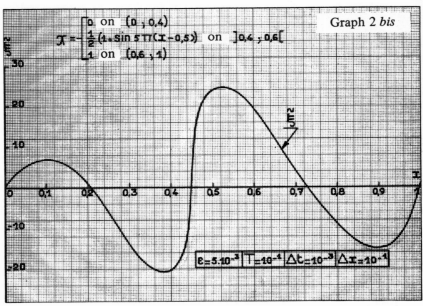

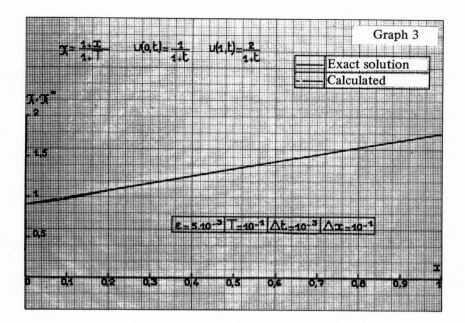

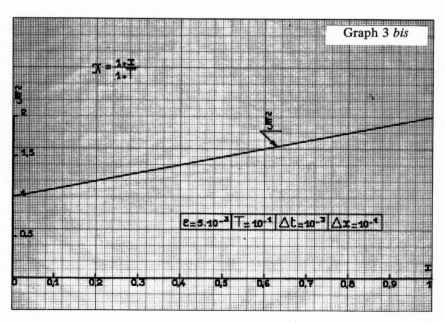

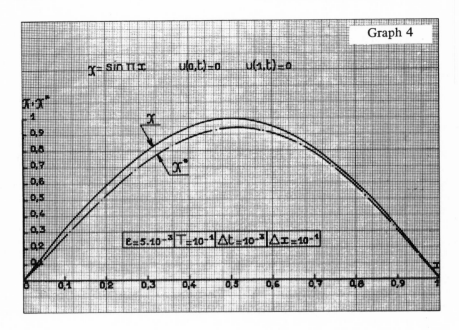

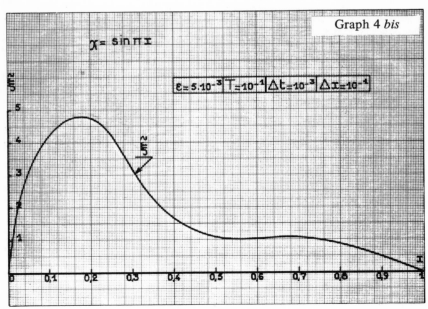

BIBLIOGRAPHY

M. S. Agranovich, I. M. Visik [1], Problems elliptiques avec parametre et problemes paraboliques generaux, *Uspekhi Mat. Nauk*, XIX (1965), pp. 53-161 (in Russian).

L. Amerio [1], Sui problemi di Cauchy e di Dirichlet per l'equazione di Laplace in due variabili, *Atti della Reale Accad. d'Italia* (1943), XXI, pp. 393-425.

Arcangeli [1], to appear.

R. Arima [1], On general boundary valur problem for parabolic equations, Journal of Math. of Kyoto Univ., vol. 4 (1964), pp. 207-244.

Artola [1], Equations d'evolution avec retard (to appear).

J.P. Aubin [1], These, Paris, 1966 (to appear).

C. Baiocchi [1], Regolarita e unicita della soluzione di una equazione differenziale astratta, *Rend. Sem. Padova*, XXXV (1965), pp. 380-417.

M. S. Baouendi [1], These, Paris (1966).

R. Bellman, R. Kalata, J. Lockett [1], *J. M. Anal. Appl.*, 12 (1965), pp. 393-400.

R. Bellman, B. Gluss, R. Roth [1], Segmential differential approximation and the "Black Box" problem, *Journal of Math. Anal. and Appl.*, 12 (1965), pp. 91-104.

B. Brelot-Collin [1], These 3e cycle, Paris (to appear).

G. Cahen [1], Determination experimentale des parametres des systemes a retard. *Revue Francaise Traitement Information* (1964) (1), pp. 15-13.

J. Cea [1], Approximation Variationnelle des problemes aux limites, *Ann. Inst. Fourier*, 14,·2 (1964), pp. 345-444.

Y. Cherruault [1], These, Paris (1966).

A. Cicurel [1], to appear.

Comminicioli [1], to appear.

R. Courant [1], Variational methods for the solution of problems of equilibrium and vibrations, *Bull. Amer. Math. Soc.*, 49 (1943), pp. 1-23.

J. Dixmier [1], *Les algebres d'operateurs dans les espaces hilbertiens*, Paris Gauthier-Villars, 1957.

J. Douglas Jr. [1], Approximate solution of physically unstable problems, Ecole C.E.A.-E.D.F., Paris, 1965.

[2], Approximate continuation of harmonic and parabolic fonctions. Dans *Numerical Solution of Partial Differential Equations*, ed. by J. H. Bramble, Acad. Press (1966).

[3], On the numerical integration of $\frac{\partial^2 u}{\partial x^2} + \frac{\partial^2 u}{\partial y^2} - \frac{\partial u}{\partial t}$ by implicit methods, *J. Siam*, III (1955), pp. 42-65.

J. Douglas-Gallie [1], *Duke Math. J.*, 26 (1959), pp. 339-348.

J. Douglas Jr., S. Jones [1], The determination of a coefficient in a Parabolic Diff. Eq. *Journal of Maths and Mechanics* (1962), vol. II, pp. 919-926.

K. O. Friedrichs [1], Asymptotic Phenomena in Mathematical Physics, *Bull. Ameri. Math. Soc.*, 61, 6 (1955), pp. 485-504.

H. Fujita [1], On the blowing up of solutions of the Cauchy Problem for $u_t = \Delta u + u^{1+\alpha}$, *Journal of the Faculty of Science*, University of Tokyo, vol. XIII, 1967.

Garabadian, Lieberstein [1], *J. Aeron. Sciences*, 25 (1958), pp. 109-118.

J. Hadamard [1], *Le probleme de Cauchy et les equations aux derivees partielles hyperboliques*, Paris, Hermann (1932).

P. Henrici [1], An algorithm for analytic continuation, *Siam J.*, vol. 3 (1966), pp. 67-78.

E. Hille, R. S. Phillips [1], Functional Analysis and Semi groups, *Colloq. Publ. Amer. Math. Soc.*, 1957.

D. Huet [1], Phenomenes de perturbation singuliere dans les problemes aux limites, *Annales Inst. Fourier* (1960), pp. 1-96.

P. Jamet, S.I. Parter [1], Numerical methods for elliptic differential operators whose coefficients are singular on a portion of the boundary (to appear).

F. John [1], *Comm. Pure Applied Math.*, XIII (1960), pp. 551-585.

K. Jorgens [1], *Comm. Pure Applied Math.*, 11 (1958), pp. 219-242.

H. Kagiwada [1], Thesis, U.C.L.A. (1964).

S. Kaplan [1], On the growth of solutions, *Comm. Pure Applied Math.*, 16 (1963), p. 305-330.

T. Kato, H. Tanabe [1], On the abstract evolution equation, *Osaka Math. Journal*, 14 (1962), p. 107-133.

J. Kohn, L. Nirenberg [1], Non coercive boundary value problems. *Comm. Pure Applied Math.*, vol. XVIII (1965), pp. 443-492.

[2], to appear in *Comm. Pure Applied Math.*).

O. A. Ladyzenskaya [1], *Theorie mathematique des fluides visqueux*, Moscow (1961) (in Russian).

B. Lago [1], Meilleures solutions de $Au = g$. *Revue Francaise Traitement Information* (1963) (1), pp. 29-58.

N. Landis [1], *Uspekhi Mat. Nauk*, XVIII (109) (1963), pp. 3-62 (in Russian).

M. M. Lavrentiev [1], Sur certains problemes mal poses de la Physique Mathematique, pp. 258-276 of the book edited by Marchouk [3] (in Russian).

[2], Idem, Novosibirsk, 1962 (in Russian).

[3], Sur le probleme de Cauchy pour l'equation de Laplace (in Russian), *Izv. Akad. Nauk*, 20 (1956), pp. 819-842.

P. D. Lax, B. Wendroff [1], On the stability of difference schemes with variable coefficients, *Comm. Pure Applied Math.*, 14 (1961), pp. 497-520.

J. Leray [1], Essai sur le mouvement plan d'un liquide visqueux que limitent des parois, *J. Math. Pures et Appl.*, XIII (1934), pp. 331-418.

[2], Sur le mouvement d'un liquide visqueux emplissant l'espace, *Acta Math.*, 63 (1934), pp. 193-248.

J. L. Lions [1], *Equations differentielles operationnelles et problemes aux limites*, Springer, Collection Jaune, 111 (1961).

[2], Cours C.I.M.E., Varenna, May 1963.

[3], Remarques sur les equations differentielles operationnelles, *Osaka Math. J.*, V. 15 (1963), pp. 131-142.

[4], Sur l'approximation des solutions de certains problemes aux limites. *Rend. Sem. Padova*, vol. XXXII (1962), pp. 3-54.

[5], Sur le controle optimale du systemes governes par des equations aux equations partielles, Paris, Dunod, 1968.

J. L. Lions, E. Magenes [1], *Problemes aux limites non homogenes et applications*, 1 (1968).

J. L. Lions, B. Malgrange [1], Sur l'unicite retrograde dans les problemes mixtes paraboliques, *Math. Scand.*, 8 (1960), pp. 277-286.

J. L. Lions, G. Prodi [1], *C. R. Acad. Sci. Paris*, 248 (1960), pp. 277-286.

J. L. Lions, P. A. Raviart [1], *I.C.C. Bull.*, vol. 5, no. 1 (1966), p. 1-20.

[2], Calcolo, 1967.

J. Loeb, G. Cahen [1], Extraction . . . des parametres dynamiques d'un systeme, *Automatisme*, 8 (1963), pp. 479-486.

[2], More about process identification, *I.E.E.E. Transactions on Automatic Control* (1965), pp. 359-361.

E. Magenes, G. Stampacchia [1], I problemi al contorno per le equazione differenziali di tipo ellittico, *Ann. Sci. Norm. Sup. Pisa*, XII (1958), pp. 247-357.

G. I. Marchouk [1], *Methodes numeriques* . . . , Novosibirsk (1965) (in Russian).

[2], Formulation of some converse problems, *Soviet Math.* (1964), pp. 675-678.

[3], *Problemes de mathematiques numeriques et appliquees*, Novosibirsk (1966), G. I. Marchouk (editor) (in Russian).

A. Mignot [1], Thesis, Paris (1967).

K. Miller [1], to appear.

S. Mizohata [1], Unicite du prolongement des solutions pour quelques operateurs differentiels paraboliques, *Mem. Coll. Sci. Univer. Kyoto*, ser. A, 31 (1958), pp. 219-239.

H. Morel [1], Introduction de poids dans l'etude de problemes aux limites, *Annales Institut Fourier*, 12 (1962), pp. 1-115.

J. Von Neumann, R. Richtmyer [1], A method for the numerical calculations of hydrodynamical shocks, *J. Appl. Phys., hydrodynamical shocks*, J. Appl. Phys., vol. 21 (1950), p. 232.

L. Nirenberg [1], Remarks on strongly elliptic differential equations, *Comm. Pure Applied Maths.*, 8 (1955), pp. 643-674.

O. A. Oleink [1], Solutions discontinues des equations aux derivees partielles non lineaires, *Uspekhi Mat. Nauk*, 12 (1957), pp. 3-73.

[2], A problem of Fichera, *Doklady Akad. Nauk*, vol. 157 (1964), pp. 1297-1301.

D. W. Peaceman, M. M. Rachford [1], The numerical solution of parabolic and elliptic differential equations, *Journal S.I.A.M. (1955), pp. 28-42.*

B. L. Phillips [1], A technique for the numerical solution of certain equations of the first kind, *Journal of the A.C.M.*, vol. 9 (1962), pp. 84-97.

C. Pucci [1], Sui problemi di Cauchy non "ben posti," *Rend. Accad. Naz. Lincei* (8), 18 (1955), pp. 473-477.

[2], Discussione del problema di Cauchy per le equazioni di tipo ellitco, *Annali Mat. Pura Appl.*, 46 (1958), pp. 131-153.

P. A. Raviart [1], These, Paris (1965), *Journal de Liouville* (1967).

[2], to appear.

G. Ribiere [1], Institut B. Pascal, Paris (1966).

R. D. Richtmyer [1], *Difference Methods for Initial-Value Problems*, Acad. Press, 1957.

J. D. Riley [1], *Math. Tables Aid Computation*, 9 (1956), p. 96-101.

D. L. Russell [1], On boundary value controllability of linear symmetric hyperbolic systems (to appear).

Sauliev [1], *Sibirsk. Mat. J.*, 4 (1963), pp. 912-925 (in Russian).

L. Schwartz [1], *Theorie des distributions*, 1 and 2, Paris (1950-1951).

    [2], Theorie des distributions a valeurs vectorielles; *Annales Institut Fourier*, 1 (1957), pp. 1-141; 2 (1958), pp. 1-209.

S. L. Sobolev [1], *Applications de l'analyse fonctionnelle aux problemes de la Physique Mathematique*, Leningrad (1950).

M. H. Stone [1], Linear transformations in Hilbert space and their applications to Analysis, *Pub. Amer. Math. Soc. Coll.*

W. Strauss [1], to appear.

L. Tantan [1], *Quelques resultats sur la penalization (to appear).*

R. Teman [1], *C. R. Acad. Sci.*, Paris, 263 (1966), p. 241-244; 265-267; 459-462.

    [2], These, Paris (1967).

    [3], Approximation numerique des solutions des equations de Navier-Stokes (to appear).

G. Torelli [1], *Rend. Sem. Padova*, XXXIV (1964), pp. 224-241.

N. Tychonoff [1], Sur la regularisation de problemes mal poses, *Doklady*, 151 (1963), pp. 501-504.

    [2], id., 153 (1963), pp. 49-52.

    [3], id., 162 (1965), pp. 763-765.

I. M. Visik, Liousternik [1], Degenerescence reguliere pour les equations differentielles avec un petit parametre, *Ospekhi. Mat. Nauk.*, XII (1957), pp. 1-121 (in Russian).

N. N. Yanenko [1], *Methode des pas fractionnaire pour la resolution numerique des problemes de la Physique Mathematique* (in Russian), Novosibirsk (1966).

K. Yosida [1], *Functional Analysis*, Springer Collection Jaune, 123 (1965).

SUBJECT INDEX